Springer Theses

Recognizing Outstanding Ph.D. Research

Aims and Scope

The series "Springer Theses" brings together a selection of the very best Ph.D. theses from around the world and across the physical sciences. Nominated and endorsed by two recognized specialists, each published volume has been selected for its scientific excellence and the high impact of its contents for the pertinent field of research. For greater accessibility to non-specialists, the published versions include an extended introduction, as well as a foreword by the student's supervisor explaining the special relevance of the work for the field. As a whole, the series will provide a valuable resource both for newcomers to the research fields described, and for other scientists seeking detailed background information on special questions. Finally, it provides an accredited documentation of the valuable contributions made by today's younger generation of scientists.

Theses may be nominated for publication in this series by heads of department at internationally leading universities or institutes and should fulfill all of the following criteria

- They must be written in good English.
- The topic should fall within the confines of Chemistry, Physics, Earth Sciences, Engineering and related interdisciplinary fields such as Materials, Nanoscience, Chemical Engineering, Complex Systems and Biophysics.
- The work reported in the thesis must represent a significant scientific advance.
- If the thesis includes previously published material, permission to reproduce this must be gained from the respective copyright holder (a maximum 30% of the thesis should be a verbatim reproduction from the author's previous publications).
- They must have been examined and passed during the 12 months prior to nomination.
- Each thesis should include a foreword by the supervisor outlining the significance of its content.
- The theses should have a clearly defined structure including an introduction accessible to new PhD students and scientists not expert in the relevant field.

Indexed by zbMATH.

Shiyuan Zhao

Noise, Dynamics and Squeezed Light in Quantum Dot and Interband Cascade Lasers

Doctoral Thesis accepted by
Institut Polytechnique de Paris, Paris, France

Springer

Author
Dr. Shiyuan Zhao
Institut Polytechnique de Paris
Paris, France

Supervisor
Prof. Frédéric Grillot
Télécom Paris
Institut Polytechnique de Paris
Paris, France

Center for High Technology Materials
University of New-Mexico
Albuquerque, NM, USA

ISSN 2190-5053 ISSN 2190-5061 (electronic)
Springer Theses
ISBN 978-3-031-66817-3 ISBN 978-3-031-66818-0 (eBook)
https://doi.org/10.1007/978-3-031-66818-0

This Springer imprint is published by the registered company Springer Nature Switzerland AG
The registered company address is: Gewerbestrasse 11, 6330 Cham, Switzerland

Supervisor's Foreword

I am very honored that Springer is publishing Zhao's thesis. His work focuses on a theoretical study of the physics of quantum dot lasers (QDs) and interband cascade lasers (ICLs), which are quantum-confined devices poised to gain importance in communications and sensing.

From a general point of view, his manuscript explores various aspects of the physics of these lasers, including noise properties from a classical and quantum mechanical point of view, nonlinear dynamics in different feedback scenarios and potential applications in chaotic communications or on-chip integration. Quantum mechanical fluctuations in semiconductor lasers are deeply examined to assess the potential of these lasers for photon number squeezing. Zhao's work not only develops theoretical frameworks but also compares them with experimental results, offering a holistic view of the field's challenges and opportunities. As mentioned above, part of his work provides a detailed analysis of the amplitude and frequency noise characterization performed with rate equations, which allow an in-depth analysis of the static and dynamic characteristics of epitaxial QD lasers on silicon, in particular by taking into account the effects of amplitude-phase coupling (alpha factor) and Shockley-Read-Hall recombination. His analysis fully confirms why and how QD devices are less sensitive to optical feedback, a critical issue for on-chip silicon integration. A similar analysis carried out for ICLs reveals substantial modulation bandwidths, making them prime candidates for free-space optical communications in the mid-infrared. The chapter also highlights their strong potential for chaos communications. Finally, the manuscript demonstrates how controlled optoelectronic feedback applied to QD lasers produces rich dynamics that can be used for optical computing applications.

Finally, the fully quantum-mechanical description aims to explore the potential for photon number squeezing in QD lasers and ICLs. The discussion is extensive and covers general aspects of squeezing to show the intrinsic advantages, particularly on the experimental and technological side, that semiconductor lasers can bring. It also highlights the reasons why little attention has been paid to their introduction into quantum communications and quantum computing, which is due to the need to work at cryogenic temperatures to achieve sufficient squeezing levels. In his thesis, Zhao showed promising properties of ICLs and QD lasers through a fully stochastic numerical investigation of noise squeezing in these devices. With the latter, he experimentally demonstrated a room temperature noise reduction close to 1 dB (without noise correction) below the quantum limit, with numerical models supporting the experimental data. This first experimental observation of photon number squeezing in QD lasers confirms the strong potential of such devices for future quantum applications.

To sum up, Zhao's thesis is a very strong theoretical contribution that clearly contains many breakthroughs in key areas of photonics. Recognized by multiple publications, his thesis significantly advances the fields of semiconductor laser and quantum theory. I am proud that Zhao's thesis will enrich Springer's prestigious series, and I am certain that it will attract international interest. Congratulations to Dr. Zhao!

Palaiseau/Paris, France
May 2024

Prof. Frédéric Grillot

Abstract

Semiconductor lasers have become ubiquitous in both scientific research and engineering applications, with significant strides in miniaturization since their initial demonstration in 1960. Notably, two prominent advancements in this field are quantum dot (QD) lasers, which operate in the near-infrared wavelength range, and interband cascade lasers (ICLs) designed for mid-infrared applications. In the current landscape of optoelectronics, photonic integrated circuits (PICs) play a pivotal and far-reaching role. They offer unparalleled scalability, reduced weight, cost-effectiveness, and energy efficiency. They achieve this by facilitating the fabrication of complete optical systems through versatile building blocks seamlessly integrated onto a single chip. In this context, direct epitaxial growth of III–V materials on silicon presents a compelling approach for coherent laser source development. QD lasers, with their ultimate three-dimensional carrier confinement, high thermal stability, and robust tolerance for epitaxial defects, emerge as promising candidates for on-chip laser sources. Similarly, ICLs also prove well suited for integration into silicon, making them ideal for compact chemical sensing systems. Noise considerations are of paramount importance when assessing the quality and reliability of this technology. Achieving the shot noise limit and the Schawlow-Townes linewidth has long been recognized as significant milestones. To address noise issues, a range of noise reduction techniques have been explored, encompassing passive optical feedback within an external cavity and active electronic feedback mechanisms to compensate for injection current fluctuations. However, while feedback systems can mitigate laser noise, they can also introduce more intricate nonlinear dynamics, giving rise to phenomena like periodic oscillation, square-wave oscillation, and chaos. The first part of this thesis involves an in-depth investigation into noise and dynamics in two distinct laser types. QD lasers are found to exhibit a high degree of robustness when exposed to parasitic optical reflections but manifest increased sensitivity to optoelectronic feedback. In contrast, ICLs display a spectrum of dynamic behaviours when subjected to optical feedback. Furthermore, recent advancements in low-noise pumping circuits for semiconductor lasers have led to generating amplitude-squeezed states of light. This represents a transition from classical noise to quantum noise, opening up new possibilities in the field of laser technology and quantum optics. The second part of this thesis delves into the phenomenon of amplitude squeezing in both QD lasers and ICLs. The findings indicate that both types of lasers can show broad-band squeezing bandwidth and a significant level of squeezing. All these outcomes in this study contribute to a deeper comprehension of the characteristics of QD lasers and ICLs, laying the groundwork for the development of high-performance classical and quantum emitters on PICs in the future.

Publications related to this thesis

Journals

[1] **Shiyuan Zhao**†, Shihao Ding†, Heming Huang, Isabelle Zaquine, Nicolas Fabre, Nadia Belabas, and Frederic Grillot, "Broadband amplitude squeezing at room temperature in electrically driven quantum dot lasers," *Physical Review Research*, May 2024.

[2] **Shiyuan Zhao** and Frédéric Grillot, "Stochastic model of sub-Poissonian quantum light in interband cascade laser," *Physical Review Applied*, December 2022.

[3] **Shiyuan Zhao** and Frédéric Grillot, "Modeling of Amplitude Squeezing in a Pump-Noise-Suppressed Interband Cascade Laser," *IEEE Photonics Journal*, June 2022.

[4] Yu Deng, Zhuo-Fei Fan, Bin-Bin Zhao, Xing-Guang Wang, **Shiyuan Zhao**, Jiagui Wu, Frédéric Grillot, and Cheng Wang, "Mid-infrared hyperchaos of interband cascade lasers," *Light: Science and Applications*, January 2022.

[5] **Shiyuan Zhao** and Frédéric Grillot, "Effect of Shockley-Read-Hall recombination on the static and dynamical characteristics of epitaxial quantum-dot lasers on silicon," *Physical Review A*, June 2021.

Acknowledgements

I would like to express my deepest gratitude to Prof. Frédéric Grillot for his invaluable guidance and unwavering support throughout my academic journey. He introduced me to the captivating and rewarding realm of the classical and quantum theory of lasers, serving as both my mentor and supervisor. Even before that, his second-year Master's lectures in laser physics were my initial formal exposure to the subject, cementing my determination to pursue a career as a semiconductor laser engineer. During the course of my Ph.D., he has consistently demonstrated his approachability and willingness to assist me in overcoming academic challenges. I hold in high regard his competence to maintain a comprehensive perspective on complex issues. When I encountered obstacles in generalizing certain results, his unbounded confidence in our capacity to find solutions proved prescient time and again. I am profoundly thankful for the priceless advice and encouragement he bestowed upon me, which greatly contributed to my growth and the successful completion of my doctoral research.

Beyond the confines of my department, I extend my heartfelt gratitude to Prof. John Bowers, Dr. Weng W. Chow, Prof. Cheng Wang, and Dr. Nadia Belabas for their enlightening discussions during our meeting in Sicily in July 2023. The ideas shared during that encounter have been a constant source of inspiration, and I value the communication we have had since then. I am also deeply indebted to the members of my supervisory committee, especially Prof. Gian Luca Lippi, Prof. Marc Vallet, and Prof. Jesper Mørk. Their critical feedback and insightful comments have played an instrumental role in shaping my research. Their guidance has been constructive in refining my research. I would like to acknowledge the useful discussions I've had with Prof. Isabelle Zaquine, Dr. Elie Awwad, and Dr. Nicolas Fabre within the GTO group of Télécom Paris.

Sincere appreciation to my senior colleagues, Dr. Jianan Duan, Dr. Heming Huang, Dr. Olivier Spitz, and Dr. Bozhang Dong for their generous allocation of time, wealth of expertise, and substantial assistance. Special thanks to individuals who have been precious contributors through their camaraderie and thought-provoking discussions. These include Shihao Ding, Pierre Didier, Thibaut Renaud, Di Cui, Janik Ehlert, Sara Zaminga, and Hyunah Kim, whose willingness to collaborate and engage has greatly enriched my experience. Furthermore, to my fellow students, Jingtian Liu, Yibo Wang, Xiaolin Wang, Yi Liu, Yao Yuan, Yinghao Wang, Louis Tomczyk, and Peter Nwakamma at Télécom Paris and Institut Polytechnique de Paris. Your diverse perspectives and interdisciplinary views have expanded my horizons.

Last but by no means least, I express my gratitude to my parents and my girlfriend, Xinyue Ma, who were always there to offer a listening ear, a word of advice, or a much-needed distraction from my work. Their firm love, unshakable belief in my abilities, patience during challenging times, and jubilant celebrations during moments of triumph have served as the sturdy pillars of strength that have sustained me.

As I draw this thesis to a close, I want to emphasize that its completion does not mark the culmination of my academic endeavours, but rather, it signifies the commencement of a new chapter. As I embark on the next phase of my career, I carry with me the wisdom, knowledge, and support that have shaped the work presented here. Thank you all for being a part of this journey.

Contents

1 Introduction 1
- 1.1 60 Years of Lasers: Pump Up the Light 1
  - 1.1.1 The Miniaturization of Semiconductor Lasers: From Bulk to Dot 2
  - 1.1.2 Mid-Infrared Cascade Lasers 6
  - 1.1.3 Towards Fully Integrated Photonics 7
  - 1.1.4 Towards Quantum Photonics 10
- 1.2 Motivation of the Dissertation 12
- 1.3 Organization of the Dissertation 12
- References 14

2 Semiconductor Lasers: Fundamentals and Theory 19
- 2.1 Fundamentals of Quantum-Confined Lasers 19
  - 2.1.1 Quantum Dot Lasers 20
  - 2.1.2 Quantum Dot Lasers on Silicon 22
  - 2.1.3 Interband Cascade Lasers 25
- 2.2 Theory of Quantum-Confined Lasers 26
  - 2.2.1 Semiclassical Description: Langevin Rate Equations 28
  - 2.2.2 Stochastic Description: Gillespie Algorithm 33
  - 2.2.3 Fully Quantum-Mechanical Description 39
- References 44

3 Classical Noise and Feedback Dynamics in Quantum-Confined Lasers 49
- 3.1 Introduction to the Classical Noise 50
  - 3.1.1 Relative Intensity Noise 52
  - 3.1.2 Frequency Noise 55
  - 3.1.3 Linewidth Enhancement Factor (α_H) 61
  - 3.1.4 Correlation Between Intensity Noise and Frequency Noise 64

3.2 Theoretical Investigation of the Classical Noise ... 66
3.2.1 Quantum Dot Lasers ... 67
3.2.2 Quantum Dot Lasers on Silicon ... 75
3.2.3 Interband Cascade Lasers ... 86
3.3 Introduction to the Time-Delayed Feedback on Semiconductor Lasers ... 91
3.3.1 External Optical Feedback: Lang–Kobayashi Model ... 93
3.3.2 Optoelectronic Feedback: Ikeda-Like Model ... 99
3.4 Investigation of the Time-Delayed Feedback on Quantum-Confined Lasers ... 101
3.4.1 Optical Feedback Insensitivity in Quantum Dot Lasers on Silicon ... 101
3.4.2 Broadband Chaos Generation in Interband Cascade Lasers ... 108
3.4.3 Optoelectronic Feedback in Quantum Dot Lasers on Silicon ... 116
References ... 124

4 Quantum Fluctuations and Amplitude Squeezing in Quantum-Confined Lasers ... 131
4.1 Introduction to the Quantum Fluctuations ... 131
4.1.1 Coherent State of Light ... 133
4.1.2 Squeezed States of Light ... 137
4.1.3 Measurement of Quantum States of Light ... 141
4.1.4 Quantum Noise from Electronics ... 145
4.1.5 Generation of Quantum States of Light ... 146
4.2 Introduction to Amplitude Squeezing in Semiconductor Lasers ... 150
4.2.1 Suppressed-Pump-Noise Mechanism ... 152
4.2.2 A Noise Equivalent Circuit of the Semiconductor Laser ... 154
4.2.3 Limitation of Squeezing Level ... 157
4.2.4 Limitation of Squeezing Bandwidth ... 158
4.3 Amplitude Squeezing in Interband Cascade Lasers ... 159
4.3.1 Semiclassical Langevin Rate Equation ... 160
4.3.2 Stochastic Gillespie Algorithm ... 171
4.4 Amplitude Squeezing in Quantum Dot Lasers ... 180
4.4.1 Experimental Evidence ... 181
4.4.2 Theoretical Investigation ... 186
4.4.3 Data Correction ... 192
References ... 194

5 Conclusion and Perspectives ... 201
References ... 211

Appendix A: Simulation Parameters 215

Appendix B: Small-Signal Analysis of Quantum-Confined Lasers 221

Appendix C: Density Operator and Phase-Space Distribution in a (Very Small) Nutshell 229

Appendix D: Balanced Homodyne Detection 235

Abbreviations

CW	Continuous-Wave
DFB	Distributed FeedBack lasers
EOF	External Optical Feedback
ES	Excited States
FN	Frequency Noise
FP	Fabry-Perot lasers
GS	Ground States
ICL	Interband Cascade Lasers
LFF	Low-Frequency Fluctuations
LK	Lang-Kobayashi
LO	Local Oscillator
LWIR	Long-Wavelength InfRared
MWIR	Mid-Wavelength InfRared
NIR	Near-InfRared
ODE	Ordinary Differential Equation
OEF	Opto-Electronic Feedback
OEO	Opto-Electronic Oscillators
PIC	Photonic Integrated Circuit
QCL	Quantum Cascade Lasers
QD	Quantum Dot
QW	Quantum Well
RIN	Relative Intensity Noise
RS	Reservoir States
SNL	Shot Noise Limit
SOI	Silicon-On-Insulator
SQL	Standard Quantum Limit
SWIR	Short-Wavelength InfRared

Chapter 1
Introduction

1.1 60 Years of Lasers: Pump Up the Light

In 2020, the laser, an acronym for "Light Amplification by Stimulated Emission of Radiation" celebrated its 60th anniversary. The concept of stimulated emission, which forms the basis of lasers, was initially proposed by Einstein [1]. Einstein's theory laid the foundation for understanding how electromagnetic fields could be amplified through population inversion. Rudolf Ladenburg's work in 1928 [2] provided indirect evidence of stimulated emission, although, at the time, physicists referred to this effect as "negative absorption." In 1951, Charles H. Townes took a significant conceptual leap by proposing that stimulated emission at microwave frequencies could lead to oscillations within a resonant cavity, ultimately producing coherent output. Then, in 1954, Charles H. Townes and his student James P. Gordon demonstrated the first microwave maser. They achieved this by directing excited ammonia molecules into a resonant cavity where these molecules oscillated at a frequency of 24 GHz [3]. The quest to create a laser was underway. However, two critical questions remained unanswered: first, how to generate a population inversion, and second, what to employ as the active medium for the laser. Then, on July 8, 1960, a brief yet historic report titled "Light Amplification Claimed by Scientist" appeared in the lower-left corner of the front page of the New York Times. This public announcement revealed that Theodore H. Maiman had successfully demonstrated the first operational Ruby laser [4]. Astonishingly, despite the groundbreaking nature of this discovery, it was summarily rejected by *Physical Review Letters* upon its initial submission. The Ruby laser left a profound impression on many laser researchers, and while it stunned some, it also served as inspiration for exploring alternative approaches. On December 12, 1960, Donald Herriott, Ali Javan, and William Bennett achieved a milestone by creating the first continuous-wave helium-neon gas laser [5]. In 1962, Elias Snitzer successfully demonstrated the first neodymium-glass fiber laser [6]. Meanwhile, Joseph E. Geusic, H. W. Marcos, and LeGrand van

S. Zhao, *Noise, Dynamics and Squeezed Light in Quantum Dot and Interband Cascade Lasers*, Springer Theses, https://doi.org/10.1007/978-3-031-66818-0_1

Uiter were actively testing Nd-YAG for lasing, which would become the dominant solid-state laser [7].

The concept of producing stimulated emission in semiconductors had been considered by John von Neumann as early as 1953. However, his proposal remained unpublished until nearly three decades after his death [8]. Independent proposals for achieving population inversions in semiconductors were also put forth by Pierre Aigrain of the École Normale Supérieure in France and Nikolai Basov of the Lebedev Physics Institute in Moscow in the late 1950s. Unfortunately, due to the lack of clarity in these early ideas, it required studies of light emission at p-n junctions to initiate the development of the semiconductor diode laser. The observation of light emission from semiconductor junctions dates back to 1907 when Henry J. Round first observed this phenomenon. However, it received little attention until the invention of the transistor, which sparked research on III–V compounds. Rubin Braunstein made an observation of light emission from junctions in gallium arsenide, indium phosphide, and indium antimonide in 1955 [9]. This discovery hinted at the potential of III–V junctions as laser candidates. However, their observed emission efficiency was relatively low at the time, and the full significance of direct bandgaps had not yet been fully understood. As a result, progress in the development of semiconductor lasers initially proceeded at a gradual pace.

In the fall of 1962, Hall [10] and other research groups reported the successful development of p-n homojunction semiconductor lasers based on GaAs, as depicted in Fig. 1.1a. Later, in 1963, Herbert Kroemer, along with Zhores Alferov and Rudolf Kazarinov (for details, refer to The Nobel Prize in Physics 2000), came up with the idea of the p-i-n double heterostructure laser diode. This innovative design involves sandwiching a thin slab of gain material between p- and n-type cladding layers, characterized by a smaller refractive index but a higher conduction-valence band gap, as illustrated in Fig. 1.1b. This arrangement forms a transverse potential well that localizes the injected electrons, thus maximizing their interaction for recombination. Moreover, the large refractive index discontinuity results in a highly focused and intense optical transverse energy profile. The advent of the bulk double heterostructure design, which harnessed both carrier and optical confinement effects, rendered semiconductor lasers truly practical from that point onward. A more detailed history of the laser can be found in Ref. [11].

1.1.1 The Miniaturization of Semiconductor Lasers: From Bulk to Dot

The introduction of double heterostructure represented a game changer in the development of semiconductor lasers. In 1970, Leo Esaki and Ray Tsu initiated research on synthesized semiconductor superlattices [12]. These superlattices are one-dimensional periodic structures composed of alternating layers with thicknesses

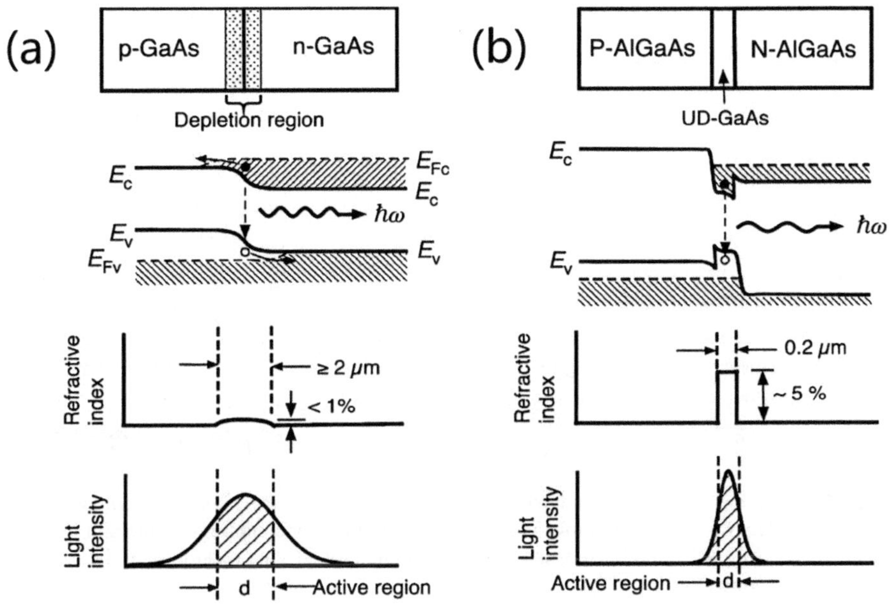

Fig. 1.1 Aspects of **a** p-n homojunction laser and **b** p-i-n double heterostructure laser. (*Source* In the textbook of *III–V Compound Semiconductors and Devices*, written by Keh Yung Cheng)

smaller than the electron mean free path. This concept allowed for the engineering of desirable electronic and optical properties through quantum confinement in narrow-gap layers, which could be as thin as a few nanometers, comparable to the de Broglie wavelength. In 1974, Raymond Dingle, W. Weigmann, and Charles H. Henry conducted a pioneering discovery on quantum wells (QW) [13], which are two-dimensional thin-film planes (less than 500 Ångströms in thickness) that effectively confine electron motion. The two-dimensional nature of QW gave rise to new features that were not observed in bulk materials. such as optical absorption and gain spectra peculiar to the step-like density of states, leading to increased efficiency and reduced threshold currents. QW lasers have since become a standard in optical telecommunication and networks. Going beyond QWs, the confinement of electron motion in an additional dimension has led to the quantum wires or dashes [14]. In 1982, Arakawa and Sakaki [15] proposed the concept of ultimate quantum confinement in all three dimensions, originally referring to the nanostructures as "three-dimensional quantum wells" or "quantum boxes," rather than quantum dots (QD). By assimilating real QD gain medium with QW laser, theories forecast low threshold, high thermal stability and enhanced dynamic properties in these lasers. As material dimensions diminish, the density of states for electrons becomes perfectly discretized in the three directions, leading to an atom-like energy spectrum shown in Fig. 1.2. Consequently, the concentration of most injected nonequilibrium carriers tends to localize in an increasingly narrow energy range near the bottom of the conduction

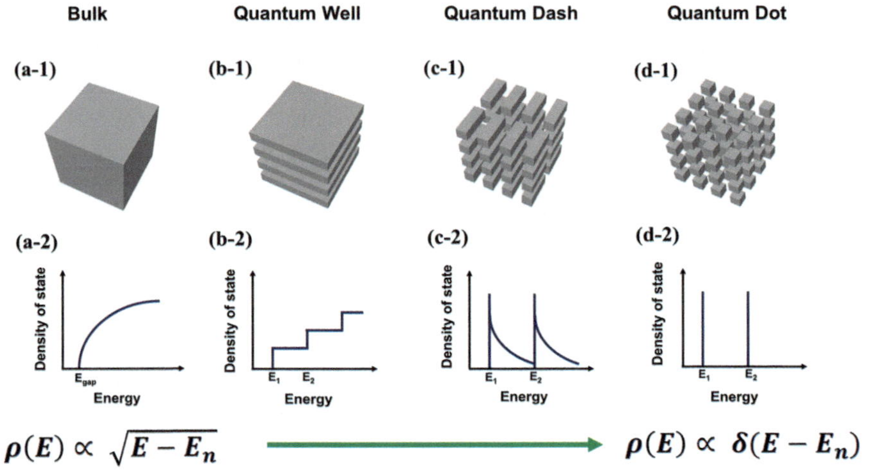

Fig. 1.2 Evolution of the semiconductor nanostructures in active layer for lasers. In bulk semiconductors, electrons are free to move in all three dimensions. By confining electrons in QD with quantized energy levels, all electrons are populated at the ground state regardless of the nature of the fermion. (*Source* In Ref. [17])

band and/or the top of the valence band. Recombination, absorption, and gain consistently exhibit excitonic (or biexcitonic) behaviour, thereby enhancing the maximum material gain. Furthermore, even if the separation between the lowest size quantization level exceeds several kT, no thermal or excitation dependent broadening of the gain function is possible, further diminishing the impact of temperature on the device's performance [16].

Despite the potential advantages of QD lasers, several negative predictions were also provided. In 1988, Vahala introduced the inhomogeneous size distribution of QDs into consideration [18]. Benisty proposed that the lack of matching energies for phonon relaxation of carriers in QDs ('the phonon bottleneck effect in a QD') will inhibit lasing in QD structures [19]. All discussions on QD lasers had merely academic character, and the practical realization of QD lasers was seen as a challenging task until the Stranski–Krastanov growth mode was used to create self-assembled dots. The first QD laser was demonstrated by teams led by D. Bimberg and Zh. I. Alferov in 1994, with the gain medium containing a single dot layer developed via molecular beam epitaxy (MBE) [20]. While strain-induced self-assembled MBE remained the mainstream growth method initially, Metal-Organic Chemical Vapor Deposit (MOCVD) has quickly proven effective in developing InAs/GaAs QD lasers as well [21, 22]. However, high-quality QD lasers are typically fabricated using the MBE system, which has been established as an ideal method for this purpose.

The fabrication of QD lasers typically requires growing multiple stacks of dots in the active region to increase material gain [23]. However, this process necessitates careful optimization because, as the number of stacks increases, strain accumulates in each dot layer. Consequently, a thicker spacer in the structure is required, which reduces the contribution to the gain and slows down carrier dynamics. For InAs/GaAs QD lasers, the maximum number of dot layers is usually limited to 10–15. In 1996, a better-optimized structure was demonstrated with three dot layers and a very low room temperature threshold current density at 62 A/cm^2, a characteristic temperature of 385 up to 300 K, and a large material gain of $10^5\,cm^{-1}$ [24]. Two years later, further breakthroughs were unveiled owing to the development of the "Dot-in-a-well" (DWELL) structure [25]. Such a structure, in which the QDs are embedded into QWs, allows for much better capture and localization of electrons into the QDs and speeds up carrier dynamics, as well as improves material gain owing to a larger dot density up to $7.5 \times 10^{10}\,cm^{-2}$. In addition, with advancements in MOCVD growth, InAs/GaAs QD lasers have been obtained with ultra-low transparency current density (<6 A/cm^2 per dot sheet) and internal losses ($\approx$1.5 cm^{-1}) [26]. Over time, more improvements have been obtained in InAs/GaAs QD lasers [27], for instance, threshold current densities down to 8.8 A/cm^2 [28], negative characteristic temperature [29], high output power [26], as well as prolonged device lifetime. Furthermore, exceptional thermal stability has been confirmed in GaAs-based QDs, experiments demonstrated 10 Gbit/s transmission from 20 to 70 °C [30], as well as sustained CW operation at an extreme temperature of 220 °C, which is the highest recorded operating temperature among any kind of semiconductor lasers [31]. In the current market, mature commercialized GaAs-based QD laser products, operating around 1.3 μm, are readily available for telecommunication applications from various companies such as Innolume GmbH in Germany, Quintessent in America, and QD Laser Inc in Japan. The first InAs/InP QD laser was demonstrated in 1998, initially with lasing operation limited to 77 K, and later achieved lasing at room temperature [32]. Although the InAs/InP and InAs/GaAs systems share the same material in the dots, they remain quite different from each other, mostly because the lattice mismatch in InAs/InP (3%) is smaller than that in InAs/GaAs (7%). Therefore, the challenge regarding dot formation on InP substrates is to address the lattice mismatch and the complex strain distribution, which tends to produce elongated dots, known as quantum dash, a class of self-assembled nanostructures that exhibits mixed properties between quantum wire and QD. Therefore, applications of InP-based QD lasers, operating around 1.5 μm, remain limited due to the complex fabrication process, which is not highly compatible with industrial standards. On the other hand, the best dynamic performance ever achieved at 1.5 μm reported a modulation bandwidth as large as 17.5 GHz on an InAs/InP(100) QD laser [33]. Instead, for 1.3 μm InAs/GaAs QD lasers grown on GaAs substrate, a record bandwidth of 13.1 GHz was reported in Ref. [34] by reducing the GaAs barrier thickness. As of today, the modulation capabilities of QD lasers still fall below expectations when compared with their QW counterparts. However, there is still ample room for progress.

The Nobel Prize in Chemistry 2023 has been awarded to Moungi G. Bawendi, Louis E. Brus, and Alexei I. Ekimov for their discovery and development of quantum dots (for details, refer to The Nobel Prize in Chemistry 2023).

1.1.2 Mid-Infrared Cascade Lasers

Considerable attention from researchers has been dedicated to the development of reliable, compact, high-power, and cost-effective light sources tailored for various applications within the mid-infrared spectrum. These applications include free-space communications, medical diagnostics, infrared countermeasures, atmospheric pollution monitoring, infrared radar, etc. Among the promising candidates for these applications, semiconductor lasers have risen to prominence due to their compact size and affordability. Moreover, they offer a wide wavelength tuning range, narrow linewidth, and high-speed response, making them exceptionally versatile. Conventional laser diodes operate through optical transitions between the conduction and valence bands of semiconductor materials. The lasing frequency ν is determined by the energy gap E_g between these two bands. Nonetheless, the attainable wavelengths of these laser diodes are constrained to the near-infrared (NIR) range, below $2\,\mu m$, due to the bandgap limitation inherent in III–V materials. To overcome this limitation and broaden the lasing spectra, covering the short-wavelength infrared (SWIR, 1.4–$3\,\mu m$), mid-wavelength infrared (MWIR, 3–$8\,\mu m$), and long-wavelength infrared (LWIR, 8–$15\,\mu m$) ranges, two possible solutions have been worked on.

The first attempt was the unipolar $Ga_xIn_{1-x}As/Al_xIn_{1-x}As$ quantum cascade lasers (QCL) that are based on intersubband transitions ($\approx$1 ps) within the conduction band of the III–V semiconductor [35]. Therefore, the emission wavelength is no longer constrained by the material bandgap but rather by the energy spacing between the subbands, which is altered through the quantum engineering of the multiple-quantum-well (MQW) active region. A cascading effect has been incorporated into this structure to improve efficiency, as depicted in Fig. 1.3a. Each electron traverses through multiple active regions via the intraband tunnelling effect in this process, leading to photon emission at each step. However, the fast nonradiative relaxation between subbands via optical-phonon scattering limits the device's performance. Recently, the research team led by Frédéric Grillot successfully established a long-range communication link employing a commercial 9-μm QCL, an external modulator based on the Stark effect, and two types of detectors: an uncooled quantum cascade detector and a nitrogen-cooled quantum well infrared photodetector. Initially, the maximum data rate of the link was evaluated in a back-to-back configuration. Subsequently, a Herriott cell was introduced to extend the optical path length to 31 m. Applying pulse shaping, pre-processing, and post-processing techniques, an impressive record-breaking data transfer rate of $30\,Gbit\,s^{-1}$ was achieved for both two-level (OOK) and four-level (PAM-4) modulation schemes [36].

To improve the performance of MWIR lasers, especially those using antimony (Sb), a second solution was proposed: bipolar $InAs/In_xGa_{1-x}Sb$ and $InAs/In_xAs_{1-x}Sb$ interband cascade lasers (ICLs), as introduced by R. Q. Yang shortly after the invention of QCL [37]. These ICLs combine a relatively long upper-level lifetime of interband transitions ($\approx$1 ns) with the voltage-efficient cascading scheme originally introduced for QCL. The advantage of cascade injection and the ability to tailor the emission wavelength is thus retained. In contrast to

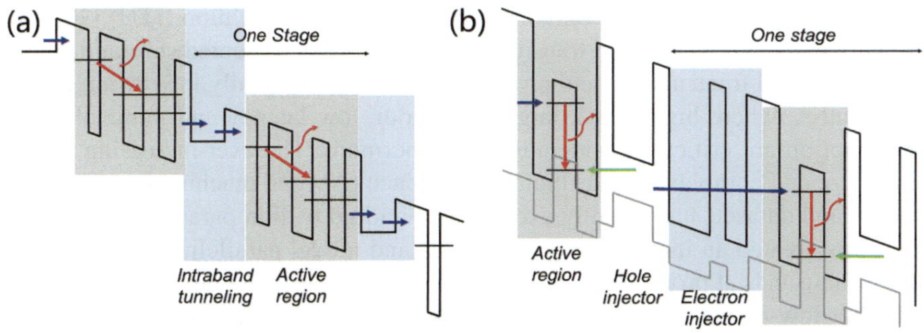

Fig. 1.3 Comparision of band structure in **a** quantum cascade lasers, and **b** interband cascade lasers

QCL, this design allows for both electrons and holes to be present in each stage of the ICL's active region [38], as depicted in Fig. 1.3b, which minimizes voltage and current requirements by reducing Auger coefficients and suppressing phonon scattering. As a result, ICLs exhibit high radiative efficiency and can achieve low threshold currents. Due to these advantages, ICLs have gained increasing popularity for portable mid-infrared applications, including their utilization in the methane detector aboard the NASA Curiosity rover. In a more recent development, Frédéric Grillot's research group demonstrated a full interband cascade system designed for high-speed data transmission at a wavelength of 4.18 μm. This proof-of-concept experiment employed a multimode ICL and an interband cascade infrared photodetector based on a type-II InAs/GaSb superlattice. The direct-modulated system's bandwidth was evaluated to be approximately 1.5 GHz, enabling the achievement of data rates up to 12 Gbit/s using on-off keying and 14 Gbit/s with a 4-level pulse amplitude modulation scheme [39].

1.1.3 Towards Fully Integrated Photonics

In the early stages of the photonics market, different material systems were used, including compound semiconductors like indium phosphide (InP) and gallium arsenide (GaAs), elementary semiconductors such as silicon (Si) and germanium (Ge), silica and rare-earth-doped glasses like glass fibre, and polymers. Each of these systems caters to specific applications or components. At that time, photonic integration was largely dependent on native III–V material substrates, which marked the first generation of commercially viable photonic technologies. However, the scalability of this technique was severely limited due to the diverse optical functionalities relying on a variety of materials, making fabrication prohibitively expensive. The rapid expansion of the commercial electronics industry, driven by the increased functionality of smartphones, cloud services, and Internet of Things devices, has ignited explosive growth in global Internet traffic over the past decades, according to recent estimates from the International Telecommunication Union (ITU) [40]. The surge in demand for increased transmission capacity has presented substantial challenges to signal transmission and processing systems, especially in very large-scale data centres, where high speed, large bandwidth, low latency and minimal power consumption are critical considerations. Furthermore, The ever-increasing growth in the size and complexity of artificial intelligence (AI) and machine learning (ML) models and datasets has recently exceeded a hundred trillion parameters [41]. This surge has created an imperative need for data and model parallelization distributed across thousands of processing and memory nodes. These nodes demand low-latency and ultra-low-power multi-terabit-per-second (Tb/s) inter/intra-rack optical I/O, as well as chip-to-chip interconnects. For instance, the latest Nvidia DGX H100 system for ML training consists of eight GPUs, each requiring a 7.2 Tb/s bidirectional off-package bandwidth [42]. To address this demand, multiple interconnect technologies such as Compute Express Link (CXL) and NVLink have emerged to deliver high bandwidth, low latency connectivity between processors, network switches, and controllers. However, the current implementation of chip-to-chip links relies on copper-based electrical links, which fall short of meeting the stringent speed, energy efficiency, and bandwidth density requirements of future distributed AI computing.

Silicon photonics, under these circumstances, has experienced enormous growth. Figure 1.4 depicts the progression of datacom transceivers across five generations of technology. Utilizing silicon photonics, datacom transceivers have swiftly transitioned from initial product introductions to achieving multi-million units per year. With data rates scaling from 100 Gbps to Tbps, integrated electronics play a critical role in reducing both system-level power consumption and device-level capacitance in connecting transmitter modulators and receiver photodetectors. Over the past decade, target power requirements have decreased from thousands of pJ/bit to sub-pJ/bit. A pivotal component of the interconnect, the semiconductor laser, is expected to transition from disaggregated architectures to integration on the photonic

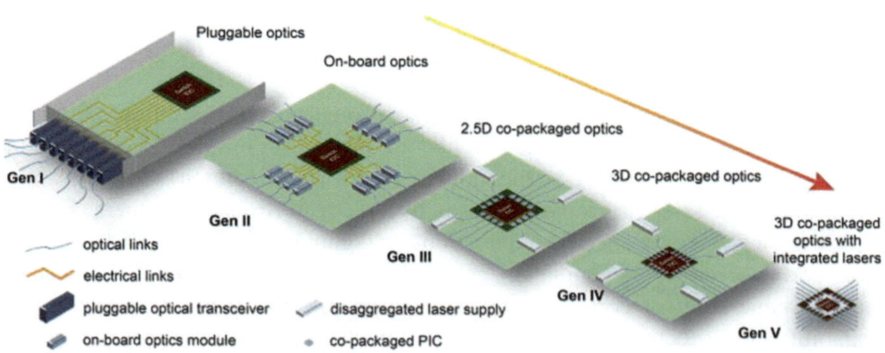

Fig. 1.4 Evolution of optical technologies across generations and their application in data centres. This progression leads to increasingly integrated network switching and optical I/O, potentially culminating in 3D co-packaged optics featuring integrated lasers. (*Source* In Ref. [43])

integrated circuit (PIC). This advancement facilitates scaling from one laser per PIC to potentially thousands of lasers per PIC with wavelength division multiplexing (WDM), enabling Pbps-scale data connections. Simultaneously, PICs are evolving from pluggable transceivers at the periphery of the board to co-packaged optics and electronics, and eventually to 3D integrated PICs and electrical integrated circuits. Esteemed electronics companies such as Intel, IBM, Hewlett Packard, STMicro-electronics, IMEC, Broadcom, Nvdiva, Huawei and Nokia have collaborated with research institutes worldwide, receiving support from government, industry, and academia to drive advancements in Silicon photonics [44–47]. This collaborative effort has contributed to the extensive adoption of silicon photonic transceivers, which offer a promising solution to this challenge by integrating optical I/O directly with high-performance CPUs and GPUs. Recently, co-packaged optics have emerged as an integral component of network top-of-the-rack (ToR) switches and GPUs, as announced by Broadcom and Nvidia. This technology integrates both optically active and passive components onto a single chip [48, 49] including lasers, semiconductor optical amplifiers, modulators, and photodetectors as illustrated in Fig. 1.5. It is anticipated to achieve bandwidths surpassing 10 Tb/s with sub-pJ/bit energy efficiency by leveraging mature complementary metal-oxide-semiconductor (CMOS) technology [50].

Considerable strides have been achieved over the past few decades in developing critical components for silicon-based PICs, including high-performance modulators, photodetectors, and waveguides. Nonetheless, creating stable light sources on silicon remains an ongoing challenge. Generally, the prerequisites for on-chip light sources are as follows [51]: (1) Electrically pumped, continuous-wave (CW) lasing; (2) Low

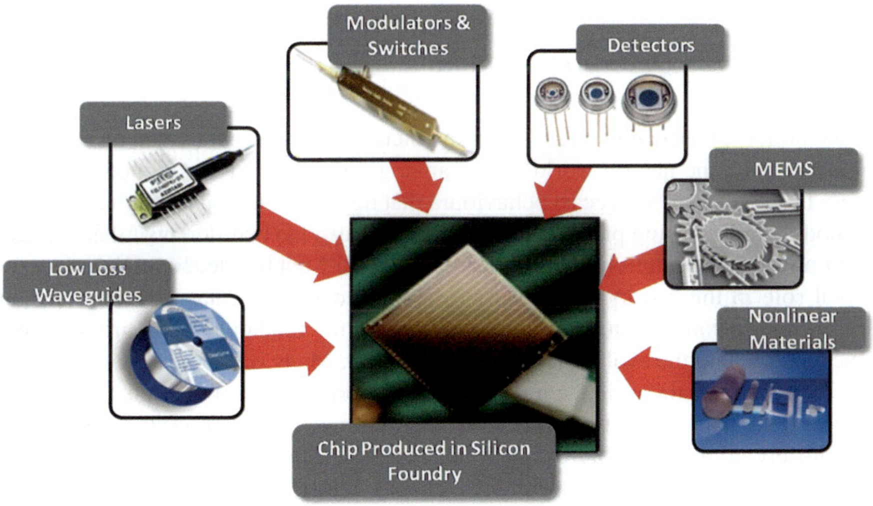

Fig. 1.5 The trend to put everything on silicon. (*Source* In the plenary talk of John E. Bowers, CLEO 2020)

Energy Consumption and High Output Power; (3) Operation in Telecom and Datacom Wavelength Bands; (4) Direct Integration onto silicon Platform. Unfortunately, directly realizing a silicon laser has long been regarded as the "Holy Grail". This challenge arises from the intrinsic properties of the two commonly used materials (Si and Ge), both of which are characterized as indirect bandgap materials [52], leading to inefficient light emission. Despite extensive efforts to achieve group-IV lasers using various methods such as Si Raman lasers, band-engineered Ge lasers, and (Si) GeSn-based lasers, they have been plagued by high threshold current and low internal quantum efficiency, rendering them impractical as on-chip light sources. These limitations have, in turn, hindered them from meeting the requirements mentioned earlier.

In stark contrast, integrating III–V materials onto silicon substrates has emerged as a transformative solution. This integration profits vastly from the respective strengths of III–V materials, characterized by their superior optical properties, and silicon, known for its large wafer size, cost-effectiveness, and well-established CMOS foundry infrastructures. Furthermore, low propagation loss has been another critical factor in propelling the evolution of integrated photonics. Silicon-on-insulator (SOI) waveguides exhibit propagation losses that are approximately an order of magnitude lower than their III–V counterparts. This property equips silicon-based PICs with the capability to accommodate a greater number of individual components and support more complex photonic systems, while also boosting the performance of passive structures and coherent light sources in the SWIR or even the MWIR or LWIR range [53]. These advantages open up a plethora of new applications such as data centres [54], neural networks [55], frequency comb [56], and quantum information [57].

1.1.4 Towards Quantum Photonics

Throughout the past century, it has become increasingly evident that, at its most fundamental level, the natural world resists analogy with human experience. Quantum theory has consistently forecast behaviours that defy explanation through any classical model, underscoring profound disparities between the operations of the quantum world and the expectations set by our daily lives. Meanwhile, light has played a pivotal role in the history of physics. It has posed fundamental questions and, in turn, furnished solutions to them. The inquiry into the fundamental nature of light is critical in shaping the evolution of quantum theory. Many of the earliest experimental tests aimed at validating the most astonishing predictions of quantum mechanics were initially conducted using photons. Furthermore, many modern technologies rely entirely upon the ability to manipulate and measure electromagnetic radiation across a broad spectrum of wavelengths. Consequently, quantum optics or quantum photonics stands as one of the most dynamic and vibrant fields of research at present with extraordinary breakthroughs. It continually provides insight into the enigmatic behaviour of the quantum world while offering practical applications across a diverse array of technologies [58].

Thanks to huge advancements in fabrication technologies, integrated photonics has evolved into a vital element of numerous classical technologies, and there is growing evidence to suggest that it will be equally important in the development of quantum photonic systems [59–61]. In pursuit of the desired functionality and performance in optical quantum technologies, it becomes imperative to have a deep understanding of semiconductor lasers. This is because semiconductor lasers are the preferred choice for their strong light-matter interaction. The main arguments in favour of this perspective can be succinctly summarized, and they all implicate the necessity of adopting a quantum approach in academic research related to semiconductor lasers.

1. The miniaturization of semiconductor laser cavities has progressed from millimetre-sized resonators to nanocavities [62], which approaches and, in some cases, surpasses the diffraction limit of light in dielectrics. At such minuscule dimensions, the effects of cavity quantum-electrodynamics become the predominant influence.
2. In the microscopic semiconductor laser cavity design, it's possible to leverage quantum effects, such as those seen in ultra-coherent Fano lasers [63, 64]. These lasers operate by using the bound state in the continuum, generated through Fano interference, to effectively suppress quantum fluctuations and enhance their coherence.
3. An alternative approach to achieving the quantum regime involves the use of a harmonic potential cavity, which can produce a large quality (Q) factor, often combined with a passive nano-resonator [65–67]. For instance, in the case of a silicon nitride (Si_3N_4) cavity ($Q > 2.5 \times 10^8$), an electrically pumped semiconductor laser was demonstrated, showing an ultra-narrow linewidth of just 1.2 Hz [68] (Fig. 1.6).

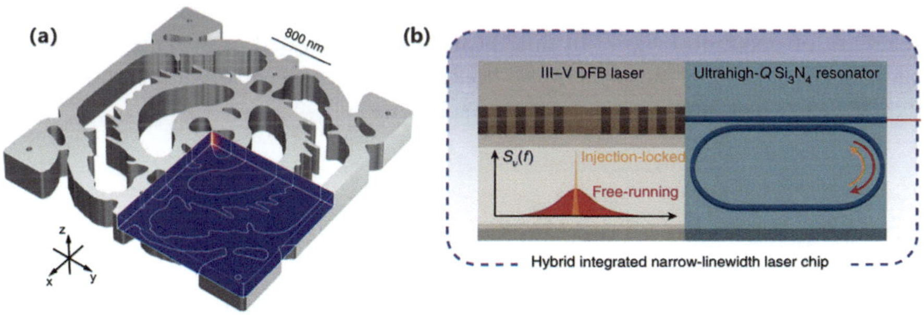

Fig. 1.6 Schematic of **a** a topology-optimized silicon dielectric bowtie cavity, **b** a hybrid integrated narrow-linewidth laser based on the ultra-high-Q Si_3N_4 microresonator. (*Source* In Refs. [62, 68])

1.2 Motivation of the Dissertation

Noise issues usually have a bearing on the quality and reliability of communication applications. Understanding noise, both from a fundamental physics perspective and in terms of practical design considerations, has been of paramount importance since the early days of laser oscillators. Some fundamental phenomena, such as the shot noise limit and the Schawlow–Townes linewidth [69], have been recognized as achievable boundaries in the evolution of semiconductor lasers. Typically, free-running semiconductor lasers exhibit linewidths ranging from several kilohertz to megahertz, largely due to spontaneous emission and phase-amplitude coupling [70]. The wide cavity bandwidths, reaching up to tens of gigahertz, are mainly determined by the dynamics of the gain medium and its interaction with the optical field.

In recent decades, great advances in the comprehension of noise issues in photonics have pushed the performance of semiconductor lasers beyond what was previously assumed to be basic limitations for both intensity noise and linewidth. Many techniques have been suggested to reduce noise, including passive optical feedback in the external cavity and active electronic feedback to compensate for injection current. Additionally, fresh insights into low-noise pumping circuits for semiconductor lasers have led to the generation of amplitude-squeezed light, representing a transition from classical noise to quantum noise. However, while feedback systems can reduce laser noise, they can also trigger more complex nonlinear dynamics, such as periodic oscillation, square-wave oscillation, and optical chaos. Given the ubiquity of semiconductor lasers in commercial settings, abundant promising applications of low-noise or dynamic devices would be highly demanding in different domains.

For this reason, the contemplation of classical and quantum noise associated with the generation and detection of light raises fundamental questions and urges for answers. To name a few, how to measure the laser intensity noise, linewidth and linewidth enhancement factor? How to leverage the nonlinear dynamics of semiconductor lasers? what causes quantum noise? How is quantum noise related to the uncertainty principle? Are vacuum fluctuations important compared with quantum noise? How does the transition to classical behaviour come about when the light fields are of quantum origin? In this thesis, I am motivated to provide a clear and rigorous description of classical noise, nonlinear dynamics and quantum noise, while clarifying my own understanding of this subject.

1.3 Organization of the Dissertation

Based on the aforementioned motivations, this dissertation is organized as follows:

Chapter 2 serves as a reminder of the fundamentals of quantum dot (QD) lasers and interband cascade lasers (ICLs). It delves into their electronic structure and elucidates carrier scattering processes. Additionally, it introduces three distinct modelling approaches: semiclassical Langevin rate equations, the stochastic Gillespie algorithm, and a fully quantum-mechanical description. These modelling techniques are essential for comprehending the underlying physical phenomena in semiconductor lasers. Therefore, this chapter sets the stage for the subsequent experimental and theoretical investigations within the thesis, providing essential background knowledge and modelling tools.

In Chap. 3, the theoretical investigation concentrates on key aspects of QD lasers and ICLs. Firstly, the chapter explores relative intensity noise (RIN), frequency noise (FN), linewidth enhancement factor (α_H), and the correlation between RIN and FN for both QD lasers and ICLs. These investigations are carried out through small-signal analysis, which accounts for contributions from both gain theory and carrier dynamics. Therefore, three well-established rate equation models are developed, namely for QD lasers, QD lasers on silicon, and ICLs, with a focus on studying their RIN and FN noise characteristics. In the second part of this chapter, optical feedback and optoelectronic feedback are considered using the Lang–Kobayashi method and the Ikeda-like method, which are solved using a fourth-order Runge–Kutta algorithm. These techniques are employed to numerically demonstrate several phenomena: optical feedback insensitivity in QD lasers on silicon, the generation of broadband chaos in ICLs, and the emergence of multiple dynamics induced by optoelectronic feedback in QD lasers on silicon. This comprehensive exploration provides valuable insights into the behaviour and characteristics of QD lasers and ICLs, both in terms of their noise properties and their responses to different feedback mechanisms.

In Chap. 4, the focus is on exploring quantum fluctuations and amplitude squeezing in QD lasers and ICLs. The chapter unfolds in several sections: The initial part of the chapter systematically introduces fundamental concepts such as coherent states of light, squeezed states of light, methods for measuring quantum states of light, quantum noise originating from electronics, and the generation of quantum states of light. Following this introduction, the chapter dives into the specifics of amplitude squeezing in semiconductor lasers, detailing the mechanisms that enable it and the constraints that influence its performance. This section covers topics such as the suppressed-pump-noise mechanism, the implementation of a noise equivalent circuit model for semiconductor lasers, and the constraints associated with squeezing level and squeezing bandwidth. Subsequently, the chapter presents a theoretical investigation of amplitude squeezing in ICLs, utilizing two different models: a semiclassical rate equation approach and a stochastic simulation. The results reveal that ICLs have the potential to achieve significant intensity noise squeezing over a broad bandwidth, extending into the gigahertz range. This characteristic holds promise for the development of ultra-low noise laser sources in the mid-infrared wavelength range. The final part of the chapter shifts its spotlight to the experimental and theoretical investigation of amplitude squeezing in QD lasers. It discusses sub-shot-noise noise spectra, sub-Poissonian photon distribution, and the second-order correlation function $g^{(2)}(0)$, all of which provide support for these experimental findings. Moreover, these experimental data are further validated through a comprehensive stochastic simulation.

Chapter 5 functions as the concluding chapter of the thesis, bringing closure to the research presented while pointing the way forward for future works within the field.

References

1. Einstein A (1917) Zur Quantentheorie der Strahlung. Physikalische Zeitschrift 18:121–128. https://doi.org/10.1515/9783112596609-016
2. Ladenburg R (1928) Untersuchungen über die anomale Dispersion angeregter Gase: I. Teil. Zur Prüfung der quantentheoretischen Dispersionsformel. Zeitschrift f ü r Physik 48(1–2):15–25. https://doi.org/10.1007/BF01351571
3. Gordon JP, Zeiger HJ, Townes CH (1954) Molecular microwave oscillator and new hyperfine structure in the microwave spectrum of NH_3. Phys Rev 95:282–284. https://doi.org/10.1103/PhysRev.95.282
4. Maiman T (1960) Stimulated optical radiation in Ruby. Nature 187(4736):493–494. https://doi.org/10.1038/187493a0
5. Javan A, Bennett WR Jr, Herriott DR (1961) Population inversion and continuous optical maser oscillation in a gas discharge containing a He-Ne mixture. Phys Rev Lett 6(3):106. https://doi.org/10.1103/PhysRevLett.6.106
6. Snitzer E (1961) Optical maser action of Nd^{+3} in a barium crown glass. Phys Rev Lett 7:444–446. https://doi.org/10.1103/PhysRevLett.7.444
7. Geusic J, Marcos H, Van Uitert L (1964) Laser oscillations in Nd-doped yttrium aluminum, yttrium gallium and gadolinium garnets. Appl Phys Lett 4(10):182–184. https://doi.org/10.1063/1.1753928
8. Von Neumann J (1953) Notes on the photon-disequilibrium-amplification scheme. An unpublished manuscript written before September **16**:659–673. https://ieeexplore.ieee.org/document/1073414
9. Braunstein R (1955) Radiative transitions in semiconductors. Phys Rev 99(6):1892–1893. https://doi.org/10.1103/PhysRev.99.1892
10. Hall RN, Fenner GE, Kingsley J, Soltys T, Carlson R (1962) Coherent light emission from GaAs junctions. Phys Rev Lett 9(9):366–368. https://doi.org/10.1103/PhysRevLett.9.366
11. Hecht J (2010) Short history of laser development. Opt Eng 49(9):091002–091002. https://doi.org/10.1117/1.3483597
12. Esaki L, Tsu R (1970) Superlattice and negative differential conductivity in semiconductors. IBM J Res Dev 14(1):61–65. https://ieeexplore.ieee.org/abstract/document/5391729
13. Dingle R, Wiegmann W, Henry CH (1974) Quantum states of confined carriers in very thin $Al_xGa_{1-x}As - GaAs - Al_xGa_{1-x}As$ heterostructures. Phys Rev Lett 33(14):827–830. https://doi.org/10.1103/PhysRevLett.33.827
14. Petroff P, Gossard A, Logan R, Wiegmann W (1982) Toward quantum well wires: fabrication and optical properties. Appl Phys Lett 41(7):635–638. https://doi.org/10.1063/1.93610
15. Arakawa Y, Sakaki H (1982) Multidimensional quantum well laser and temperature dependence of its threshold current. Appl Phys Lett 40(11):939–941. https://doi.org/10.1063/1.92959
16. Bimberg D, Grundmann M, Ledentsov NN (1999) Quantum dot heterostructures. Wiley
17. Grillot F, Duan J, Dong B, Huang H (2021) Uncovering recent progress in nanostructured light-emitters for information and communication technologies. Light: Sci & Appl 10(1):156. https://doi.org/10.1038/s41377-021-00598-3
18. Vahala KJ (1988) Quantum box fabrication tolerance and size limits in semiconductors and their effect on optical gain. IEEE J Quantum Electron 24(3):523–530. https://doi.org/10.1109/3.157

19. Benisty H, Sotomayor-Torres C, Weisbuch C (1991) Intrinsic mechanism for the poor luminescence properties of quantum-box systems. Phys Rev B 44(19):10945. https://doi.org/10.1103/PhysRevB.44.10945
20. Kirstaedter N, Ledentsov N, Grundmann M, Bimberg D, Ustinov V, Ruvimov S, Maximov M, Kopév PS, Alferov ZI, Richter U et al (1994) Low threshold, large T_0 injection laser emission from (InGa) as quantum dots. Electron Lett 30(17):1416–1417. https://doi.org/10.1049/el:19940939
21. Oshinowo J, Nishioka M, Ishida S, Arakawa Y (1994) Highly uniform InGaAs/GaAs quantum dots (15 nm) by metalorganic chemical vapour deposition. Appl Phys Lett 65(11):1421–1423. https://doi.org/10.1063/1.112070
22. Arakawa Y, Nakamura T, Kwoen J (2019) Quantum dot lasers for silicon photonics. Semiconductors and Semimetals, vol 101. Elsevier, pp 91–138. https://doi.org/10.1016/bs.semsem.2019.07.007
23. Grundmann MGM, Bimberg DBD (1997) Gain and threshold of quantum dot lasers: theory and comparison to experiments. Jpn J Appl Phys 36(6S):4181. https://iopscience.iop.org/article/10.1143/JJAP.36.4181
24. Ledentsov N, Shchukin V, Grundmann ME, Kirstaedter N, Böhrer J, Schmidt O, Bimberg D, Ustinov V, Egorov AY, Zhukov A et al (1996) Direct formation of vertically coupled quantum dots in Stranski-Krastanow growth. Phys Rev B 54(12):8743. https://doi.org/10.1103/PhysRevB.54.8743
25. Maximov MV, Kochnev IV, Shernyakov YM, Zaitsev SV, Gordeev NY, Tsatsul'nikov AF, Sakharov AV, Krestnikov IL, Petr S, Alferov ZI et al (1997) InGaAs/GaAs quantum dot lasers with ultrahigh characteristic temperature. Jpn J Appl Phys 36:4221–4223
26. Sellin R, Ribbat C, Grundmann M, Ledentsov N, Bimberg D (2001) Close-to-ideal device characteristics of high-power InGaAs/GaAs quantum dot lasers. Appl Phys Lett 78(9):1207–1209. https://doi.org/10.1063/1.1350596
27. Mirin R, Gossard A, Bowers J (1996) Room temperature lasing from InGaAs quantum dots. Electron Lett 32(18):1732–1734. https://doi.org/10.1049/el_19961147
28. Deppe DG, Shavritranuruk K, Ozgur G, Chen H, Freisem S (2009) Quantum dot laser diode with low threshold and low internal loss. Electron Lett 45(1):54–56. https://doi.org/10.1049/el_20092873
29. Badcock T, Royce R, Mowbray D, Skolnick M, Liu H, Hopkinson M, Groom K, Jiang Q (2007) Low threshold current density and negative characteristic temperature 1.3 μm InAs self-assembled quantum dot lasers. Appl Phys Lett 90(11). https://doi.org/10.1063/1.2713136
30. Otsubo K, Hatori N, Ishida M, Okumura S, Akiyama T, Nakata Y, Ebe H, Sugawara M, Arakawa Y (2004) Temperature-insensitive eye-opening under 10-Gb/s modulation of 1.3-μm p-doped quantum-dot lasers without current adjustments. Jpn J Appl Phys 43(8B):L1124
31. Kageyama T, Nishi K, Yamaguchi M, Mochida R, Maeda Y, Takemasa K, Tanaka Y, Yamamoto T, Sugawara M, Arakawa Y (2011) Extremely high temperature (220 C) continuous-wave operation of 1300-nm-range quantum-dot lasers. In: The European conference on lasers and electro-optics. Optica Publishing Group, PDA_1
32. Nishi K, Yamada M, Anan T, Gomyo A, Sugou S (1998) Long-wavelength lasing from InAs self-assembled quantum dots on (311) B InP. Appl Phys Lett 73(4):526–528. https://doi.org/10.1063/1.121922
33. Abdollahinia A, Banyoudeh S, Rippien A, Schnabel F, Eyal O, Cestier I, Kalifa I, Mentovich E, Eisenstein G, Reithmaier JÁ (2018) Temperature stability of static and dynamic properties of 1.55 μm quantum dot lasers. Opt Express 26(5):6056–6066. https://doi.org/10.1364/OE.26.006056
34. Kageyama T, Vo Q, Watanabe K, Takemasa K, Sugawara M, Iwamoto S, Arakawa Y (2016) Large modulation bandwidth (13.1 GHz) of 1.3 μm-range quantum dot lasers with high dot density and thin barrier layer. In: 2016 compound semiconductor week (CSW) [Includes 28th international conference on indium phosphide & related materials (IPRM) & 43rd international symposium on compound semiconductors (ISCS)]. IEEE, pp 1–2

35. Faist J, Capasso F, Sivco DL, Sirtori C, Hutchinson AL, Cho AY (1994) Quantum cascade laser. Science 264(5158):553–556. https://www.science.org/doi/abs/10.1126/science.264.5158.553
36. Didier P, Dely H, Bonazzi T, Spitz O, Awwad E, Rodriguez É, Vasanelli A, Sirtori C, Grillot F (2022) High-capacity free-space optical link in the midinfrared thermal atmospheric windows using unipolar quantum devices. Adv Photonics 4(5):056004–056004. https://doi.org/10.1117/1.AP.4.5.056004
37. Yang RQ (1995) Infrared laser based on intersubband transitions in quantum wells. Superlattices Microstruct 17(1):77–83. https://doi.org/10.1006/spmi.1995.1017
38. Vurgaftman I, Bewley W, Canedy C, Kim C, Kim M, Merritt C, Abell J, Lindle J, Meyer J (2011) Rebalancing of internally generated carriers for mid-infrared interband cascade lasers with very low power consumption. Nat Commun 2(1):585. https://doi.org/10.1038/ncomms1595
39. Didier P, Knötig H, Spitz O, Cerutti L, Lardschneider A, Awwad E, Diaz-Thomas D, Baranov A, Weih R, Koeth J et al (2023) Interband cascade technology for energy-efficient mid-infrared free-space communication. Photonics Res 11(4):582–590. https://doi.org/10.1364/PRJ.478776
40. Tariq F, Khandaker MR, Wong K-K, Imran MA, Bennis M, Debbah M (2020) A speculative study on 6G. IEEE Wirel Commun 27(4):118–125. https://doi.org/10.1109/MWC.001.1900488
41. Zhang C, Mezzanotte C (2022) Training Deep Learning-based recommender models of 100 trillion parameters over Google Cloud. Google
42. Elster AC, Haugdahl TA (2022) Nvidia hopper gpu and grace cpu highlights. Comput Sci & Eng 24(2):95–100. https://doi.org/10.1109/MCSE.2022.3163817
43. Xiang C, Bowers SM, Bjorlin A, Blum R, Bowers JE (2021) Perspective on the future of silicon photonics and electronics. Appl Phys Lett 118(22). https://doi.org/10.1063/5.0050117
44. Rahim A, Goyvaerts J, Szelag B, Fedeli J-M, Absil P, Aalto T, Harjanne M, Littlejohns C, Reed G, Winzer G et al (2019) Open-access silicon photonics platforms in Europe. IEEE J Sel Topics Quantum Electron 25(5):1–18. https://doi.org/10.1109/JSTQE.2019.2915949
45. Arakawa Y, Nakamura T, Kurata K (2022) Highlights of 10-years of research in a Japanese Si photonics project. In: Optical fiber communication conference. Optica Publishing Group, Th3C–6
46. Liang D, Srinivasan S, Kurczveil G, Tossoun B, Cheung S, Yuan Y, Descos A, Hu Y, Huang Z, Sun P et al (2022) An energy-efficient and bandwidth-scalable DWDM heterogeneous silicon photonics integration platform. IEEE J Sel Topics Quantum Electron 28(6):1–19. https://doi.org/10.1109/JSTQE.2022.3181939
47. Zhou Z, Chen W, He X, Ma D (2023) Photonics in a time of rapid growth: silicon-based optoelectronics in China. IEEE Photonics J 1–10. https://doi.org/10.1109/JPHOT.2023.3324756
48. Tran MA, Zhang C, Morin TJ, Chang L, Barik S, Yuan Z, Lee W, Kim G, Malik A, Zhang Z et al (2022) Extending the spectrum of fully integrated photonics to submicrometre wavelengths. Nature 610(7930):54–60. https://doi.org/10.1038/s41586-022-05119-9
49. Xiang C, Jin W, Terra O, Dong B, Wang H, Wu L, Guo J, Morin TJ, Hughes E, Peters J et al (2023) 3D integration enables ultralow-noise isolator-free lasers in silicon photonics. Nature 620(7972):78–85. https://doi.org/10.1038/s41586-023-06251-w
50. Stojanović V, Ram RJ, Popović M, Lin S, Moazeni S, Wade M, Sun C, Alloatti L, Atabaki A, Pavanello F et al (2018) Monolithic silicon-photonic platforms in state-of-the-art CMOS SOI processes. Optics Express 26(10):13106–13121. https://doi.org/10.1364/OE.26.013106
51. Zhou Z, Ou X, Fang Y, Alkhazraji E, Xu R, Wan Y, Bowers JE (2023) Prospects and applications of on-chip lasers. Elight 3(1):1–25. https://doi.org/10.1186/s43593-022-00027-x
52. Liang D, Bowers JE (2010) Recent progress in lasers on silicon. Nat Photonics 4(8):511–517. https://www.nature.com/articles/nphoton.2010.167
53. Tournié E, Monge Bartolome L, Rio Calvo M, Loghmari Z, Díaz-Thomas DA, Teissier R, Baranov AN, Cerutti L, Rodriguez J-B (2022) Mid-infrared III-V semiconductor lasers epitaxially grown on Si substrates. Light: Sci & Appl 11(1):165. https://www.nature.com/articles/s41377-022-00850-4

54. Shi Y, Zhang Y, Wan Y, Yu Y, Zhang Y, Hu X, Xiao X, Xu H, Zhang L, Pan B (2022) Silicon photonics for high-capacity data communications. Photonics Res 10(9):A106–A134. https://doi.org/10.1364/PRJ.456772
55. Shastri BJ, Tait AN, Ferreira de Lima T, Pernice WH, Bhaskaran H, Wright CD, Prucnal PR (2021) Photonics for artificial intelligence and neuromorphic computing. Nat Photonics 15(2):102–114. https://www.nature.com/articles/s41566-020-00754-y
56. Chang L, Liu S, Bowers JE (2022) Integrated optical frequency comb technologies. Nat Photonics 16(2):95–108. https://doi.org/10.1038/s41566-021-00945-1
57. Zhang G, Haw JY, Cai H, Xu F, Assad S, Fitzsimons JF, Zhou X, Zhang Y, Yu S, Wu J et al (2019) An integrated silicon photonic chip platform for continuous-variable quantum key distribution. Nat Photonics 13(12):839–842. https://www.nature.com/articles/s41566-019-0504-5
58. Bremner M, Clark R, Bartlett S, Lam PK (2021) The impact of quantum technologies on secure communications
59. Moody G, Sorger VJ, Blumenthal DJ, Juodawlkis PW, Loh W, Sorace-Agaskar C, Jones AE, Balram KC, Matthews JC, Laing A et al (2022) 2022 Roadmap on integrated quantum photonics. J Phys: Photonics 4(1):012501
60. Moody G, Chang L, Steiner TJ, Bowers JE (2020) Chip-scale nonlinear photonics for quantum light generation. AVS Quantum Sci 2(4). https://doi.org/10.1116/5.0020684
61. Wang J, Sciarrino F, Laing A, Thompson MG (2020) Integrated photonic quantum technologies. Nat Photonics 14(5):273–284. https://www.nature.com/articles/s41566-019-0532-1
62. Albrechtsen M, Vosoughi Lahijani B, Christiansen RE, Nguyen VTH, Casses LN, Hansen SE, Stenger N, Sigmund O, Jansen H, Mørk J et al (2022) Nanometer-scale photon confinement in topology-optimized dielectric cavities. Nat Commun 13(1):6281. https://www.nature.com/articles/s41467-022-33874-w
63. Yu Y, Sakanas A, Zali AR, Semenova E, Yvind K, Mørk J (2021) Ultra-coherent Fano laser based on a bound state in the continuum. Nat Photonics 15(10):758–764. https://www.nature.com/articles/s41566-021-00860-5
64. Zhong H, Yu Y, Zheng Z, Ding Z, Zhao X, Yang J, Wei Y, Chen Y, Yu S (2023) Ultra-low threshold continuous-wave quantum dot mini-BIC lasers. Light: Sci & Appl 12(1):100. https://www.nature.com/articles/s41377-023-01130-5
65. Gomez S, Huang H, Duan J, Combrié S, Shen A, Baili G, de Rossi A, Grillot F (2020) High coherence collapse of a hybrid III-V/Si semiconductor laser with a large quality factor. J Phys: Photonics 2(2):025005. https://iopscience.iop.org/article/10.1088/2515-7647/ab6a74
66. Malik A, Xiang C, Chang L, Jin W, Guo J, Tran M, Bowers J (2021) Low noise, tunable silicon photonic lasers. Appl Phys Rev 8(3):031306. https://doi.org/10.1063/5.0046183
67. Alkhazraji E, Chow WW, Grillot F, Bowers JE, Wan Y (2023) Linewidth narrowing in self-injection-locked on-chip lasers. Light: Sci & Appl 12(1):162. https://www.nature.com/articles/s41377-023-01172-9
68. Jin W, Yang Q-F, Chang L, Shen B, Wang H, Leal MA, Wu L, Gao M, Feshali A, Paniccia M et al (2021) Hertz-linewidth semiconductor lasers using CMOS-ready ultra-high-Q microresonators. Nat Photonics 15(5):346–353. https://www.nature.com/articles/s41566-021-00761-7
69. Schawlow AL, Townes CH (1958) Infrared and optical masers. Phys Rev 112(6):1940. https://doi.org/10.1103/PhysRev.112.1940
70. Henry C (1982) Theory of the linewidth of semiconductor lasers. IEEE J Quantum Electron 18(2):259–264. https://doi.org/10.1109/JQE.1982.1071522

Chapter 2
Semiconductor Lasers: Fundamentals and Theory

This chapter will introduce the fundamentals and theory of semiconductor lasers. To this end, it is important to understand that a semiconductor laser is a non-equilibrium system that undergoes a second-order phase transition at the oscillation threshold. It consists of many electrically- or optically-pumped emitters, ranging from a few to over trillions (10^{12}), which interact with different optical modes within a cavity. Consequently, this presents a highly complex problem that necessitates a diverse range of physical concepts, including semiconductor physics, quantum optics, nonlinear dynamics, and others. In the following sections, I will briefly introduce the principles of quantum-confined lasers, followed by an overview of different theoretical models that address specific aspects of laser behaviour.

2.1 Fundamentals of Quantum-Confined Lasers

The objective of this section is to investigate diverse physical processes occurring within the quantum-confined structures of semiconductor lasers and to elucidate their fundamental properties across electronic, optical, and optoelectronic domains. In its simplest form, a laser diode comprises a single pn-junction within a cavity formed by two mirrors. The cleaved facets of the laser diode possess sufficient reflectivity to enable lasing, owing to the refractive-index difference between semiconductor materials and air. To trigger stimulated emission of light, population inversion of the gain medium is required, often achieved through electrical or optical pumping. In a state of thermodynamic equilibrium, the electron distribution in the gain material would be characterized by a Fermi energy E_F, with the occupation probability described by a Fermi function:

$$\rho(\mathfrak{E}) = F(\mathfrak{E}, E_F, T) \equiv \left[1 + \exp\left(\frac{\mathfrak{E} - \mathrm{E_F}}{\mathrm{k_B T}}\right)\right]^{-1} \tag{2.1}$$

S. Zhao, *Noise, Dynamics and Squeezed Light in Quantum Dot and Interband Cascade Lasers*, Springer Theses, https://doi.org/10.1007/978-3-031-66818-0_2

with the energy of the electronic state $\mathfrak{E}$, the equilibrium temperature T, and Boltzmann's constant k_B. In equilibrium conditions, realizing the population inversion, which entails a higher occupation of the energetically higher conduction band than the valence band, is unattainable. To induce this inversion, an external pump source is employed to drive the system away from equilibrium and into a non-equilibrium state. This process involves injecting electrons into the conduction band and removing them from the valence band. As a result, stimulated emission of photons can take place through the recombination of a conduction-band electron with an available valence-band state, commonly referred to as a "hole". In the next sections, I will introduce two typical types of quantum-confined lasers, namely quantum dot lasers and interband cascade lasers.

2.1.1 Quantum Dot Lasers

The quantum wells (QW) and quantum dots (QD) lasers have garnered consistent attention in the scientific community due to their potential to enhance the performance of semiconductor lasers at the fundamental physics level. Moving from QW to QD, there is a transition from a quasi-continuous density of states reminiscent of a solid-state system to an atom-like structure characterized by discrete states, as explained in Fig. 1.2. This shift principally alters the nature of carrier transport and has a profound impact on the optical properties of the semiconductor material.

During the Stranski–Krastanov growth process of InAs QD on a GaAs substrate [1], several key steps come about as shown in Fig. 2.1. Initially, a thin layer of InAs, referred to as the wetting layer, forms on top of the substrate. The subsequent formation of pyramidal structures is a consequence of the strain induced by the lattice constant mismatch between InAs and GaAs. These pyramidal structures are then further grown over with InGaAs, which flattens the tops of the pyramids and creates an InGaAs QW in which the InAs QD are embedded. This resulting structure is known as a dot-in-a-well (DWELL) and is the primary focus of investigation [2].

A schematic of a QD laser and its band structure is provided in Fig. 2.2a and b, respectively. The band structure of QD devices is strongly influenced by the coupling of QD states to other states [3, 4]. In general, the active region of QD lasers comprises a three-dimensional separate confinement heterostructure (SCH states), a two-dimensional wetting layer serves as a charge-carrier reservoir (RS states),

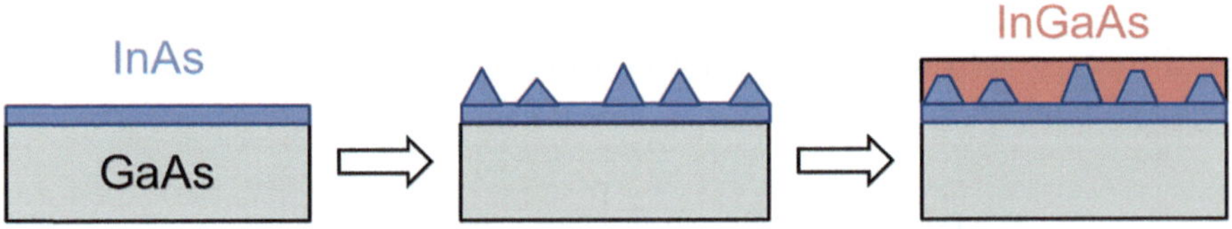

Fig. 2.1 Sketch of the epitaxial Stranski–Krastanov growth process of a DWELL structure

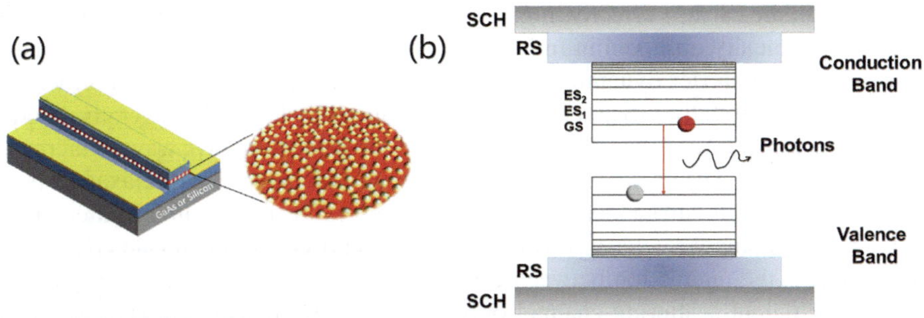

Fig. 2.2 **a** Schematic illustration of the QD laser. **b** The band structure of the QD laser. The localized InAs QD states, namely a ground state (GS), and two degenerate excited states (ES_1 and ES_2) lie within the bandgap of RS states inside the InGaAs QW (light blue). The whole structure is embedded within the SCH states of the surrounding GaAs substrate (light grey)

and QD states that can be approximately displayed as a three-dimensional potential well with discretized energy levels, including excited states (ES) and ground states (GS). In this structure, carriers are typically treated as quasi-free particles within the quasi-continuum SCH and RS charge-carrier reservoir. However, the RS states in the wetting layer, which incorporate the localized QD states, lead to smaller energy separations within the QD states, causing them to overlap at higher energies [5], as illustrated in Fig. 2.2b.

The presence of Coulomb interaction between electrons and holes in the semiconductor active region results in the formation of excitons, which are bound electron-hole pairs. The size of the exciton, known as the Bohr radius, is determined by the separation between the electron and hole within the exciton and mostly ranges from a few to several nanometers [6]. The properties of the exciton, which can be modified by the confinement structure, are central in determining the optical characteristics of the system. The exciton approximation is habitually employed in the semi-classical modelling of QD lasers, as discussed in Sect. 3.2.1. This approximation simplifies the modelling process by reducing the number of equations related to the carrier populations. Moreover, it provides an intuitive understanding of the physical processes, allowing direct input from experimental data. However, it should be noted that this approximation holds true as long as there are no significant deviations in the electron and hole populations. In certain cases, more sophisticated approaches may be necessary to accurately capture the system dynamics [7–9].

In a QD laser, the pumping process creates a high-density carrier plasma in the delocalized states, traversing through the 3D SCH before reaching the 2D RS. Because the carrier transport in the SCH is a fast process (1–5 ps) according to the time-resolved photoluminescence (TPRL) [10, 11], its impact on the overall carrier scattering process is minimal. When the carriers are captured by the QD from RS, two kinds of relaxation processes can take place. In the first case, the carriers relax through the higher energy levels of ES and eventually settle into the GS. At this stage, recombination between electrons and holes triggers the onset of lasing. In the second case, the carriers directly move from the RS to the GS. This direct relaxation channel is typically observed in InP-based QD lasers rather than GaAs-based ones. It speeds up the carrier dynamics and leads to a dual-state lasing operation, where emissions from both the GS and ES occur simultaneously [12]. Moreover, the lateral coupling among QD and the assistance of defects are also alternative ways to increase carrier dynamics. The carrier dynamics thus depend on capture and relaxation processes [3]. At high carrier densities in the RS reservoir, the Coulomb interaction (carrier-carrier Auger scattering) will dominate the carrier capture into and out of the QD, whereas the interaction between carrier and longitudinal-optical (LO) phonon offers an efficient scattering channel only at a low carrier densities. Hence, it can be inferred that LO phonon-assisted capture stands for a widely applicable capture mechanism that relies on both carrier density and temperature, whereas the Auger process exhibits temperature-independent behaviour while being dependent on carrier density. Upon analyzing the photoluminescence rise time at room temperature, it is observed that the capture times exhibit a wide range of variation, spanning from 1 to 100 ps [13, 14]. Specifically, for moderate RS carrier density ranging from 10^{11} to $10^{12}\,cm^{-2}$, the typical carrier scattering times are in the order of 1–10 ps [11]. At the very beginning, the relaxation process within QD was believed to be relatively slow, lasting a few nanoseconds due to the phonon bottleneck effect [15]. However, experimental findings have consistently demonstrated carrier relaxation times on the picosecond scale, suggesting the involvement of multiple mechanisms such as carrier-carrier interactions, carrier-phonon interactions, and direct relaxation from continuum states. Notably, the relaxation process within QD occurs at a faster timescale compared to the carrier capture from the RS into the QD [3].

2.1.2 *Quantum Dot Lasers on Silicon*

In the early 1960s, diode lasers made from Group III–V compound semiconductors were successfully demonstrated, coinciding with the widespread popularity of Si-based transistor radios. Since then, numerous scientists and engineers have dedicated their efforts to exploring the possibility of photonic integrated circuits (PICs). The initial concept of PICs was introduced by Miller [16], during a period when the integrated electronic industry was dawning. In the early stages of PIC development, the designs were mainly centred around the use of complementary metal-oxide-semiconductor (CMOS) wafers. However, to create a complete on-chip optical communication system, a more extensive range of photonic building blocks was required. It was not until the early 1990s that most of these essential optical components became available for integration [17]. The benefits of photonic integration through silicon photonics can be observed in Fig. 2.3, which enables the miniaturization of high-speed transceivers over several product generations. By leveraging these techniques, each new generation of transceivers was able to enhance capacity while simultaneously reducing power consumption and physical size.

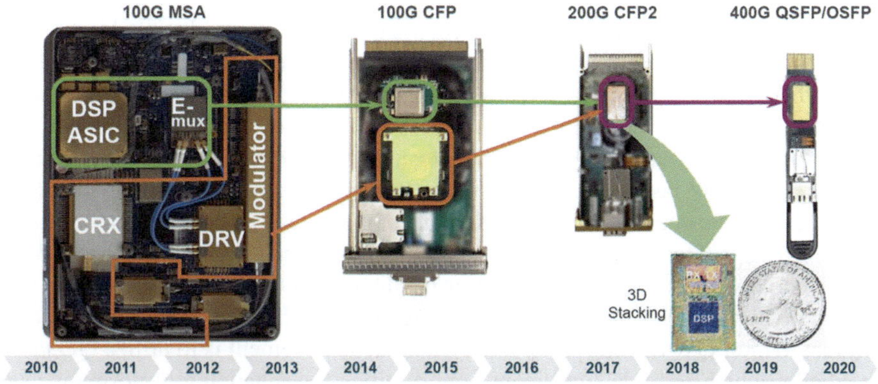

Fig. 2.3 PICs have enabled coherent transceivers to become smaller over the last decade. (*Source* In the OFC 2019 Plenary talk from Benny P. Mikkelsen in Acacia)

The fundamental hurdle to achieving lasing directly with silicon (Si) was identified early on. The issue stemmed from the fact that optical transitions must conform to the principles of both energy and momentum conservation, and this dual requirement was not simultaneously satisfied in Si, which is characterized as an indirect bandgap material [18]. Despite its inherent limitations as a gain medium, Si possesses several crucial properties that make it an excellent substrate for diode lasers. Firstly, Si wafers are renowned for their exceptional purity, low defect density, and larger size, typically reaching 12 inches or 300 mm in diameter. In comparison, GaAs $\leq$ 6 inches, InP $\leq$ 4 inches. Secondly, the state-of-the-art 45 and 32 nm Si-based CMOS technologies offer a "sweet spot" for incorporating photonic capabilities and expanding integrated system applications beyond Moore's Law scaling [19]. This means that leveraging highly optimized CMOS processes and materials allows for high-volume industrial manufacturing, low-cost production, excellent material compatibility (supporting integration of Ge, $LiNbO_3$, Si_3N_4, SiC, etc.), and most importantly, high integration densities. A recent development involving the three-dimensional integration of ultralow-loss photonic integrated circuits represents a significant milestone in the progression toward complex systems and networks on Si [20]. Thirdly, the high-quality native oxide of Si, SiO_2, serves as an effective protective layer and an excellent optical waveguide cladding, because of its substantial refractive index difference from Si ($\delta n \sim 2.1$). This factor contributes to the development of Si waveguides with propagation losses typically one order of magnitude lower than those of compound semiconductor waveguides. Additionally, Si boasts high thermal conductivity, a highly advantageous characteristic for active device substrates. These unparalleled advantages offered by Si have prompted researchers to shift their focus towards the integration of III–V semiconductors onto Si substrate platforms.

The main challenge in silicon photonics is the engineering of active building blocks such as laser diodes and optical amplifiers. There are generally two main approaches to realize semiconductor lasers integrated into Si, as depicted in Fig. 2.4. The first approach is heterogeneous integration. The initial work was demonstrated in 2006, where a multi-quantum-well laser was packaged on silicon via a wafer bonding technique [21]. Nevertheless, this approach often requires complicated and precise alignment in the co-packaging process that can potentially compromise the scalability. Moreover, QW lasers do not exhibit an intrinsic threshold or thermal steadiness. Since the ultimate goal of silicon photonics is to integrate PICs close to heat-radiating CMOS-based digital circuits, device cooling can become exceedingly challenging. To this end, owing to the ultimate carrier localization, QD laser displays more reliable characteristics and, thus can be considered a much better-suited candidate in contrast to their QW counterparts. Lately, a great achievement is to utilize QD lasers on silicon, which provide significant opportunities for enhancing performance and reducing costs [22]. The second approach is monolithic integration, which involves the direct epitaxial growth of QD lasers on Si substrate [23, 24]. Recent advancement has demonstrated the feasibility of electrically pumped QD lasers on a 300 mm patterned (001) Si wafer [25], which unleashes more potential of silicon photonics into industrial applications.

However, the formation of inevitable crystalline defects occurs during the merging of III–V and Si substrate via epitaxial growth, although QD lasers exhibit greater tolerance to threading dislocations compared to QW structures [28]. Threading dislocations, which nucleate as the III–V film relaxes due to lattice mismatch, are considered the most detrimental to QD laser performance as they penetrate through all layers. These epitaxial defects lead to nonradiative recombination of carriers via the Shockley–Read–Hall process, impeding efficient radiative recombination for light emission. Further details of these investigations are presented in Sect. 3.2.2. Considerable resources are dedicated to improving epitaxial growth in this field, resulting in a consequent reduction of defects. This advancement has elevated the performance of QD devices to a level comparable to those on their native substrates [29]. In

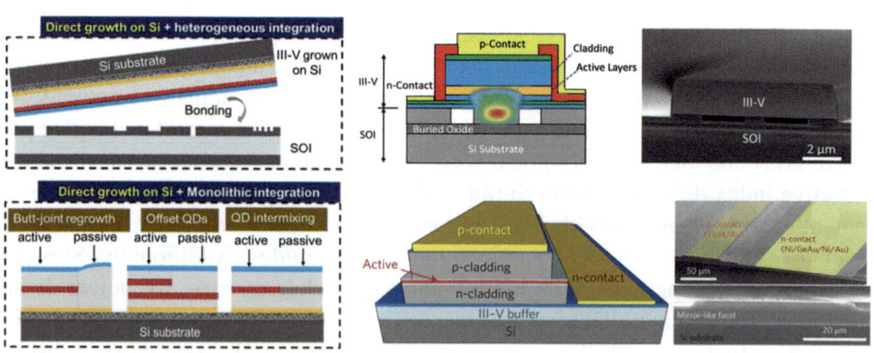

Fig. 2.4 Top panel: Heterogeneous integration of QD devices. Bottom panel: Monolithic integration of QD devices. (*Source* In Refs. [26, 27])

conclusion, the eventual realization of large-scale, low-cost, and high-density monolithically integrated lasers and amplifiers on Si through direct epitaxial growth would pave the way for the next generation of photonics integrated circuit platforms.

2.1.3 Interband Cascade Lasers

To enable continuous-wave (CW) operation of mid-infrared (MIR) semiconductor lasers well above the threshold at room temperature, multiple quantum wells (QWs) are often required to provide sufficient gain. The challenge lies in optimizing the connection between these QWs. Historically, electrically pumped MIR lasers have adopted a conventional diode geometry with multiple QWs connected in parallel. However, this approach has a drawback: if the electrons and holes starting at opposite ends of the line of QWs fail to populate each well uniformly, the generated gain for a given injected current density will degrade. Cascading offers an efficient and practical means. By arranging them in series with the same current flowing through each stage, and by having each carrier traverse every QW in turn, the interband cascade laser (ICL) [30] can significantly reduce the threshold density. In an ICL, as depicted in Fig. 1.3c, electrons and holes are generated internally at an interface between the narrow-gap semiconductors InAs and Ga(In)Sb, and this interface exhibits energy alignment that changes from semiconducting to semimetallic with an applied bias voltage V.

Figure 2.5a illustrates how this internal generation is facilitated by the alignment of the conduction band minimum of InAs, which is located approximately 0.2 eV below the valence band maximum of GaSb [31]. When an applied electric field induces a transition from a positive energy gap caused by quantum confinement to a semimetallic alignment with a voltage-dependent band overlap of approximately $E_{SM}(V) \approx 100\,\text{meV}$ as shown in Fig. 2.5b, both electron states in the InAs quantum well and hole states in the GaSb quantum well can be populated in thermal quasi-equilibrium. As the external electric field sweeps the generated carriers away from the junction, holes flow to the left and electrons to the right. To maintain quasi-equilibrium, equal numbers of additional electrons and holes must be continuously regenerated at the semimetallic interface (SMIF) to replenish the carriers removed by the field.

The active region of an ICL comprises multiple repeated stages, each of which consists of (1) the "W"-shaped QWs, usually constructed by two InAS QWs sandwiching a layer of GaInSb; (2) a hole injector, made by coupled GaSb/AlSb QWs; and (3) an electron injector, made by coupled InAs/AlSb QWs. A band structure of an ICL is presented in Fig. 2.6. The semimetallic interface, depicted in Fig. 2.5, separates the hole injector from the electron injector. The field-dependent band overlap between states in the hole injector and active hole QW, and those in the electron injector and active electron QW in the next stage, determines the carrier densities throughout the active region. Assuming a common quasi-Fermi level (black dashed line) on both sides of the semimetallic interface, due to the rapid carrier transport times via

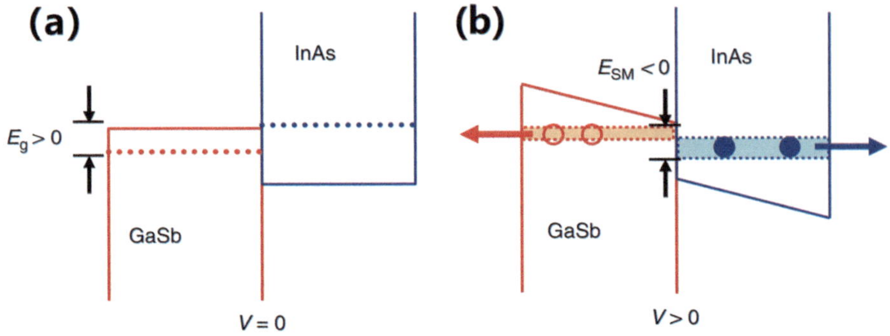

Fig. 2.5 The energy alignment of adjacent InAs and GaSb QWs in both **a** equilibrium and **b** under bias. The solid blue and red lines correspond to the conduction and valence band edges, respectively. In the absence of bias, quantum confinement leads to an energy gap (E_g) between the lowest conduction and highest valence states, as represented by the blue and red dotted lines, respectively. Under bias conditions, a semimetallic overlap (E_{SM}) is induced, which generates equal densities of electrons and holes that are depicted by the solid blue and open red circles, respectively. Furthermore, the applied field drives both carrier types away from the interface, as indicated by the arrows. (*Source* In Ref. [31])

direct, phonon-assisted, and other tunnelling mechanisms, the quasi-Fermi level is discontinuous across the active region of each stage, owing to the much longer carrier lifetime in the active QWs (~1 ns). In the ideal ICL design, the applied bias must be sufficient to achieve two objectives simultaneously: (1) produce enough optical gain to compensate for the photon loss in the cavity, and (2) generate a quasi-equilibrium carrier density consistent with the voltage drop. The first condition depends only on the active QWs, while the second condition depends on the design and doping of all electron and hole QWs on both sides of the semimetallic interface. For instance, if the QWs in the electron injector are thicker than optimal, excess carriers will be generated, leading to unnecessary free-carrier absorption [32]. Finally, the optical transitions of carriers within the band structure create several photons, which will be discussed in Sect. 3.2.3.

2.2 Theory of Quantum-Confined Lasers

In this section, I shall emphasize the theoretical treatment of the laser theory. In recent years, numerous textbooks have been devoted to the physics of semiconductor lasers [9, 33–36]. Basically, there are three defining ingredients of a semiconductor laser: a resonant optical cavity, a gain medium, and a pumping mechanism. The gain medium is responsible for creating an inverted electron distribution, deviating from thermal equilibrium. The operation of most semiconductor lasers can be described in this manner: it begins with spontaneous emission into the lasing mode, and this initial light is then amplified through stimulated emission within the gain medium.

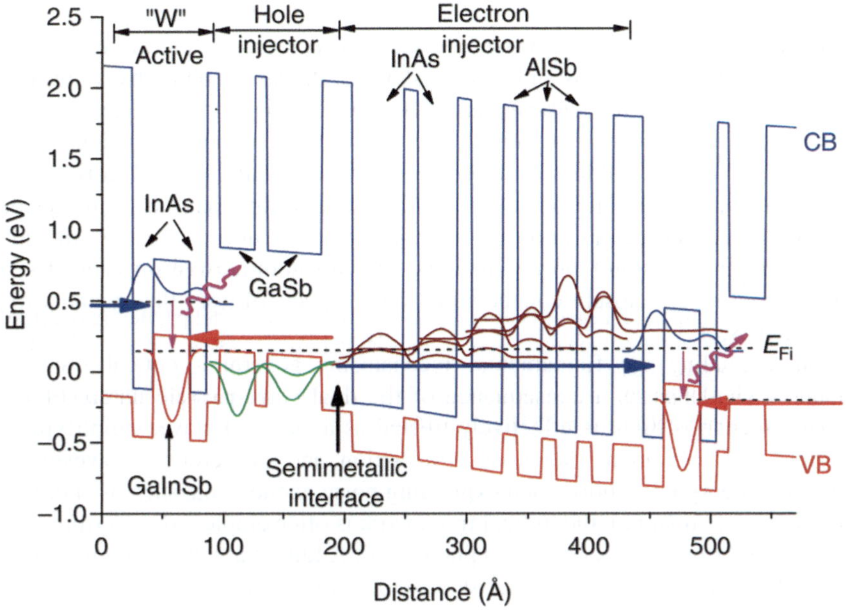

Fig. 2.6 Band diagram for $1\frac{1}{2}$ stages of the ICL active region. The dashed lines indicate the position of the quasi-Fermi levels in each stage. (*Source* In Ref. [31])

This process generates light that matches the incoming light in terms of direction, wavelength, and phase. Subsequently, this coherent beam of light bounces back and forth inside the optical resonator until it reaches an equilibrium point, maintained by a balance between gain and loss. The minimum amount of pump power required to initiate lasing action is referred to as the laser threshold. A laser is essentially an open system that continuously exchanges energy with its surroundings. It emits photons through one of its mirrors, which has a certain degree of transmissivity. On the other hand, to sustain the laser process, energy must be continually supplied. A variety of interesting processes are going on inside the laser cavity that necessitate a detailed analysis.

In the 1970s, a valuable discovery was made from a profound physical perspective, suggesting that the transition from light generated by thermal sources to laser light bears a striking resemblance to phase transitions in thermodynamics [37, 38]. The laser is, in fact, a *trailblazer of "synergetics"* [34], creating macroscopic order when driven far from thermal equilibrium. This laid the foundation for deep-rooted analogies between quite different systems in physics, chemistry, biology, and other fields. For instance, the connections between fluid dynamics and laser light led to the prediction and experimental observation of chaotic laser light. These intriguing topics will be explored in Sect. 3.3. The macroscopic order, achieved through a second-order phase transition, is governed by the interplay of two opposing forces: a stabilizing force within the system and a fluctuating force from the external reservoir. In semiconductor lasers, the stabilizing force is provided by stimulated emission, which maintains the laser's amplitude around its steady-state value. Furthermore, relaxation oscillations between photons and carriers stabilize the laser's phase by ensuring the emission of photons with identical phases. However, both photons and inverted carriers continuously dissipate into external reservoirs. Photon decay occurs due to output coupling loss and internal losses caused by absorption, while inverted carriers decay due to spontaneous emission and other non-radiative recombinations. Consequently, these dissipation processes introduce fluctuating forces into the system, including pump noise and vacuum field fluctuations. The quantum noise properties of semiconductor lasers are closely connected to these aspects of laser theory and will be explored in Sect. 4.1.

Comprehending the fundamental mechanisms governing optical transitions remains a challenge as a consequence of the dual wave-particle nature of light. In many scenarios, light is effectively treated as a classical wave, simplifying the description of phenomena like modes, diffraction, and interference. However, classical wave theory falls short when explaining noise phenomena, such as shot noise. In these cases, a quantum theoretical framework is often employed, describing these phenomena through the concept of a photon gas. Additionally, the discrete nature of spectral lines can be directly attributed to the principles of quantum theory. To tackle this challenge, several theoretical frameworks have been developed, each with its strengths and limitations. In the following sections, I shall introduce three theoretical descriptions of semiconductor lasers: Langevin rate equations, Gillespie stochastic algorithm, and fully quantum-mechanical treatments. The choice of method depends on the interpretive and solvable difficulty posed by the specific phenomenon under investigation.

2.2.1 Semiclassical Description: Langevin Rate Equations

The rate equation model is the simplest way to describe a laser. It is based on ordinary differential equations (ODEs) that track the temporal change of the number of particles, such as photons and carriers, occupying individual cavity modes. Albert Einstein used these equations to analyze optical transitions of carriers, as illustrated in Fig. 2.7, with phenomenological rate coefficients [39]. In doing so, he postulated the existence of stimulated emission, which, like absorption, is induced by a photon. The typical rate equations for a QW laser with the carrier number N and the photon number S are of the form:

$$\begin{aligned} \frac{\mathrm{d}N}{\mathrm{d}t} &= G_n - R_n \\ \frac{\mathrm{d}S}{\mathrm{d}t} &= G_s - A_s \end{aligned} \tag{2.2}$$

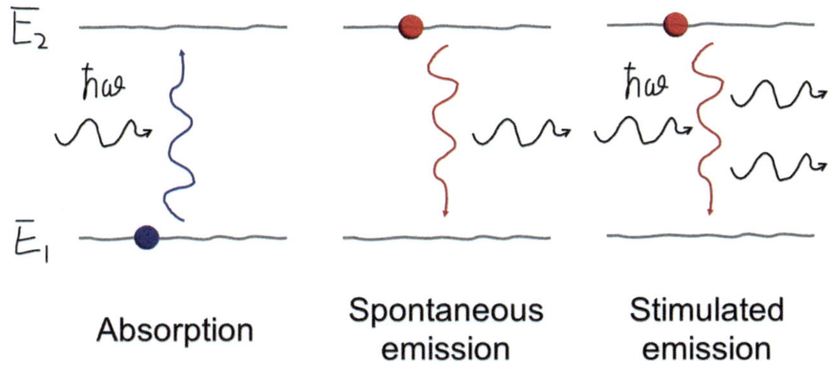

Fig. 2.7 Illustration of the processes of absorption, spontaneous emission, and stimulated emission

where G_n and R_n stand for carrier generation rate and recombination rate, and G_s and A_s stand for photon generation rate and annihilation rate, respectively. This approach is still relevant today when examining global phenomena such as laser threshold, single mode or multimode laser action [36].

The QW laser reservoir model, as summarized in Fig. 2.8, is used to derive the rate equations and helps ensure particle conservation in the system. Each arrow in the flowchart represents the rate at which particles flow into or out of the system per unit of time. To understand the rate equations, it's essential to grasp the inputs and outputs of this flowchart. A current density I/q is driven into the top contact, and a fraction η of it enters the carrier reservoir. In this context, the carrier density rate equation is formulated in terms of the remaining current, $J = \eta I/q$. Electrons are lost from the active region due to net stimulated emission, $R_{stim}^{net}(N, S)$, spontaneous emission, $R_{spon}(N)$, and nonradiative recombination, $R_{nr}(N)$. The carrier density in the well, represented as a reservoir, is determined by the competition among these rates. Therefore, the net rate of change of the area density of electrons in the active region can be expressed as:

$$\frac{\mathrm{d}N}{\mathrm{d}t} = J - R_{spon}(N) - R_{nr}(N) - R_{stim}^{net}(N, S) \tag{2.3}$$

Photons can be lost due to internal scattering, which is represented by the internal mode loss parameter α_i, as well as through the end mirrors, which is represented by the mirror loss parameter α_m. The photon lifetime τ_p is an important property of the cavity:

$$\frac{1}{\tau_p} = v_g(\alpha_i + \alpha_m) \tag{2.4}$$

The loss mentioned earlier pertains to the loss that occurs in the absence of any amplification, known as a "cold cavity" loss. However, when a laser operates above its threshold and amplifies the light, the average lifetime of photons within the cavity is

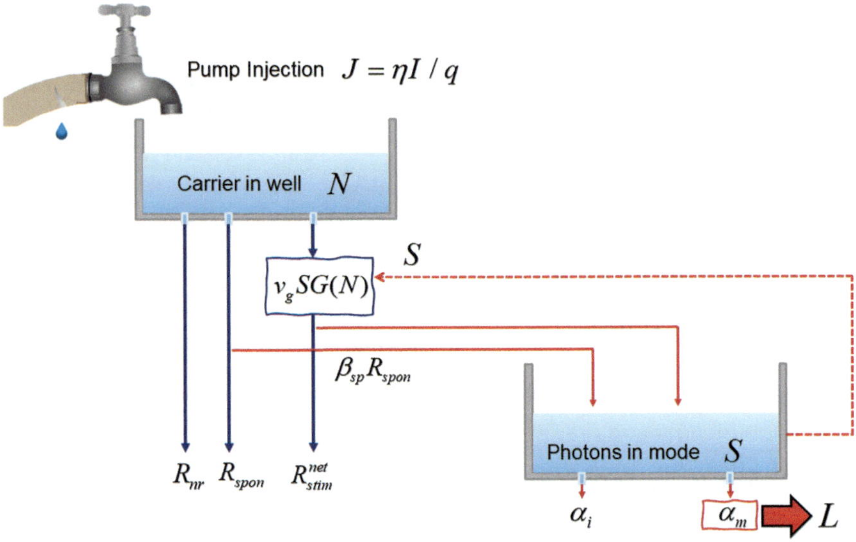

Fig. 2.8 Schematic diagram of the coupled carrier and photon reservoirs in a QW laser. Carriers are injected into the device from the contact through J and lost through nonradiative, spontaneous, and net-stimulated recombination processes represented by blue lines. Photons, represented by red lines, are gained by the fraction of spontaneous emission in the mode, $\beta_{sp} R_{spon}$, and net stimulated emission. They are lost through internal loss, α_i, and through the end mirrors, α_m, which represents the laser light output of the device, denoted as L. The stimulated rate is influenced by gain $G(N)$ and the photon density, which is indicated by the dashed red line

increased. This increase results from the addition of photons through both stimulated emission and the fraction of spontaneous emission, denoted as β_{sp}, that enters the mode. The temporal evolution of the photon density S is:

$$\frac{\mathrm{d}S}{\mathrm{d}t} = R_{stim}^{net}(N, S) + \beta_{sp} R_{spon}(N) - \frac{S}{\tau_p} \tag{2.5}$$

For a monochromatic beam, the increase in photon density per unit length is $G(N)S$ and $G(N) = a(N - N_0)$ is the material gain with a the differential gain and N_0 the transparency carrier density. Therefore, the net stimulated emission rate per unit of time $R_{stim}^{net}(N, S)$ is:

$$R_{stim}^{net}(N, S) = v_g SG(N) = \frac{S}{\tau_p^{stim}} \tag{2.6}$$

The quantity $v_g G(n)$ represents the average net stimulated rate per photon, with units of $[s]^{-1}$. This quantity is equivalent to $1/\tau_p^{stim}$, where τ_p^{stim} is the photon lifetime due to stimulated emission. Similarly, the nonradiative recombination rate is typically expressed in terms of a nonradiative lifetime τ_{nr}, such that $R_{nr}(N) = N/\tau_{nr}$. And, the spontaneous recombination rate is given by $R_{spon}(N) = N/\tau_{spon}$, with a spontaneous recombination lifetime τ_{spon}. Equation 2.6 highlights the interdependence of photons and carriers, as the gain is dependent on the carrier density. However, it's essential to note that this treatment considers all laser emissions to be in a single mode. To fully understand the laser emission spectrum, one must formulate multimode equations, which are coupled through the homogeneous linewidth.

Figure 2.9 illustrates the emission rates into the laser mode resulting from spontaneous emission and net stimulated emission, and their sum produces the light output. At first, the net stimulated rate R^{net}_{stim} is negative, which indicates the absorption of photons when the population is not inverted. The rate becomes positive once the transparency current density N_0 of about 240 A cm^{-2} is exceeded. The spontaneous emission rate increases with current as the carrier density builds up and finally pins just above the threshold owing to the rapid increase in photon density, which causes a corresponding increase in the stimulated rate. In a more comprehensive laser theory, noise must be considered. The practical way of dealing with noise is to add random *Langevin forces* to a set of rate equations [36]. The concept of Langevin forces or Langevin noise sources was originally inspired by the random motion of small macroscopic particles immersed in a liquid, known as *Brownian motion*, which clearly illustrates the statistical fluctuations occurring in a system in thermal equilibrium [41]. Brownian motion is a prototype problem that provides insights into the mechanisms responsible for the existence of fluctuations and dissipation of energy,

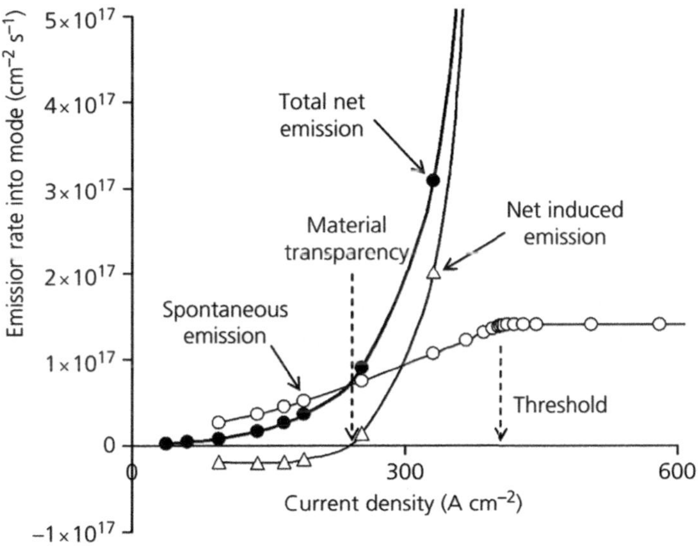

Fig. 2.9 An example of the dynamic interplay between spontaneous and stimulated emission rates with laser mode and light output variation, as a function of current density. The transparency current density N_0 is approximately 240 A/cm^{-2}. (*Source* On page 219 in Ref. [40])

making it of great interest to investigate such fluctuations that constitute a background of noise limiting the accuracy of detailed physical measurements [42].

A Langevin noise source, $F(t)$, is characterized by its rapidly fluctuating nature, which varies irregularly over time. The exact functional dependence of F on t cannot be specified precisely. To address this, the problem is often formulated in statistical terms. The one-time average, denoted as $\langle F(t)\rangle$, is defined as the statistical average over many similarly prepared systems at the same time. However, in cases where the statistical processes involved are both *stationary* and *ergodic*, the same result can be obtained by averaging a single system over extended time intervals. This is the assumption made here. Therefore, whether $\langle\ \rangle$ is defined as a statistical average or a time average depends on conceptual convenience. A time average is typically more convenient in the time domain, while a statistical average is often more appropriate in the frequency domain. Therefore, $\langle F(t)\rangle$ does not depend on t but vanishes over time:

$$\langle F(t)\rangle = 0 \tag{2.7}$$

Mathematically, a key characteristic of such a process is referred to as "*memoryless*." This means that the value of $F(t)$ at time t is completely uncorrelated with any previous value $F(t')$, even as $\left|t - t'\right| \to 0^+$:

$$\left\langle F(t)F(t')\right\rangle = 0 \tag{2.8}$$

Suppose that the characteristic time in which $F(t)$ appreciably changes is the correlation time τ_c, the average of the product of two Langevin noise sources, $F_i(t)$ and $F_j(t')$, will become negligible for $\left|t - t'\right| > \tau_c$. This is because, beyond τ_c, the product is just as likely to be negative as positive. Hence, the correlation function $\left\langle F_i(t)F_j(t')\right\rangle$ is concentrated at $t = t'$ and falls off to zero as $\left|t - t'\right|$ exceeds τ_c. When τ_c is much shorter than all other relevant timescales, one has:

$$\left\langle F_i(t)F_j(t')\right\rangle = 2D_{ij}\delta(t - t') \tag{2.9}$$

The diffusion coefficients D_{ij} are quantities that represent the proportionality constants between the magnitude of fluctuating forces [36, 43]. These coefficients have units of (seconds) $\times$ (fluctuating variable units)^2. There are two important types of strength to study: The *autocorrelation strength* is defined by $2D_{ij}$ when $i = j$, while the *cross-correlation strength* is determined by $2D_{ij}$ when $i \neq j$. The cross-correlation strength is only non-zero if the fluctuations of one noise source are correlated with the fluctuations of another noise source. The system is considered Markovian when Eq. 2.9 provides a good approximation. There are various methods to determine these diffusion coefficients. One approach, common among engineers familiar with noise in microwave circuits, is to draw an analogy between lasers and transmission lines with Nyquist noise sources, as described on page 77 of Ref. [35]. Another method, which I will introduce, involves adapting the diffusion coefficients by examining the correlation function in the frequency domain [36].

The Wiener–Khinchin theorem, which applies to all types of stationary processes, states that the spectral density $S_{ij}(\omega)$ and the correlation function $\langle F_i(t)F_j(t')\rangle$ are related as a Fourier transform pair. Therefore, assuming $t' = t - \tau$ and considering Eq. 2.9 the spectral density $S_{ij}(\omega)$ is:

$$S_{ij}(\omega) = \int \langle F_i(t)F_j(t-\tau)\rangle e^{-j\omega\tau} d\tau = 2D_{ij} \int \delta(\tau)e^{-j\omega\tau} d\tau = 2D_{ij} \quad (2.10)$$

The Langevin noise spectral density is equal to the correlation strength and is independent of frequency, making it a "white" noise source. The McCumber model [44] assumes that laser noise arises from shot noise associated with the discrete random flow of particles into and out of the carrier and photon reservoirs. Noise spectral density is constant and proportional to the average rate of particle flow. With this in mind, consider a reservoir where particles flow into and out of it via a variety of discrete random processes. In the Langevin formalism, each of these discrete processes contributes shot noise to the overall noise in the reservoir. To determine the total Langevin noise spectral density or correlation strength $\langle F_i F_i\rangle$, one needs to sum over simply all rates of particle flow into and out of reservoir i. To determine the cross-correlation strength $\langle F_i F_j\rangle$ between two reservoirs i and j, one needs to sum only over particle flows that affect both reservoirs simultaneously. However, in this case, when one reservoir gains a particle ($F_i > 0$), the other loses a particle ($F_j < 0$). Therefore, the product $\langle F_i F_j\rangle$ is always negative, and the noise between the two reservoirs is said to be "*negatively correlated*" [36].

$$\begin{aligned} \langle F_i F_i\rangle &= \sum R_i^+ + \sum R_i^- \\ \langle F_i F_j\rangle &= -\left[\sum R_{ij} + \sum R_{ji}\right] \end{aligned} \quad (2.11)$$

The Langevin noise sources for the reservoirs, F_i and F_j, as well as the rates of particle flow into (R_i^+), out of (R_i^-), and between the two reservoirs (R_{ij} and R_{ji}), are all expressed in units of numbers per unit time. Note that for single-sided spectral densities, an additional factor of 2 would be required on the right-hand side of these definitions.

2.2.2 *Stochastic Description: Gillespie Algorithm*

The Gillespie algorithms, also known as the stochastic simulation algorithm (SSA), were introduced by American physicist Gillespie [45]. These algorithms were developed as a numerical method to address the limitations of deterministic approaches when simulating chemical kinetics. They are practical algorithms for simulating "*coupled*"[1] stochastic processes exactly, meaning they provide a precise simulation without approximation errors.

The Gillespie algorithms offer several advantages over ordinary differential equations (ODEs) such as rate equations:

1. **Accurate Representation of Small Populations**: When dealing with systems with small populations, the ODEs description may not be accurate. ODEs assume real and continuous variables over time, but if the variables are quantized and can only take integer values, Gillespie algorithms can accurately simulate the discrete, quantized nature of these systems.
2. **Simulation of Rare Events**: Gillespie algorithms can efficiently simulate rare events with low probabilities. In contrast, ODEs, being deterministic, may not accurately represent the occurrence of such rare events.
3. **Accurate Description Near Critical Points**: ODEs may not provide an accurate description when a system is close to a critical point or a threshold. In these cases, the fluctuations of variables become significant compared to their mean values, even for large populations. Gillespie algorithms can capture these fluctuations more accurately.

While Gillespie's algorithm was initially developed for simulating chemical reactions, it relies on the general principles of stochastic processes and is applicable in various fields, including biology, finance, and epidemiology [46].

The Gillespie algorithms have found applications in semiconductor lasers, particularly in the study of "nanolasers" when the modal volume is reduced [47, 48]. These lasers exhibit interesting dynamical and statistical behaviours as the parameter β_{sp} (representing the fraction of spontaneous emission coupled into the lasing mode) varies from 0 to 1. At the threshold ($\sim\beta_{sp}^{-1/2}$) and saturation ($\sim\beta_{sp}^{-1}$) conditions, where the number of photons is low, fluctuations become increasingly significant, rendering semiclassical descriptions questionable [47, 49]. The discreteness of processes involving the exchange of photons and carriers in integer numbers becomes crucial [50, 51], making rate equations unsuitable for accurate predictions.[2] Recent developments, like the Stochastic Simulator [52], use a first-principle approach to simulate the light-matter interaction and provide efficient numerical simulations for traditional macroscopic lasers (with low β_{sp} values) to nanolasers (with high β_{sp} values), as shown in Fig. 2.10. These algorithms are especially accurate when dealing with low particle numbers but remain compatible with ODEs when dealing with large populations.

Prior to diving into the nitty-gritty of the Gillespie algorithms, it is imperative to elucidate some fundamental concepts.

---

[1] The word "coupled" means that an event that occurs somewhere in the system potentially influences the likelihood of future events' occurrences in different parts of the same system.

[2] In this case, the magnitude of the fluctuations is of the same order as the mean values and thus linearization is not an adequate approximation. At the same time, the small-population statistics deviate significantly from the Gaussian statistics of the Langevin forces.

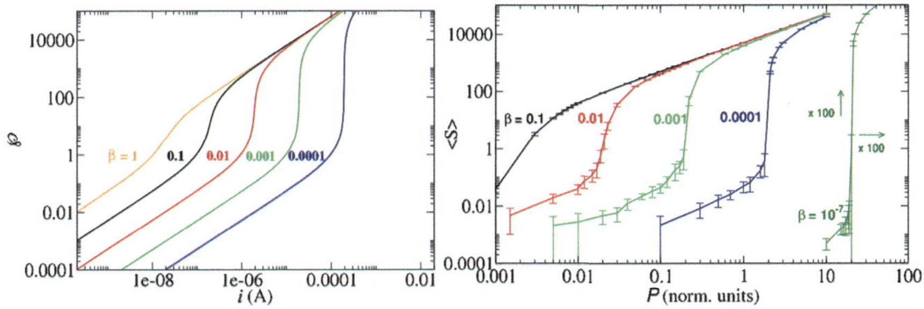

Fig. 2.10 The example of two methods in different types of semiconductor lasers. Left panel: Photon number $\wp$ versus injection current i computed from the steady states of the rate equations. Right panel: Average photon number $\langle S \rangle$ as a function of the pump P obtained from the stochastic simulation. $\langle S \rangle$ is computed as the average of the temporal average photon number $\bar{S} = \frac{1}{N}\sum_{N}^{l=1} S_l$ over ten series of events. (*Source* In Ref. [52])

- To use the Gillespie algorithms, one needs a predefined mathematical model that describes the system being simulated. This model should include parameters and rules governing how the system or its constituents change states over time.
- The Gillespie algorithms are designed for simulating stochastic processes. This means that each time one runs the same model with the same initial conditions, the results can vary due to the inherent randomness introduced by a random number generator (RNG). Usually, running a series of simulations is an effective approach to capturing the full range of possible outcomes and understanding the system's behaviour.
- The Gillespie algorithms simulate processes where changes in the system occur through discrete events taking place in continuous time. While these events are discrete (countable) and can be identified when they happen, they can occur at any point in time.

A recent review article [46] outlines two different but equivalent approaches; the direct method algorithm and the approximate explicit tau-leaping algorithm. The latter is a specific case of the former and offers an approximation to reduce computational complexity. The details of these approaches will be discussed as follows.

The direct method algorithm: The concept stems from two fundamental questions that require answers: *(1) how much time τ will elapse before the next event occurs?* and *(2) which event μ will occur at that time?* These questions apply to various event types in semiconductor lasers, including stimulated emission, absorption, spontaneous emission, internal loss, mirror loss, and pumping. Addressing these questions is not an easy task due to the immediate impact of each event on the present photon and/or carrier count, along with its influence on the probability of subsequent events. Therefore, in accordance with Gillespie algorithms [45], in order to simulate the time evolution, the formulation of this method centres on the so called "*fundamental premise of stochastic chemical kinetics*":

$$P(\tau, \mu|\mathbf{x}, t) = a_\mu(\mathbf{x})\exp\left[-\tau \sum_{\mu}^{M} a_\mu(\mathbf{x})\right] \tag{2.12}$$

where the a_μ terms represent *propensity functions* or event rates for the μth event among the M different event types, with their argument being the vector $\mathbf{x}$, denoting the counts of photons or carriers. The parameter τ signifies the time until the occurrence of the next event (sojourn time) while t is the current time. This expression can be interpreted as "the probability, given $\mathbf{x}$, that the next event in the system will occur within the infinitesimal time interval $[t + \tau, t + \tau + d\tau]$, and will correspond to the stoichiometry of the μth event." This formulation provides insight into the method by suggesting that τ is an exponentially distributed random variable with a mean (and standard deviation) of $1/\sum_{\mu}^{M} a_\mu(\mathbf{x})$, and μ is "a statistically independent integer random variable with point probabilities $a_\mu(\mathbf{x})/\sum_{\mu}^{M} a_\mu(\mathbf{x})$". In mathematical terms, this expression is also a joint probability density function in the space of the continuous variable τ ($0 \leqslant \tau < \infty$) and the discrete variable μ ($\mu = 1, 2, \ldots, M$). Notice that the variables τ and μ are quantities whose values would give the answers to the two aforementioned questions. Therefore, finding a legitime way for assigning numerical values to τ and μ in Eq. 2.12 is simply to draw two pseudorandom numbers based on the Monte-Carlo simulation, r_1 and r_2 from [0, 1], then compute [45]

$$\begin{aligned} \tau &= \frac{1}{\sum_{\mu'=1}^{M} a_{\mu'}(\mathbf{x})}\log\left(\frac{1}{r_1}\right) \\ \mu &= \text{the smallest integer satisfying} \sum_{\mu'=1}^{\mu} a_{\mu'}(\mathbf{x}) > r_2 \sum_{\mu'=1}^{M} a_{\mu'}(\mathbf{x}) \end{aligned} \tag{2.13}$$

Utilizing this generating method for the sojourn time and the type of next event, the direct method algorithm or the exact stochastic simulation algorithm is stated as

Algorithm 2.1: The direct method algorithm

1: Initialize: Set the initial number of photons and carriers $\mathbf{x}_0$ and set $t = 0$.
2: Calculate the propensity function or event rate, a_μ, for each event.
3: Set $a_0 = \sum_{\mu=1}^{M} a_\mu$.
4: Generate two independent uniform (0, 1) random numbers r_1 and r_2.
5: Set $\tau = \frac{1}{a_0} \ln \frac{1}{r_1}$, equivalent to drawing an exponential random variable with parameter a_0.
6: Find $\mu \in [1, \ldots, M]$ such that $\sum_{\mu'=1}^{\mu} a_{\mu'} < r_2 a_0 \leq \sum_{\mu'=1}^{\mu+1} a_{\mu'}$.
7: Set $t = t + \tau$ and update the number of photons and carriers according to event μ.
8: Return to step 2 or quit.

A significant characteristic of this method is its lack of numerical parameters, such as the fixed time increment. Consequently, there is no requirement for supplementary convergence checks concerning these parameters. In this context, this method stands out as an exact method for the stochastic simulation of laser dynamics. There have been some advancements over the last years, as reported in Ref. [46]. While all these improvements are valuable, simulating each event one at a time, regardless of efficiency, may still be too slow for many real-world applications. As a result, researchers have explored alternative strategies that trade-off some of the accuracies of the exact simulation for faster simulation times. One such strategy is known as the tau-leaping algorithm.

The approximate explicit tau-leaping algorithm: In pursuit of a faster algorithm, the fundamental questions posed at each iteration are altered to inquire, "How many events of type μ occurred during the time increment τ?" for M different event types. Assuming that the system is in state $\mathbf{x}$ at time t, a time increment $\tau > 0$ that satisfies the leap condition can be defined as the period of time $[t, t+\tau)$ during which no propensity function $a_\mu(\mathbf{x})$ is expected to change hugely. Since $a_\mu(\mathbf{x})$ remains nearly constant, the fundamental premise in Eq. 2.12 implies that the number of μth event fires within $[t, t+\tau)$ is a Poisson random variable $\mathbb{P}_\mu\left[a_\mu(\mathbf{x})\tau\right]$ with a mean of $a_\mu(\mathbf{x})\tau$. Hence, if the leap condition is satisfied, the system can be approximately leapt forward in time by τ [53]:

$$\mathbf{x}(t+\tau) = \mathbf{x}(t) + \sum_{\mu=1}^{M} \mathbb{P}_\mu\left[a_\mu(\mathbf{x})\tau\right]\mathbf{v}_\mu \tag{2.14}$$

The system's state-change vector is $\mathbf{v}_\mu \equiv (v_{1\mu}, v_{2\mu}, \ldots)$, where $v_{i\mu}$ represents the change in the number of photons and carriers caused by the μth event. When the system is in state $\mathbf{x}$ and the μth event fires, the system instantaneously jumps to a new state $\mathbf{x} + \mathbf{v}_\mu$. The core tau-leaping formulation is thus given in Eq. 2.14. While satisfying the leap condition is desirable, it can sometimes demand an exceedingly small τ, resulting in only a minimal number of events being leapt over. In such cases, it might be more efficient to bypass the tau-leaping algorithm and resort to the direct method algorithm instead.[3] To this end, it is crucial to carefully choose the tau-leap step τ, ensuring it is sufficiently large to achieve statistical convergence while preventing unrealistic population values:

$$a_\mu(\mathbf{x})\tau \gg 1 \text{ for all } \mu = 1, \ldots, M \tag{2.15}$$

[3] For instance, if one selects a relatively small value for τ, such as $1/a_0$ as demonstrated in Algorithm 2.1, the ensuing leap would be approximately equivalent to the expected size of the next time increment in the direct method algorithm. However, it's very likely that only one of the generated Poisson variables would be 1, with all the others being 0. Opting for even smaller values for τ would result in leaps where all the Poisson variables are expected to be 0, offering no discernible advantage.

The tau-leap step can be determined using an algorithm reliant on the event rates [54, 55]. Nevertheless, in the context of the physical domain, a fixed time step is a necessity for the Fourier transform, making the adaptive direct method algorithm less favourable. The explicit tau-leaping algorithm can be presented as follows (Fig. 2.11):

Algorithm 2.2: The tau-Leaping algorithm

Initialize: Set the initial number of photons and carriers $\mathbf{x}_0$ and set $t = 0$.
Choose a time increment, τ.
Calculate the propensity function, a_μ, for each event.
while $t < T$ **do**
    Generate M independent Poisson random variables, $\Delta x_\mu \sim \mathbb{P}(a_\mu \tau)$, which is the number of each event occurring during the time interval $[t, t + \tau)$.
    Update the number of photons and carriers as $\mathbf{x}(t + \Delta t) = \mathbf{x}(t) + \sum_{\mu=1}^{M} \Delta x_\mu$.
    Calculate the new propensity functions, $a_\mu(t + \tau)$, using the updated photon and carrier numbers.
    Set $t = t + \tau$.
end while

Next, by using the notation $\mathbb{N}(m, \sigma^2)$ to represent a Gaussian random variable with mean m and variance σ^2, and considering the mathematical approximation that a Poisson random variable with a large mean and variance can be approximated as a Gaussian random variable with the same mean and variance, one can further approximate Eq. (2.14) as follows:

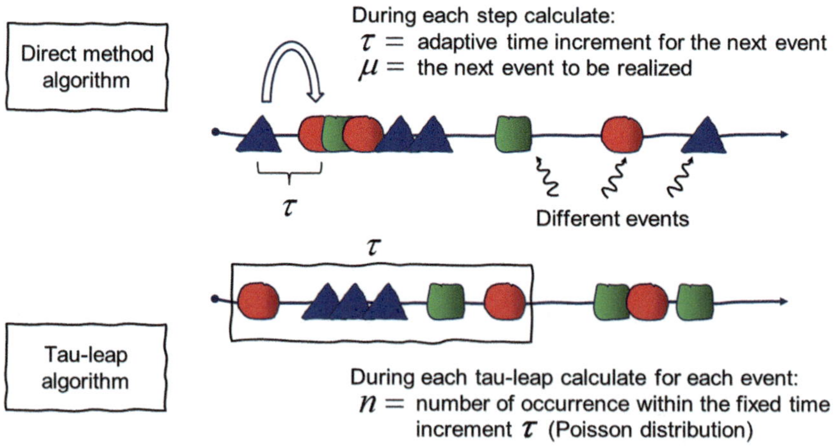

Fig. 2.11 Schematic illustration of the two Gillespie algorithms

$$\begin{aligned}\mathbf{x}(t+\tau) &= \mathbf{x}(t) + \sum_{\mu=1}^{M} \mathbb{N}\left[a_\mu(\mathbf{x})\tau, a_\mu(\mathbf{x})\tau\right]\mathbf{v}_\mu \\ &= \mathbf{x}(t) + \sum_{\mu=1}^{M} \mathbf{v}_\mu a_\mu(\mathbf{x})\tau + \sum_{\mu=1}^{M} \mathbf{v}_\mu \sqrt{a_\mu(\mathbf{x})}\sqrt{\tau}\mathbb{N}(0,1)\end{aligned} \tag{2.16}$$

The final step in this derivation makes use of a property of the Gaussian random variable, which states that $\mathbb{N}(m, \sigma^2) = m + \sigma\mathbb{N}(0, 1)$. In the realm of continuous Markov processes, it has been established that these equations can also be represented in the form of white noise:

$$\frac{\mathrm{d}\mathbf{x}(t)}{\mathrm{d}t} = \sum_{\mu=1}^{M} \mathbf{v}_\mu a_\mu(\mathbf{x}) + \sum_{\mu=1}^{M} \mathbf{v}_\mu \sqrt{a_\mu(\mathbf{x})}\Gamma_\mu(t) \tag{2.17}$$

Here, the $\Gamma_\mu(t)$ represents statistically independent Gaussian white-noise processes that fulfill the condition $\langle\Gamma_\mu(t)\Gamma_{\mu'}(t')\rangle = \delta_{\mu\mu'}\delta(t - t')$. In the *thermodynamic limit*,[4] the expression of (2.17), where the right-hand side is the combination of a deterministic drift term and a stochastic diffusion term proportional to Gaussian white noise, is commonly referred to as Langevin equations in Sect. 2.2.1.

2.2.3 Fully Quantum-Mechanical Description

Despite the successes of the semiclassical theory, there are compelling reasons to pursue a more refined treatment. Firstly, the semiclassical theory implies that laser radiation in an ideal steady state should be perfectly monochromatic. However, practical lasers are subject to various disturbances, leading to a finite radiation bandwidth. This intrinsic linewidth includes the combined effects of thermal noise, zero-point fluctuation fields, and spontaneous emission, making an accurate calculation a challenging task in nonequilibrium statistical mechanics. Secondly, the semiclassical theory falls short in explaining the spontaneous onset of oscillations, which necessitate an initial optical frequency field. Understanding how oscillations can emerge from a state without any initial radiation is a topic of significant interest, requiring the consideration of the Weisskopf–Wigner theory of spontaneous emission [56]. Thirdly, a fully quantum-mechanical approach can give the statistical distribution of energy stored in the laser cavity, referred to as "photon" statistics. This information is vital for a comprehensive analysis of the statistical distribution of photoelectrons generated by a laser.

---

[4] In a general sense, the thermodynamic limit refers to the scenario where the populations X and the volume of the system V both tend to infinity while maintaining constant species concentrations X/V. In the context of semiconductor lasers, the thermodynamic limit is specifically defined as the limit where $\beta^{-1} \rightarrow \infty$ [47].

The quantum theory of lasers has been a subject of extensive research since its emergence in the early 1960s, with contributions from prominent physicists. In the domain of semiconductor lasers, progress in all semiconductor approaches heavily depends on advancements in atomic-molecular-optics (AMO) treatments. The foundation of microscopic semiconductor laser theory can be traced back to the semi-classical gas laser theory. In this context, velocity-resolved atom populations are replaced by momentum-resolved carrier populations, and Maxwell–Boltzmann distributions are supplanted by Fermi-Dirac distributions. The active medium is treated with quantum-mechanical methods, while the radiation field can be described either semiclassically using Maxwell–Bloch equations or quantum-mechanically through quantum optics. In this overview, I will present the key concepts without delving into excessive technical details.

In the semi-classical approach, also known as Lamb laser theory [58], the radiation field is modelled using Maxwell's equations, while the response of matter is described by the optical Bloch equations. The fundamental wave equation governing the dynamics inside the laser cavity can be expressed as:

$$-\nabla^2\mathbf{E} + \mu_0\varepsilon_{bg}\frac{\partial^2\mathbf{E}}{\partial t^2} = -\mu_0\frac{\partial^2\mathbf{P}}{\partial t^2} \tag{2.18}$$

where μ_0 is the magnetic permeability of the vacuum and $\varepsilon_{bg} = n_0^2\varepsilon_0$ is the background permittivity, which includes the vacuum contribution ε_0 and the background refractive index n_0. The macroscopic polarization quantity $\mathbf{P}$ can be ascertained by solving the equations of motion for the microscopic dipole moments $\mathbf{p_i}$ within the quantum gain medium, as illustrated in Fig. 2.12. These equations should be derived from the Hamiltonian of the full system, which encompasses the contributions arising from the electric-dipole interaction between the driving field and the gain medium composed of homogeneously broadened two-level atoms. More comprehensive information on this topic can be found in Refs. [9, 33, 59–61]. Last but not least, it is important to mention that the frequency-pulling contribution in Lamb laser theory is the precursor to the carrier-induced refractive index and linewidth enhancement factor (α_H) observed in semiconductor lasers [62, 63], as discussed in

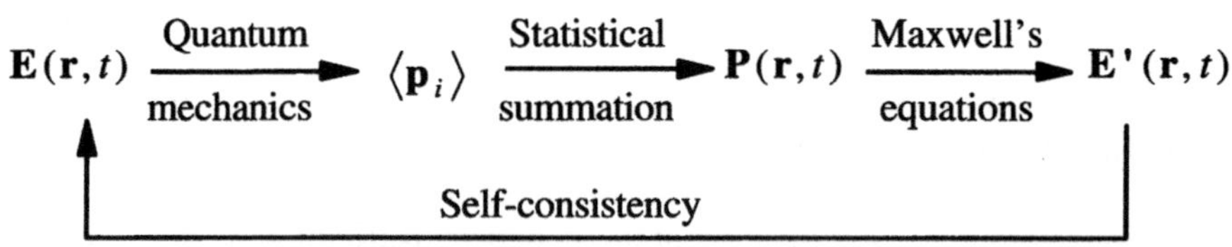

Fig. 2.12 The effect of the electromagnetic field $\mathbf{E}(\mathbf{r}, t)$ is to induce microscopic dipole moments $\mathbf{p_i}$ in the active medium according to the laws of quantum mechanics, Then the density matrix is used to facilitate the statistical summations involved in obtaining a macroscopic polarization $\mathbf{P}(\mathbf{r}, t)$ of the medium from the individual dipole moment, which acts as a source for the electromagnetic field $\mathbf{E}'(\mathbf{r}, t)$ in accordance with Maxwell's equations. The condition of *self-consistency* then requires that the resulting field $\mathbf{E}'$ should be equal to the inducing field $\mathbf{E}$. (*Source* In Page 96 of the Ref. [57])

Sect. 3.1.3. Taking a different path with the optical Bloch equations, one assumes that the dephasing process occurs much more rapidly than changes in laser intensity and carrier transport processes. This assumption gives rise to the rate equation approximation, as outlined in Sect. 3.2.

Hermann Haken and his colleagues [34] made great advancements by developing a theoretical description in the Heisenberg picture. In this picture, operators are treated as time-dependent quantities, while wave functions remain time-independent. Within this framework, they devised techniques to transform master equations into Langevin operator equations. Linearizing these equations enabled the calculation of the internal noise spectral density, which, in turn, allowed the examination of laser light properties such as coherence and quantum fluctuations. This approach provides a straightforward understanding and facilitates direct comparisons with experimental power spectra through the Fourier transform. However, accurately obtaining photon-number statistics remains a challenge in this approach. It's worth noting that the early papers primarily focused on gas lasers with closed cavities, limiting their applicability. Yamamoto [64] highlighted the importance of considering the coupling of the laser cavity to external optical modes for a correct description of shot noise in the intensity noise spectrum of lasers. Additionally, the introduction of input/output boundary conditions [65] further enhanced the understanding of laser theory by enabling the calculation of output noise spectra based on internal noise spectra. An example of this approach applied to semiconductor lasers can be found in [64]. As illustrated in Fig. 2.13, the laser system can be described by employing three quantum Langevin equations. These equations are derived after the adiabatic elimination of the dipole-moment operator, and they govern the dynamics of the inversion operator $\hat{N}_c(t)$ and the slowly varying internal optical field annihilation operator $\hat{A}(t)$, alongside the external field $\hat{r}(t)$.

$$
\begin{aligned}
\frac{\mathrm{d}}{\mathrm{d}t}\hat{N}_c(t) &= P - (1-\beta)\frac{\hat{N}_c(t)}{\tau_{sp}} - (\hat{E}_{cv} - \hat{E}_{vc})\hat{n}(t) - \left\langle \hat{E}_{cv} \right\rangle + \hat{\Gamma}_p(t) + \hat{\Gamma}_{sp}(t) + \hat{\Gamma}(t) \\
\frac{\mathrm{d}}{\mathrm{d}t}\hat{A}(t) &= \frac{1}{2}\left[\frac{\omega}{Q_i} + \frac{\omega}{Q_o} + 2j(\omega-\omega_0) - \frac{\omega}{\mu^2}(\hat{\chi}_i - j\hat{\chi}_r)\right]\hat{A}(t) + \hat{G}(t) + \hat{g}(t) + \sqrt{\frac{\omega}{Q_e}}\hat{f}_e(t) \\
\hat{r} &= -\hat{f}_e(t) + \frac{\hat{A}(t)}{\sqrt{\omega/Q_e}}
\end{aligned}
\tag{2.19}
$$

For a detailed explanation of all the terms and their significance, please consult Ref. [64]. Here, to maintain simplicity, I refrain from providing an exhaustive description of each term. Lastly, it is noteworthy that the quantum mechanical laser equations bear a striking resemblance to their semiclassical counterparts in Sect. 2.2.1. Despite the use of operators, their physical interpretation can still be understood at a classical level.

Another alternative approach utilizes the Fokker–Planck equation in the P representation (see Appendix C) or the density matrix in the Schrödinger picture [57, 66]. In this picture, the state vectors or wave functions evolve in time, while the operators remain mostly constant with respect to time. Therefore, the time evolution of

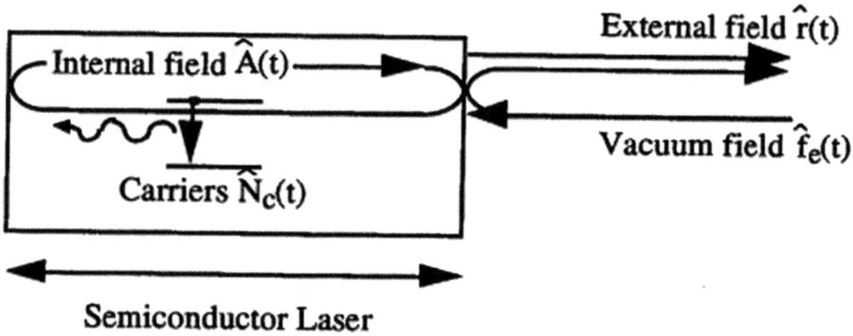

Fig. 2.13 The basic quantum Langevin model for the semiconductor lasers. (*Source* In Ref. [64])

the reduced density matrix ρ is traced. The statistics of photon numbers are connected to the diagonal elements $\rho_{n,n}$, and their master equation in the photon number representation is given by [57]:

$$\dot{\rho}_{n,n} = -\left[A - B(n+1)\right](n+1)\rho_{n,n} + (A - Bn)n\rho_{n-1,n-1} - \frac{\omega}{Q}n\rho_{n,n} + \frac{\omega}{Q}(n+1)\rho_{n+1,n+1} \tag{2.20}$$

The physical interpretation of (2.20) is succinctly depicted in Fig. 2.14. This approach offers computational convenience for calculating photon statistics while capturing the time evolution with full information but it is somehow difficult to determine power spectra. In the steady-state condition, the net flow between adjacent diagonal elements $\rho_{n,n}$ must be precisely zero to maintain time invariance. This condition, known as detailed balance, is expressed as:

$$\frac{\omega}{Q}(n+1)\rho_{n+1,n+1} - \left[A - (n+1)B\right](n+1)\rho_{n,n} = 0 \tag{2.21}$$

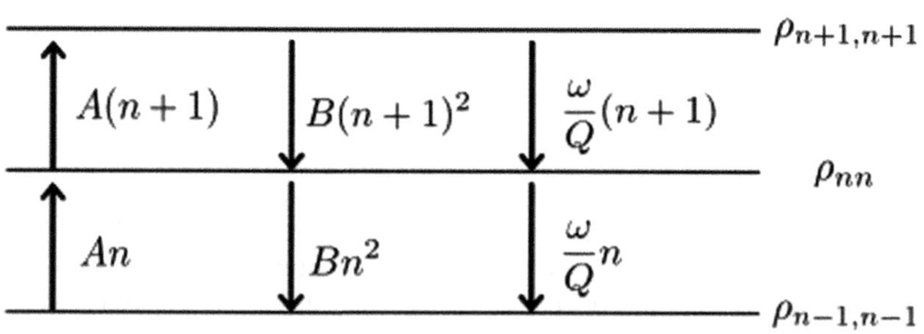

Fig. 2.14 The flow of the diagonal elements $\rho_{n,n}$ to the neighboring $\rho_{n-1,n-1}$ and $\rho_{n+1,n+1}$. An upward flow proportional to $A(n+1)$ represents stimulated and spontaneous emission of photons, while two downward flows proportional to Bn^2 and $\frac{\omega}{Q}n$ correspond to nonlinear atomic absorption (gain saturation) and linear photon decay from the cavity, respectively. (*Source* In Ref. [57])

By induction and also the conservation of probability ($\sum_{n=0}^{\infty} \rho_{n,n} = 1$), one can get

$$\rho_{n,n} = \rho_{0,0} \prod_{k=1}^{n} \frac{A - Bk}{\omega/Q}, \quad \text{with } \rho_{0,0} = \left(1 + \sum_{n=0}^{\infty} \prod_{k=1}^{n} \frac{A - Bk}{\omega/Q}\right)^{-1} \tag{2.22}$$

For operation below the laser threshold ($\langle n \rangle = \sum n\rho_n = \frac{A}{(\omega/Q)} \ll 1$), the fraction $\frac{(A-Bk)}{(\omega/Q)}$ can be approximated as $\frac{A}{\omega/Q}$. Therefore, $\rho_{n,n}$ exhibits an exponentially decreasing value:

$$\rho_{n,n} \approx \left(\frac{A}{\omega/Q}\right)^n \rho_{0,0} = \langle n \rangle^n \rho_{0,0} = \frac{1}{1 + \langle n \rangle} \left(\frac{\langle n \rangle}{1 + \langle n \rangle}\right)^n \tag{2.23}$$

At the laser threshold ($\langle n \rangle \sim 1$), the fraction $\frac{(A-Bk)}{\omega/Q}$ is less than unity for all $k > 0$. As a result, the largest probability corresponds to $\rho_{0,0}$ and the probabilities of photon number decrease monotonically $\rho_{0,0} = (\frac{1}{2})^{n+1}$. In the case of very high excitation ($\langle n \rangle \gg 1$), the average photon number $\langle n \rangle$ tends to $\frac{A^2}{B(\omega/Q)}$, and the photon numbers around $\langle n \rangle$ exceed $\frac{A}{B}$ by the ratio $\frac{A}{\omega/Q}$. As a consequence, the photon statistics follow the Poissonian distribution and approach a *coherent states*, which will be discussed in Sect. 4.1.

$$\rho_{n,n} \approx \frac{\rho_{0,0}^{-1} \langle n \rangle^n}{n!} = \frac{e^{-\langle n \rangle} \langle n \rangle^n}{n!} \tag{2.24}$$

Finally, as the diagonal elements $\rho_{n,n}$ of the density operator become nonzero, the off-diagonal elements $\rho_{n,n+1}$ decay to zero. This decay signifies the simultaneous decay of the field, ultimately leading to the generation of the laser linewidth. For a more comprehensive understanding, please refer to Chap. 17 of Ref. [57]. After that, Melvin Lax established that these two formulations are equivalent when the photon number inside the cavity is sufficiently large [67]. Moreover, this theory determines the photocount distributions of laser light from modes both below and above the threshold, and theoretical results align excellently with experiments conducted on gas lasers [55, 68].

In conclusion, a bird's eye view of modelling semiconductor lasers is provided in Fig. 2.15. In the classical treatment of radiation, the optical Bloch equations are habitually used in conjunction with Maxwell's equations. The Lamb laser theory [58] allows for the elimination of the active-medium equations when perturbation theory is applicable. The rate equation approximation can be employed when the polarization dynamics are eliminated [36]. In the quantized treatment of radiation, there are two categories of approaches. The first involves using a density matrix approach [57], while the second employs the Langevin operator approach [34, 64]. More accurate treatments, such as the quantum luminescence theory and the Lindblad approach, can be applied subsequently [61, 69].

The active gain medium in semiconductor lasers comprises charge carriers that interact through the Coulomb potential. When coupled with quantum statistics, this

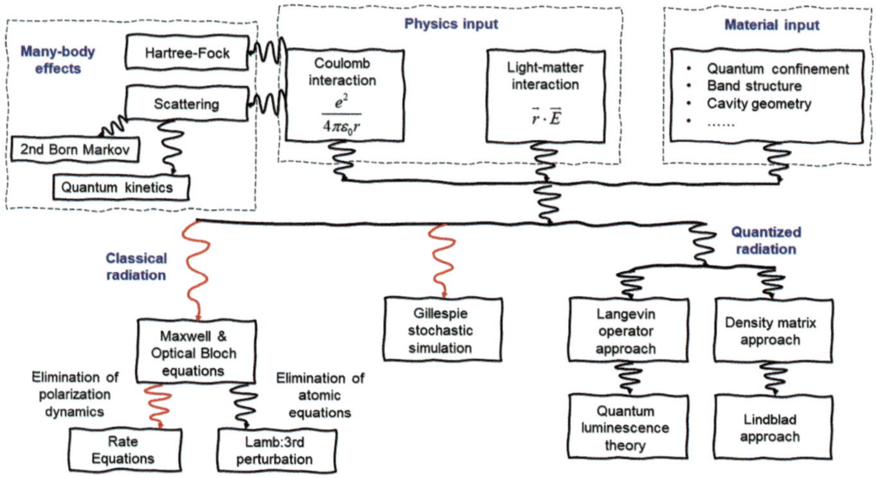

Fig. 2.15 The development of the semiclassical and quantum optical theories for semiconductors, with red lines indicating the methodologies employed in this thesis. (*Source* In Ref. [61])

leads to many-body effects [3] (left side of Fig. 2.15). In most modelling, the Hartree–Fock effects are addressed within the screened Hartree–Fock approximation, and scattering effects are described using effective relaxation rates. However, a rigorous treatment is warranted when a more in-depth examination of the active material is needed. This improved treatment involves delaying the truncation of the expansion in Coulomb correlations, extending it beyond the Hartree–Fock level when utilizing two different approaches. By using the second Born and Markov approximations to describe collisions, researchers have achieved very good agreement with experimental gain spectra for bulk and QW systems [9]. However, for QD systems, a more complex approach involving quantum kinetic equations is required due to their unique characteristics [69].

References

1. Kirstaedter N, Ledentsov N, Grundmann M, Bimberg D, Ustinov V, Ruvimov S, Maximov M, Kopév PS, Alferov ZI, Richter U et al (1994) Low threshold, large T_0 injection laser emission from (InGa) as quantum dots. Electron Lett 30(17):1416–1417. https://doi.org/10.1049/el:19940939
2. Grundmann M, Stier O, Bimberg D (1995) InAs/GaAs pyramidal quantum dots: strain distribution, optical phonons, and electronic structure. Phys Rev B 52(16):11969. https://doi.org/10.1103/PhysRevB.52.11969
3. Nielsen TR, Gartner P, Jahnke F (2004) Many-body theory of carrier capture and relaxation in semiconductor quantum-dot lasers. Phys Rev B 69(23):235314. https://doi.org/10.1103/PhysRevB.69.235314

4. Gioannini M (2006) Analysis of the optical gain characteristics of semiconductor quantum-dash materials including the band structure modifications due to the wetting layer. IEEE J Quantum Electron 42(3):331–340. https://doi.org/10.1109/JQE.2006.869811
5. Cornet C, Platz C, Caroff P, Even J, Labbé C, Folliot H, Le Corre A, Loualiche S (2005) Approach to wetting-layer-assisted lateral coupling of InAs/InP quantum dots. Phys Rev B 72(3):035342. https://doi.org/10.1103/PhysRevB.72.035342
6. Nozik AJ (2008) Multiple exciton generation in semiconductor quantum dots. Chem Phys Lett 457(1–3):3–11. https://doi.org/10.1021/jz200166y
7. Chow WW, Koch SW (2005) Theory of semiconductor quantum-dot laser dynamics. IEEE J Quantum Electron 41(4):495–505
8. Ludge K, Scholl E (2009) Quantum-dot lasers–desynchronized nonlinear dynamics of electrons and holes. IEEE J Quantum Electron 45(11):1396–1403. https://doi.org/10.1109/JQE.2009.2028159
9. Chow WW, Koch SW, Sargent MI (2012) Semiconductor-laser physics. Springer Science & Business Media
10. Marcinkevičius S, Leon R (1999) Carrier capture and escape in $In_xGa_{1-x}As$/GaAs quantum dots: effects of intermixing. Phys Rev B 59(7):4630
11. Siegert J, Marcinkevičius S, Zhao QX (2005) Carrier dynamics in modulation-doped InAs/GaAs quantum dots. Phys Rev B 72(8):085316
12. Veselinov K, Grillot F, Cornet C, Even J, Bekiarski A, Gioannini M, Loualiche S (2007) Analysis of the double laser emission occurring in 1.55-μm InAs-InP (113) b quantum-dot lasers. IEEE J Quantum Electron 43(9):810–816. https://doi.org/10.1109/JQE.2007.902386
13. Ohnesorge B, Albrecht M, Oshinowo J, Forchel A, Arakawa Y (1996) Rapid carrier relaxation in self-assembled $In_xGa_{1-x}As$/GaAs quantum dots. Phys Rev B 54(16):11532
14. Uskov A, McInerney J, Adler F, Schweizer H, Pilkuhn M (1998) Auger carrier capture kinetics in self-assembled quantum dot structures. Appl Phys Lett 72(1):58–60
15. Benisty H, Sotomayor-Torres C, Weisbuch C (1991) Intrinsic mechanism for the poor luminescence properties of quantum-box systems. Phys Rev B 44(19):10945. https://doi.org/10.1103/PhysRevB.44.10945
16. Miller SE (1969) Integrated optics: an introduction. Bell Syst Techn J 48(7):2059–2069
17. Kish F, Lal V, Evans P, Corzine SW, Ziari M, Butrie T, Reffle M, Tsai H-S, Dentai A, Pleumeekers J et al (2017) System-on-chip photonic integrated circuits. IEEE J Sel Topics Quantum Electron 24(1):1–20
18. Liang D, Bowers JE (2010) Recent progress in lasers on silicon. Nat Photonics 4(8):511–517. https://www.nature.com/articles/nphoton.2010.167
19. Stojanović V, Ram RJ, Popović M, Lin S, Moazeni S, Wade M, Sun C, Alloatti L, Atabaki A, Pavanello F et al (2018) Monolithic silicon-photonic platforms in state-of-the-art CMOS SOI processes. Optics Express 26(10):13106–13121. https://doi.org/10.1364/OE.26.013106
20. Xiang C, Jin W, Terra O, Dong B, Wang H, Wu L, Guo J, Morin TJ, Hughes E, Peters J et al (2023) 3D integration enables ultralow-noise isolator-free lasers in silicon photonics. Nature 620(7972):78–85. https://doi.org/10.1038/s41586-023-06251-w
21. Fang AW, Park H, Cohen O, Jones R, Paniccia MJ, Bowers JE (2006) Electrically pumped hybrid AlGaInAs-silicon evanescent laser. Optics Express 14(20):9203–9210
22. Wan Y, Xiang C, Guo J, Koscica R, Kennedy M, Selvidge J, Zhang Z, Chang L, Xie W, Huang D et al (2021) High speed evanescent quantum-dot lasers on Si. Laser & Photonics Rev 15(8):2100057
23. Liu AY, Bowers J (2018) Photonic integration with epitaxial III-V on silicon. IEEE J Sel Topics Quantum Electron 24(6):1–12. https://doi.org/10.1109/JSTQE.2018.2854542
24. Wei W-Q, He A, Yang B, Wang Z-H, Huang J-Z, Han D, Ming M, Guo X, Su Y, Zhang J-J et al (2023) Monolithic integration of embedded III-V lasers on SOI. Light: Sci & Appl 12(1):84
25. Shang C, Feng K, Hughes ET, Clark A, Debnath M, Koscica R, Leake G, Herman J, Harame D, Ludewig P et al (2022) Electrically pumped quantum-dot lasers grown on 300 mm patterned Si photonic wafers. Light: Sci & Appl 11(1):299

26. Chen S, Li W, Wu J, Jiang Q, Tang M, Shutts S, Elliott SN, Sobiesierski A, Seeds AJ, Ross I et al (2016) Electrically pumped continuous-wave III-V quantum dot lasers on silicon. Nat Photonics 10(5):307–311
27. Shang C, Wan Y, Selvidge J, Hughes E, Herrick R, Mukherjee K, Duan J, Grillot F, Chow WW, Bowers JE (2021) Perspectives on advances in quantum dot lasers and integration with Si photonic integrated circuits. ACS Photonics 8(9):2555–2566
28. Gérard J, Cabrol O, Sermage B (1996) InAs quantum boxes: highly efficient radiative traps for light emitting devices on Si. Appl Phys Lett 68(22):3123–3125
29. Norman JC, Jung D, Wan Y, Bowers JE (2018) Perspective: the future of quantum dot photonic integrated circuits. APL Photonics 3(3)
30. Yang RQ (1995) Infrared laser based on intersubband transitions in quantum wells. Superlattices Microstruct 17(1):77–83. https://doi.org/10.1006/spmi.1995.1017
31. Vurgaftman I, Bewley W, Canedy C, Kim C, Kim M, Merritt C, Abell J, Lindle J, Meyer J (2011) Rebalancing of internally generated carriers for mid-infrared interband cascade lasers with very low power consumption. Nat Commun 2(1):585. https://doi.org/10.1038/ncomms1595
32. Meyer JR, Bewley WW, Canedy CL, Kim CS, Kim M, Merritt CD, Vurgaftman I (2020) The interband cascade laser. Photonics 7(3):75. MDPI. https://doi.org/10.3390/photonics7030075
33. Louisell WH (1973) Quantum statistical properties of radiation
34. Haken H (1985) Laser light dynamics, vol 2. North-Holland Amsterdam
35. Yamamoto Y (1991) Coherence, amplification, and quantum effects in semiconductor lasers. Wiley series in pure and applied optics. Wiley
36. Coldren LA, Corzine SW, Mashanovitch ML (2012) Diode lasers and photonic integrated circuits. Wiley
37. DeGiorgio V, Scully MO (1970) Analogy between the laser threshold region and a second-order phase transition. Phys Rev A 2(4):1170. https://doi.org/10.1103/PhysRevA.2.1170
38. Graham R, Haken H (1970) Laserlight–first example of a second-order phase transition far away from thermal equilibrium. Zeitschrift f ü r Physik 237(1):31–46. https://doi.org/10.1007/BF01400474
39. Einstein A (1917) Zur Quantentheorie der Strahlung. Physikalische Zeitschrift 18:121–128. https://doi.org/10.1515/9783112596609-016
40. Blood P (2015) Quantum confined laser devices: optical gain and recombination in semiconductors. OUP Oxford
41. Langevin P (1908) Sur la théorie du mouvement brownien. Compt Rendus 146:530–533
42. Gardiner CW et al (1985) Handbook of stochastic methods. Springer, Berlin
43. Lax M (1960) Fluctuations from the nonequilibrium steady state. Rev Modern Phys 32(1):25. https://doi.org/10.1103/RevModPhys.32.25
44. McCumber D (1966) Intensity fluctuations in the output of CW laser oscillators. I. Phys Rev 141(1):306. https://doi.org/10.1103/PhysRev.141.306
45. Gillespie DT (1977) Exact stochastic simulation of coupled chemical reactions. J Phys Chem 81(25):2340–2361. https://doi.org/10.1021/j100540a008
46. Gillespie DT (2007) Stochastic simulation of chemical kinetics. Annu Rev Phys Chem 58:35–55. https://doi.org/10.1146/annurev.physchem.58.032806.104637
47. Rice PR, Carmichael H (1994) Photon statistics of a cavity-QED laser: a comment on the laser-phase-transition analogy. Phys Rev A 50(5):4318. https://doi.org/10.1103/PhysRevA.50.4318
48. Mørk J, Lippi G (2018) Rate equation description of quantum noise in nanolasers with few emitters. Appl Phys Lett 112(14)
49. Roy-Choudhury K, Haas S, Levi A (2009) Quantum fluctuations in small lasers. Phys Rev Lett 102(5):053902
50. Lebreton A, Abram I, Takemura N, Kuwata-Gonokami M, Robert-Philip I, Beveratos A (2013) Stochastically sustained population oscillations in high-β nanolasers. J Phys 15(3):033039. https://doi.org/10.1088/1367-2630/15/3/033039
51. Mørk J, Yvind K (2020) Squeezing of intensity noise in nanolasers and nanoLEDs with extreme dielectric confinement. Optica 7(11):1641–1644

52. Puccioni G, Lippi G (2015) Stochastic Simulator for modeling the transition to lasing. Optics Express 23(3):2369–2374. https://doi.org/10.1364/OE.23.002369
53. Gillespie DT (2001) Approximate accelerated stochastic simulation of chemically reacting systems. J Chem Phys 115(4):1716–1733
54. Cao Y, Gillespie DT, Petzold LR (2006) Efficient step size selection for the tau-leaping simulation method. J Chem Phys 124(4):044109
55. André EC, Mørk J, Wubs M (2020) Efficient stochastic simulation of rate equations and photon statistics of nanolasers. Optics Express 28(22):32632–32646
56. Weisskopf V, Wigner E (1930) Berechnung der natürlichen linienbreite auf grund der diracschen lichttheorie. Zeitschrift f ü r Physik 63:54–73
57. Sargent M, Scully M, Lamb W (1977) Laser physics. Addison-Wesley
58. Lamb WE Jr (1964) Theory of an optical maser. Phys Rev 134(6A):A1429
59. Tromborg B, Lassen HE, Olesen H (1994) Traveling wave analysis of semiconductor lasers: modulation responses, mode stability and quantum mechanical treatment of noise spectra. IEEE J Quantum Electron 30(4):939–956
60. Vey J-L, Gallion P (1997) Semiclassical model of semiconductor laser noise and amplitude noise squeezing. I. Description and application to Fabry-Perot laser. IEEE J Quantum Electron 33(11):2097–2104
61. Chow WW, Jahnke F (2013) On the physics of semiconductor quantum dots for applications in lasers and quantum optics. Prog Quantum Electron 37(3):109–184. https://doi.org/10.1016/j.pquantelec.2013.04.001
62. Lax M (1967) Classical noise. V. Noise in self-sustained oscillators. Phys Rev 160(2):290
63. Henry C (1982) Theory of the linewidth of semiconductor lasers. IEEE J Quantum Electron 18(2):259–264. https://doi.org/10.1109/JQE.1982.1071522
64. Yamamoto Y, Machida S, Nilsson O (1986) Amplitude squeezing in a pump-noise-suppressed laser oscillator. Phys Rev A 34(5):4025
65. Gardiner CW, Collett MJ (1985) Input and output in damped quantum systems: quantum stochastic differential equations and the master equation. Phys Rev A 31(6):3761
66. Scully MO, Lamb WE Jr (1967) Quantum theory of an optical maser. I. General theory. Phys Rev 159(2):208
67. Lax M, Louisell W (1969) Quantum noise. XII. Density-operator treatment of field and population fluctuations. Phys Rev 185(2):568
68. Arecchi F, Berne A, Sona A (1966) Measurement of the time evolution of a radiation field by joint photocount distributions. Phys Rev Lett 17(5):260
69. Carroll MA, D'alessandro G, Lippi GL, Oppo G-L, Papoff F (2021) Thermal, quantum antibunching and lasing thresholds from single emitters to macroscopic devices. Phys Rev Lett 126(6):063902

Chapter 3
Classical Noise and Feedback Dynamics in Quantum-Confined Lasers

This chapter focuses on exploring classical noise and feedback dynamics in quantum-confined lasers. In semiconductor lasers, the presence of classical noise and fluctuations associated with the lasing process can significantly affect both the intensity and phase of the optical field, leading to frequency and intensity noises. Additionally, understanding the dynamic behaviour of laser diodes within optical or optoelectronic feedback loops is of great importance.

The chapter begins by emphasizing the significance of low relative intensity noise (RIN) in optical communication systems. Furthermore, optical sources with low-frequency noise (FN) are highly sought after for applications requiring narrow linewidth operation, and state-of-the-art results in this regard will be exhibited. The subsequent investigation involves numerically analyzing the RIN and FN characteristics of quantum dot (QD) lasers and interband cascade lasers (ICLs) using the Langevin rate equation approach, which considers both spontaneous emission and carrier contributions. To study the FN spectrum, the rate equation model incorporates the phase of the electric field. It is worth noting that carrier fluctuations introduce additional phase fluctuations due to the coupling between carrier density and refractive index, as described by the *linewidth enhancement factor* (α_H-factor).

Time-delayed optical or optoelectronic feedback is a well-known mechanism capable of modifying laser dynamics. Even feedback strengths below 1% can trigger complex nonlinear dynamics in certain laser systems, such as periodic or chaotic oscillations. Optoelectronic feedback, on the other hand, is known to generate square-wave oscillations. While these complex dynamics are advantageous in many applications, they can be detrimental to the operational stability in the context of photonic integrated circuits (PICs) used in high-speed optical interconnects. Thus, at the end of this chapter, I also delve into investigating the impact of time-delayed feedback on QD lasers and ICLs, aiming to gain deeper insights into the nonlinear dynamics of semiconductor lasers.

S. Zhao, *Noise, Dynamics and Squeezed Light in Quantum Dot and Interband Cascade Lasers*, Springer Theses, https://doi.org/10.1007/978-3-031-66818-0_3

3.1 Introduction to the Classical Noise

Classical noise exerts a significant influence on the performance and capabilities of photonics applications, presenting limitations and impacting various aspects. In coherent optical communications and interference-based sensors, semiconductor laser frequency noise (FN) or phase noise stands out as a primary source of noise [1]. This noise originates from inherent phase fluctuations within the laser itself, setting a fundamental sensitivity limit for these systems. Furthermore, the spectral linewidth of semiconductor lasers surpasses the Schawlow-Townes limit due to amplitude-phase coupling mediated by excited carriers. This coupling not only affects the linewidth but also determines the characteristics of direct frequency modulation in semiconductor lasers. In optical data transmission, minimizing laser intensity noise is critical for effective analogue modulation [2]. This necessity ensures the faithful and high-quality transmission of signals, guaranteeing clear and accurate representation upon reception. For high-speed long-distance digital transmission systems, maintaining low bit-error rates is of utmost importance. Achieving stable single-mode operation, free from sporadic mode jumps, is essential to reduce noise. This phenomenon, known as mode-partition noise, must be mitigated to ensure reliable and error-free transmission.

Within the classical framework of the dual wave-particle nature, light is frequently understood as an electromagnetic wave. The following wave equation serves as the mathematical description for a single-mode electromagnetic field in an isotropic and insulating medium:

$$\nabla^2 \mathbf{E}(\mathbf{r}, t) - \frac{1}{(c/n)^2} \frac{\partial^2}{\partial t^2} \mathbf{E}(\mathbf{r}, t) = 0 \tag{3.1}$$

with a solution for the electric field vector $\mathbf{E}(\mathbf{r}, t)$, given the dimensionless complex amplitude α

$$\mathbf{E}(\mathbf{r}, t) = E_0 \cos(\omega t - \phi) u(\mathbf{r}) = (X_1 \cos \omega t + X_2 \sin \omega t) u(\mathbf{r}) = \frac{1}{2}\left[\alpha e^{-i\omega t} + \alpha^* e^{i\omega t}\right] u(\mathbf{r}) \tag{3.2}$$

the variables $X_1 = E_0 \cos \phi = \frac{1}{2}(\alpha + \alpha^*)$ and $X_2 = E_0 \sin \phi = \frac{1}{2i}(\alpha - \alpha^*)$ signify the in-phase and quadrature components in Cartesian coordinates, respectively. The electric field amplitude E_0 is related to the photon number S by $E_0 = \xi_0 \sqrt{S}$, where $\xi_0 = \sqrt{\frac{\hbar \omega}{\varepsilon_0 V}}$ denotes the electric field generated by a single photon within the volume V. The spatial mode function $u(\mathbf{r})$ is determined by boundary conditions and satisfies the normalization condition $\int |u(\mathbf{r})|^2 dr^3 = V$. When describing interference or diffraction in classical wave scenarios, a commonly used tool is the *phasor diagram*, depicted in the left side of Fig. 3.1, which represents the field at a specific point in space and time. In this diagram, each wave corresponds to a specific vector originating from the origin and extending to the point (X_1, X_2).

The classical model has traditionally assumed that light is composed of a continuous succession of electromagnetic waves. However, this idealization deviates from reality. Atoms that emit light possess finite lifetimes, resulting in bursts of light

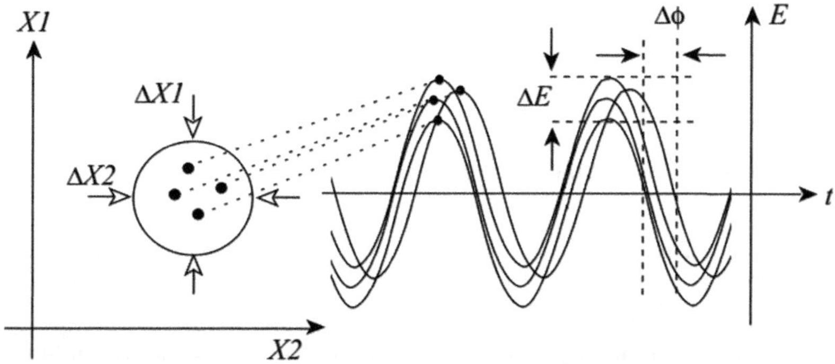

Fig. 3.1 Fluctuations depicted in the phasor diagram illustrating the classical limit. The wave can exhibit fluctuations in both its in-phase and quadrature components, which correspond to variations in the temporal traces. (*Source* On page 18 in Ref. [3])

emission. Given the brief nature of these lifetimes compared to the detection interval, these light bursts often elude individual detection. Consequently, what emerges is a well-defined average intensity that oscillates on time scales exceeding the detection interval. Similarly, most classical or thermal light sources consist of a large assembly of atoms, each independently emitting light. Additionally, these atoms are subject to motion, engendering frequency shifts, and they can interact through collisions, which disrupt the emission process and introduce phase changes.

These phenomena give rise to two implications. **Intensity Noise**: Every light source inevitably exhibits some degree of intensity noise. The intensity of light isn't constant; it experiences fluctuations over time. These fluctuations arise from variations in the number of emitting atoms, as well as sudden phase shifts or discontinuities in the light emitted by individual atoms. **Spectral Lineshape**: Real-world light sources exhibit a spectral line shape. In reality, the frequency of light doesn't remain perfectly constant; instead, it undergoes rapid changes over time. To accurately depict any light source, a spectral distribution encompassing an extended spectrum, rather than a single delta function, is indispensable.

The assessment of the statistical properties of light can be carried out in two alternative ways. The first entails measuring the fluctuations directly, either in the configuration of temporal traces or noise power spectra. The second method involves evaluating coherence or correlation functions. Despite requiring different equipment and methodologies, these two concepts are interconnected. In the upcoming Chap. 4, I shall delve into how these statistical properties can be practically modelled and how they have been instrumental in distinguishing between the classical and quantum nature of light. This chapter primarily focuses on the first method. Moreover, investigating spectral lineshape and frequency noise is crucial as they play a significant role in various applications. In applications such as spectroscopy, where high-resolution measurements are required, a well-defined spectral lineshape is essential. In coherent optical communication, the spectral purity and stability of laser sources are vital for maintaining the integrity of transmitted information. Frequency noise can lead to phase fluctuations and signal degradation, affecting the system's ability to accurately detect and decode data.

3.1.1 Relative Intensity Noise

The relative intensity noise (RIN) has a key role in determining the signal-to-noise ratio (SNR) and the bit error rate (BER) in signal transmission, respectively. Figure 3.2 provides a visual representation of the noisy intensity-modulated laser output signals in both analogue and digital applications. In analogue applications, the noise is quantified through the SNR, which is defined as follows:

$$\mathrm{SNR} = \frac{\langle i_S^2 \rangle}{\langle i_N^2 \rangle} = \frac{\langle (P_1 \sin \omega t)^2 \rangle}{\langle \delta P(t)^2 \rangle} = \frac{m^2}{2} \frac{P_0^2}{\langle \delta P(t)^2 \rangle} \tag{3.3}$$

where the intensity modulation index is given by $m = P_1/P_0$. In digital applications, a decision level is set at the midpoint to distinguish between "0" and "1" signals. However, false recordings can occur if the magnitude of the noisy term $|\delta P(t)|$ exceeds half of the signal power $P_0/2$. Assuming the noise follows a Gaussian distribution around the mean power level, in order to achieve a BER lower than 10^{-9}, the probability of encountering $|\delta P(t)| > P_0/2$ needs to be less than 1 in 10^9, requiring [4]

$$\frac{1}{2}\mathrm{erfc}\left[\frac{1}{2\sqrt{2}} \frac{P_0}{\langle \delta P(t) \rangle}\right] < 10^{-9} \Rightarrow \frac{P_0^2}{\langle \delta P(t)^2 \rangle} > (11.89)^2 \tag{3.4}$$

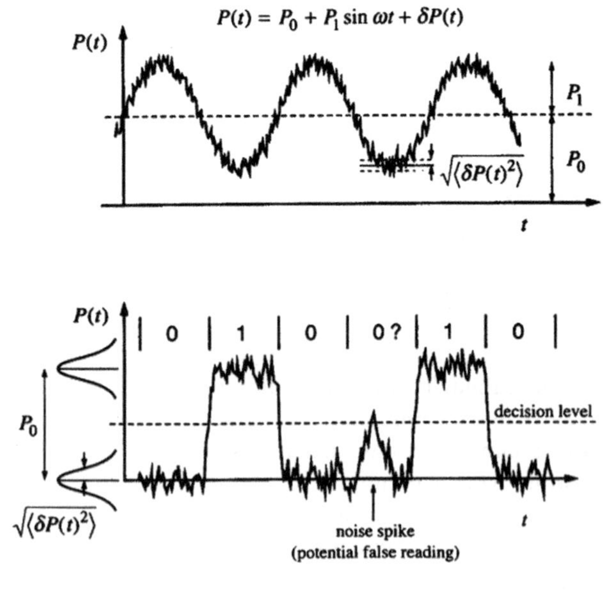

Fig. 3.2 Noise in modulated laser signals for both analogue and digital applications. (*Source* On page 222 in Ref. [5])

The RIN of a laser is commonly defined as the ratio of the power spectral density of the laser intensity fluctuations to the square of the average laser output power [5]:

$$\mathrm{RIN} \equiv \frac{\langle \delta P(t)^2 \rangle}{P_0^2} \tag{3.5}$$

The RIN, often expressed in decibels (dB) as $10\log_{10}(\mathrm{RIN})$, is represented in this manner because using a logarithmic scale allows for a more convenient comparison and characterization of noise levels over a wide range.

To characterize the output power fluctuations and thus determine the RIN, it is valuable to represent the RIN in the frequency domain. This can be achieved by employing the Fourier transform as the initial step:

$$\langle \delta P(t)^2 \rangle = \frac{1}{2\pi} \int_{-\infty}^{+\infty} S_{\delta P}(\omega)\, |F(\omega)|^2\, d\omega$$

Here, $S_{\delta P}$ represents the spectral density of $|\delta P(t)|$, while $F(\omega)$ is the passband of the spectrum analyzer's filter. Assuming that the measurement filter is centred at ω_0 and is narrowband compared to variations in the spectral density, one can simplify the analysis with $F(\omega_0) = 1$,

$$\langle \delta P(t)^2 \rangle \approx S_{\delta P}(\omega_0) \int_{-\infty}^{+\infty} |F(\omega_0)|^2\, df = S_{\delta P}(\omega_0)\dot{2}\Delta f \tag{3.6}$$

This relation is graphically illustrated in Fig. 3.3 for an arbitrary noise spectrum. The effective measurement bandwidth is $2\Delta f$ since both positive and negative frequencies must be considered. Alternatively, one could define the spectral density as "single-sided," existing only in the positive frequency domain, as indicated by the dashed line in Fig. 3.3. In this case, the measurement bandwidth would simply be Δf, and the factor of 2 would be lumped into the single-sided spectral density. The choice of using a single-sided or double-sided spectral density is academic, as long

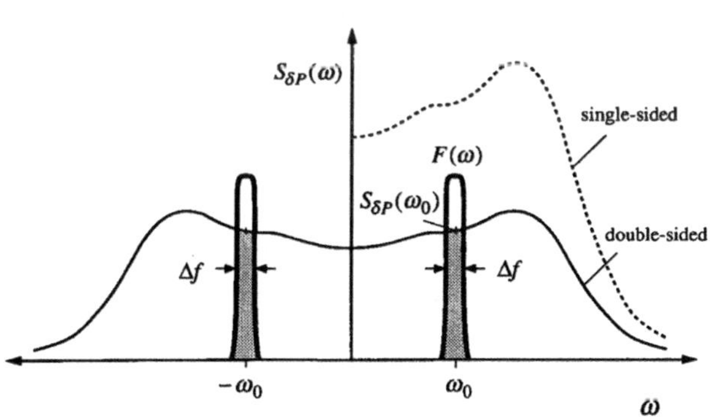

Fig. 3.3 Measured noise using a narrowband filter. One can fold the spectrum in half, if desired, and define a *single-sided* spectrum existing only in the positive frequency domain that is a factor of two larger than the *double-sided* spectrum. (*Source* On page 224 in Ref. [5])

as it is consistently applied. Finally, one defines the RIN in the frequency domain as follows:

$$\frac{\text{RIN}}{\Delta f} = \frac{2S_{\delta P}(\omega)}{P_0^2} \tag{3.7}$$

Note that this spectral density is double-sided (if it is single-sided, then the factor of 2 should be removed in (3.7)). As the measurement bandwidth can vary across different applications, it is customary to express the quantity on the left as RIN in dB/Hz or RIN per unit bandwidth. The full RIN is then determined by integrating the RIN per unit bandwidth over the detection bandwidth relevant to the practical system. When designing a communication system, the desired SNR or BER sets an upper limit on the total RIN of the laser. If the RIN spectrum is flat, then the necessary RIN per unit bandwidth of the laser can be determined from:

$$\text{RIN(dB/Hz)} = 10\log_{10}\left(\frac{\text{RIN}}{\Delta f}\right)$$

Hence, for instance, in a digital transmission link with a 2 Gbits/s (1 GHz) system bandwidth and a required BER of less than 10^{-9} (i.e., RIN < 1/11.89 or −21.5 dB as indicated in (3.4)), the laser's average RIN (dB/Hz) must be lower than −21.5 dB − 90 = −111.5 dB/Hz. If the system bandwidth is increased, the laser's RIN per unit bandwidth must be reduced to maintain the same total RIN level.

In the realm of QD lasers, many advancements have been made in minimizing the RIN level. Noteworthy experimental investigations on InAs/GaAs and InAs/InP QD devices have demonstrated impressively low RIN levels, reaching as low as −160 dB/Hz [6, 7]. Moreover, recent studies have revealed that QD lasers epitaxially grown on silicon, with or without p-doping, manifest RIN levels spanning from −140 to −150 dB/Hz, as exemplified in Fig. 3.4b and c. Both the relaxation oscillation frequency (ROF) and damping factor are extracted from the curve-fitting of the RIN spectrum through the expression [5]:

$$\text{RIN}(\omega) = \frac{a + b\omega^2}{(\omega^2 - \omega_{RO}^2)^2 + \gamma^2\omega^2} \tag{3.8}$$

with ω the angular frequency, ω_{RO} the angular ROF, γ the damping factor and a and b are used for the curve fitting. For QD lasers, the damping factor γ has rapid growth with the squared ROF f_{RO}^2 according to the relationship $\gamma = Kf_{RO}^2 + \gamma_0$. Here, the K-factor represents the slope, and the damping factor offset γ_0 is inversely related to the differential carrier lifetime τ_c, as demonstrated in Fig. 3.4e. Additionally, the K-factor is commonly used to estimate the maximum 3-dB bandwidth, given by the expression $f_{3\text{dB,max}} = 2\sqrt{2}\pi/K$.

Building upon the same principle, a recent experimental investigation focused on the RIN of an interband cascade laser (ICL) operating at 4.1 μm. The study revealed that the laser exhibited a relaxation frequency in the GHz range at room temperature

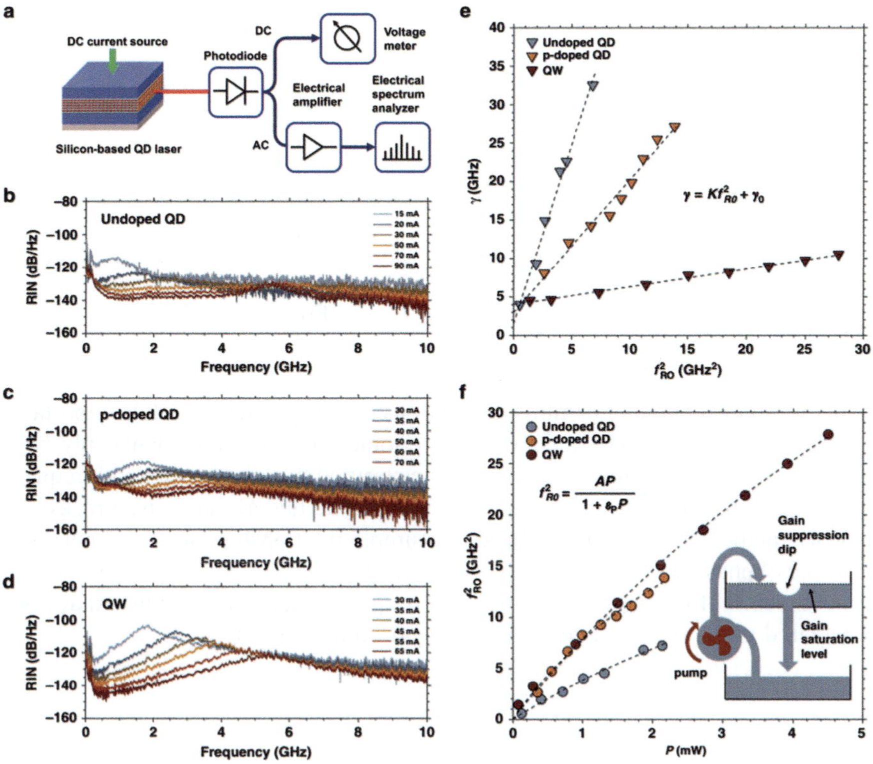

Fig. 3.4 Dynamic and RIN properties. **a** Experimental setup used for investigating the RIN of QD lasers. RIN spectra up to 10 GHz for **b** undoped QD laser, **c** p-doped QD laser and **d** QW laser. The observed high levels of low-frequency RIN can be attributed to the contributions from current source noise, thermal noise, and mode partition. **e** Damping factor (γ) as a function of the squared ROF (f_{RO}^2). **f** Squared ROF (f_{RO}^2) versus the output power (P). (*Source* In Ref. [8])

as shown in Fig. 3.5. In particular, at low frequencies ranging from 10 to 100 MHz, the RIN level was considerably high, measuring around −95 dB/Hz, which can be attributed to various noise sources including technical, thermal, and partition noises. However, the observed relaxation frequency was sufficiently high to realize multi-Gbit/s data transmission and anticipate a remarkably broad dynamic bandwidth in free space [10], which will be elaborated upon in Sect. 3.2.3.

3.1.2 Frequency Noise

In addition to intensity noise, semiconductor lasers also have frequency noise, often referred to as phase noise. This phenomenon can have a detrimental impact on the laser's spectral purity, as the output of a single-frequency laser is not perfectly

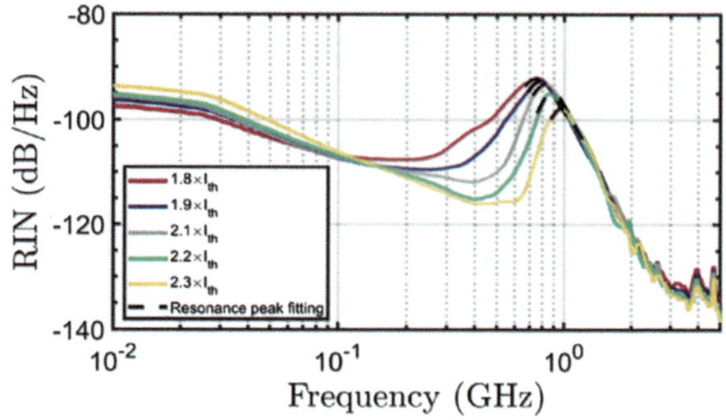

Fig. 3.5 The RIN of an ICL at 295 K for various bias currents. Dashed black lines correspond to the fitting of the resonance associated with the relaxation oscillation. It leads to the determination of the relaxation frequency around the GHz. (*Source* In Ref. [9])

monochromatic but rather exhibits fluctuations in the optical phase. It is important to note that phase noise and frequency noise are sometimes used interchangeably, which can lead to confusion. However, it should be clarified that phase noise and frequency noise represent different representations of the same underlying noise process, and they are mathematically related through a simple relationship, $\omega(t) = 2\pi\nu(t)$. The presence of phase noise can cause the laser's instantaneous frequency, $\nu(t)$, to deviate from its central frequency, ν_0, resulting in frequency fluctuations $\Delta\nu(t)$ that can be considered as the temporal derivative of the oscillation phase, $\phi(t)$:

$$\Delta\nu(t) = \nu(t) - \nu_0 = \frac{1}{2\pi}\frac{\mathrm{d}\phi(t)}{\mathrm{d}t} \tag{3.9}$$

Phase fluctuations in semiconductor lasers arise from two essential sources: (1) spontaneous emission and (2) carrier density fluctuations. Spontaneous emission, which occurs in all types of lasers, arises from the random addition of photons emitted spontaneously to the coherent resonant cavity mode. On the other hand, carrier density fluctuations primarily affect semiconductor lasers. These fluctuations result from variations in the carrier density, leading to changes in the refractive index [1]. This phenomenon, known as *frequency chirp*, becomes particularly significant during direct modulation operations. The spectral linewidth of the lasing mode serves as a well-known parameter to gauge the level of frequency noise. While single-mode lasers, such as distributed feedback (DFB) and distributed Bragg reflector (DBR) types, manage to reduce the spectral width of the laser output to quasi-single mode, the linewidth remains finite due to phase noise. In practice, narrow linewidth semiconductor lasers find indispensable applications in high-performance coherent communication systems, optical atomic clocks, frequency synthesis, spectroscopy, and distributed sensing systems. Besides, these lasers are expected to be single-frequency, energy-efficient, and amenable to monolithic integration, too. In pursuit of these objectives, semiconductor DFB lasers with strained quantum well (QW) layers have emerged as highly reliable single-wavelength sources. However, QW DFB lasers fabricated using simple grating designs often exhibit spectral linewidths on the order of a few megahertz. For a dense wavelength division multiplexing (DWDM) system with 96 channels and 50 GHz channel spacing, a 38 nm wavelength tuning range is needed and the linewidth is in the order of 100 kHz [11]. Therefore, gaining a complete understanding of the inherent linewidth is needed.

The linewidth originating from spontaneous emission is referred to as the Schawlow-Townes linewidth $(\Delta\nu)_{ST}$ [12]. In this linewidth broadening process, the full-width half maximum (FWHM) linewidth of the cold cavity is inversely proportional to the photon lifetime τ_{p0} in the absence of stimulated or spontaneous emission rate, i.e., $\Delta\nu_0 = 1/2\pi\tau_{p0}$. However, as the optical field propagates through the gain medium, the effective photon lifetime τ_p is extended due to gain compensation for cavity losses, and it can be expressed as [5]:

$$\frac{1}{\tau_p} = \frac{1}{\tau_{p0}} - \Gamma_p v_g g$$

where Γ_p is the optical confinement factor, v_g represents the group velocity in the gain medium, and g corresponds to the material gain. Consequently, the photon density at steady-state N_p is amplified due to the prolonged photon lifetime τ_p:

$$N_p = \frac{R_{sp}}{1/\tau_{p0} - \Gamma_p v_g g} = \tau_p R_{sp}$$

with $R_{sp} = v_g g n_{sp}/V$ representing the spontaneous emission rate into one optical mode, n_{sp} being the population inversion factor, and V being the volume of active region. Therefore, the correlated Schawlow-Townes linewidth $(\Delta\nu)_{ST}$ is determined by the following relationship:

$$(\Delta\nu)_{ST} = \frac{1}{2\pi\tau_p} = \frac{R_{sp}}{2\pi N_p}$$

Unfortunately, this intuitive derivation has its limitations. While the previous equation can describe the linewidth below the threshold caused by the amplified spontaneous emission, it fails to capture the behaviour above the threshold, where the intensity fluctuations are effectively suppressed, resulting in a reduction of the linewidth by a factor of 2 [13]:

$$(\Delta\nu)_{ST} = \frac{R_{sp}}{4\pi N_p} \tag{3.10}$$

To provide design guidelines for lasers with low frequency noise, an alternative definition of the Schawlow-Townes linewidth, which takes into account the laser output power $P_0 = v_g \alpha_m N_p h\nu V$ and the gain steady condition $g = \alpha$, is expressed as follows [1]:

$$(\Delta\nu)_{ST} = \frac{v_g^2 \alpha \alpha_m n_{sp} h\nu}{4\pi P_0} \tag{3.11}$$

where α and α_m are the total cavity losses[1] and the mirror losses, respectively.

Fleming and Mooradian [14] conducted careful measurements of semiconductor laser linewidth, revealing a Lorentzian lineshape and an inverse relationship between linewidth and output power, which was consistent with theoretical predictions. Surprisingly, their results showed that the linewidth was 50 times wider than what was expected based on the modified Schawlow-Townes formula in (3.11). Charles H. Henry provided an explanation for this discrepancy [1]. This explanation involves two parts. First, there's the instantaneous phase change caused by spontaneous emission. Second, there's a delayed phase change resulting from the instantaneous change in field intensity. When the laser aims to restore its steady-state field intensity, it goes through relaxation oscillations. During this period, there is a net gain change in the laser, and this phenomenon accounts for the 2-fold reduction in linewidth as predicted by the modified Schawlow-Townes formula in the above-threshold regime. It's important to note that the gain variation also induces changes in the refractive index, leading to additional phase fluctuations. To incorporate this effect, the modified Schawlow-Townes-Henry formula for the optical linewidth $(\Delta\nu)_{OL}$ needs to be corrected by a factor of $(1 + \alpha_H^2)$ [1]:

$$(\Delta\nu)_{OL} = (\Delta\nu)_{ST}(1 + \alpha_H^2) = \frac{v_g^2 \alpha \alpha_m n_{sp} h\nu}{4\pi P_0}(1 + \alpha_H^2) \tag{3.12}$$

with α_H the linewidth enhancement factor. In conventional QW lasers, this factor falls within the range of 4–7, causing huge linewidth increases. A similar correction was proposed by Lax [13] for gas or solid-state lasers, where the cavity resonance is tuned away from the centre of the gain spectrum. The detuning in these systems is small, and experimental verification is limited. In semiconductor lasers, however, the laser line is located in the low-energy tail of the optical absorption edge, leading to substantial detuning and consequently larger values of α_H. This parameter will be further discussed in the Sect. 3.1.3. Figure 3.6 illustrates a frequency noise spectrum. In the low-frequency range, the dominant contribution comes from carrier noise, which gradually diminishes as the modulation frequency exceeds the relaxation resonance frequency. In the high-frequency range, carrier noise becomes negligible, resulting in frequency noise being determined by the white noise background originating from spontaneous emission noise. The FWHM linewidth of the laser can be determined through the self-heterodyne method in the experimental setup. This involves introducing a time delay in the optical field before correlating it with itself.

To achieve spectral linewidth in the kHz or even sub-Hz range, extensive efforts have been dedicated to enhancing the quality factor (Q-factor) of the cavity, thereby increasing the number of photons stored within it. In this regard, an alternative

---

[1] The total cavity losses α include the mirror losses α_m and the internal loss α_i.

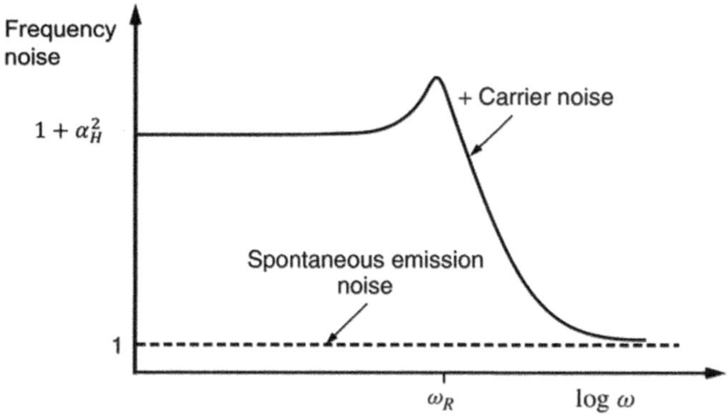

Fig. 3.6 Frequency noise spectrum, illustrating contributions from both carrier and spontaneous emission noise. The values of $(\Delta\nu)_{OL}$ and $(\Delta\nu)_{ST}$ can be found at the y-intercepts of this spectrum for low frequencies and high frequencies, respectively. (*Source* On page 236 in Ref. [5])

formulation of (3.12) formula is proposed, with the goal of further refining the understanding of laser linewidth [15]:

$$(\Delta\nu)_{OL} = (\Delta\nu)_{ST}(1+\alpha_H^2) = \frac{\pi h\nu^3 n_{sp}}{P_0 Q_E Q}(1+\alpha_H^2) \tag{3.13}$$

with the total Q-factor $Q = 2\pi\nu\tau_p = \frac{2\pi\nu}{v_g(\alpha_i+\alpha_m)}$, the external Q-factor $Q_E = \frac{2\pi\nu}{v_g\alpha_m}$ and the internal Q-factor $Q_I = \frac{2\pi\nu}{v_g\alpha_i}$. And these three quality factor are linked $\frac{1}{Q} = \frac{1}{Q_E} + \frac{1}{Q_I}$. By referring to (3.13), it becomes evident that the spectral linewidth can be reduced by increasing the Q-factor, achieved through an elongation of the photon lifetime (τ_p) and a reduction in the losses (α_i and α_m). Recent advancements in heterogeneous integration have explored the utilization of high-Q Si/Si_3N_4 microresonators together with III–V lasers on silicon as a possible solution for this objective [15, 16]. These integrated systems bridge the gap between semiconductor lasers and coherent optical systems by leveraging CMOS-foundry-fabricated microresonators with exceptional performance metrics, including an impressive high Q factor exceeding 260 million, enabling the achievement of an exceptionally low short-term linewidth of only 1.2 Hz, and a finesse over 42,000, as clearly depicted in Fig. 3.7.

Additional advancements can be achieved by incorporating a gain medium composed of self-assembled quantum dots (QDs), leading to great improvements in laser performance through enhanced carrier confinement and a modified density of states. This approach results in substantially reduced values of the linewidth enhancement factor (α_H) (approaching zero) and the population inversion parameter (n_{sp}) (approaching unity), thereby yielding superior laser characteristics. Very recently, a milestone has been accomplished with an InAs/InP QD distributed feedback (DFB) laser, which demonstrated a record intrinsic spectral linewidth as low as 30 kHz at 20 °C [17], as depicted on the left side of Fig. 3.8. Moreover, a record-narrow intrinsic linewidth of only 1.62 kHz is attained, accompanied by a white noise level of

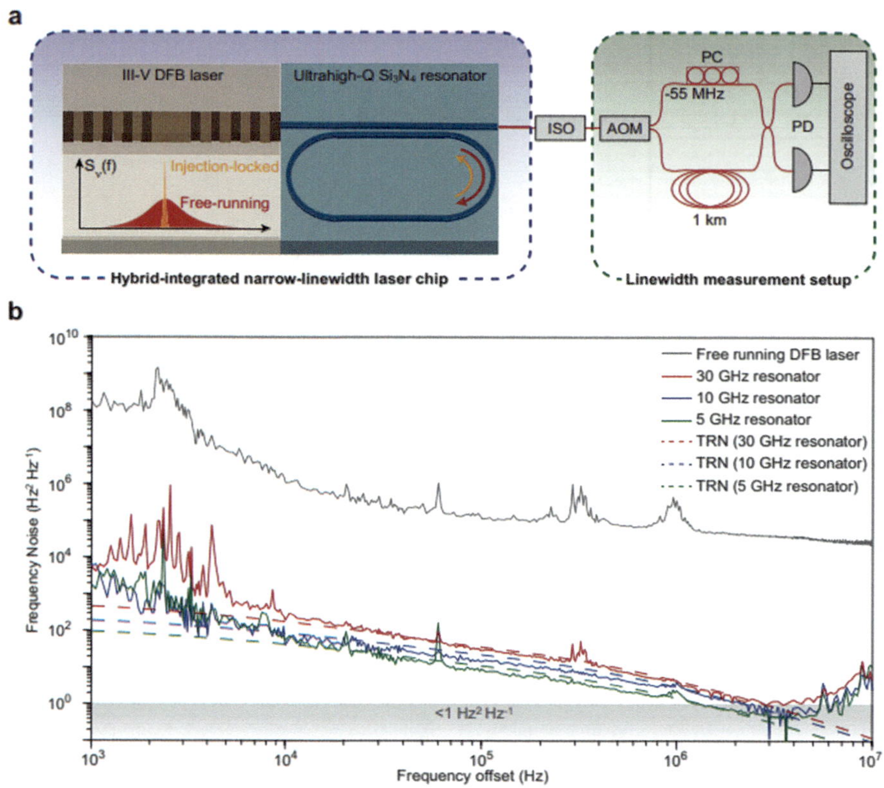

Fig. 3.7 Hybrid-integrated narrow-linewidth laser based on ultra-high-Q Si_3N_4 microresonator. **a** Schematic of the hybrid laser design and the setup used for linewidth testing. **b** Measurement of single-sideband frequency noise for both the free-running and self-injection locked DFB laser. The white-frequency-noise levels are 1 Hz^2Hz^{-1}, 0.8 Hz^2Hz^{-1}, and 0.5 Hz^2Hz^{-1} for resonators with a free spectral range (FSR) of 20 GHz, 10 GHz, and 5 GHz, respectively. (*Source* In Ref. [16])

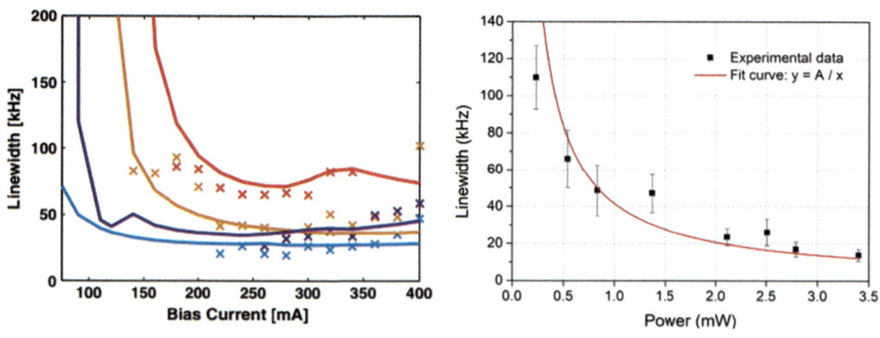

Fig. 3.8 Left panel: Extracted QD laser linewidth shown in marks. The simulated values are shown in solid lines: 20 °C (blue), 40 °C (purple), 60 °C (orange), and 80 °C (red). Right panel: Extracted ICL linewidth is shown in marks. (*Source* In Refs. [17, 18])

just 515 Hz^2/Hz^{-1} [19]. Similarly, in interband cascade lasers (ICLs), the linewidth of a 4.6 μm DFB ICL can be reduced to approximately 10 kHz by increasing the pump current [18], as shown on the right side of Fig. 3.8. Consequently, the quantity $n_{sp} \times (1 + \alpha_H^2)$ can be utilized as a figure of merit for narrow linewidth operation.

3.1.3 Linewidth Enhancement Factor (α_H)

In semiconductor lasers, the coupling between the laser field and the gain medium is mediated by the gain and the carrier-induced refractive index through the *Kramers-Kronig relation* or, equivalently, by the complex optical susceptibility χ. The interaction between the carrier-induced variations of the real and imaginary parts of the susceptibility χ is described by the linewidth enhancement factor (α_H-factor) [1, 20]. To investigate the α_H-factor theoretically, it is essential to solve the quantum mechanical equations of motion for the microscopic polarization of the gain medium, as discussed in Sect. 2.2.3. These equations include various contributions such as kinetic energies, many-body Coulomb interactions, electric-dipole interaction between carriers and the laser field, carrier-phonon interactions, and the effects of the injection current [21–23]. The macroscopic polarization **P** can be expressed as the sum of the microscopic polarizations $\mathbf{p_i}$:

$$\mathbf{P} = \frac{1}{V}\sum_i \mu_i \mathbf{p_i}$$

with V the volume of the active region volume, μ_i the dipole matrix element for the i transitions. Assuming the gain medium is isotropic, the complex optical susceptibility χ is connected with the polarization via

$$\mathbf{P} = \varepsilon_0 n_0^2 \chi \mathbf{E} = \varepsilon_0 n_0^2 (\Re\mathrm{e}\lfloor\chi\rfloor + j\Im\mathrm{m}\lfloor\chi\rfloor)\mathbf{E}$$

where ε_0 is the vacuum permittivity and **E** the electric field amplitude. To facilitate the calculation of the material gain g and the carrier-induced refractive index δn, a simplified scenario is adopted, wherein a monochromatic plane wave propagating along the z-direction is considered. This choice enables the illustration of the essential physics of the semiconductor gain medium with minimal algebraic complexity [21]:

$$\mathbf{E}(z, t) = \frac{1}{2}\hat{\mathbf{x}}E(z)e^{j(Kz-\omega t-\phi(z))} + c.c.$$

where $\hat{\mathbf{x}}$ is the unit vector in the x direction, $E(z)$ and $\phi(z)$ are the real field amplitude and phase shift that vary little in an optical wavelength, ω is the lasing angular frequency in radians/second and $\exp(jKz)$ accounts for most of the spatial variation in the laser field with $K = \omega n_0/c$ the wave number. In the same spirit, the laser field induces a polarization in the medium

$$\mathbf{P}(z,t) = \frac{1}{2}\hat{\mathbf{x}}P(z)e^{j(Kz-\omega t-\phi(z))} + c.c.$$

where $P(z)$ is a complex polarization amplitude that varies little in a wavelength. Then substituting these two ansatzes into Eq. 2.18:

$$-\frac{\mathrm{d}^2E}{\mathrm{d}z^2} - 2j\left(K - \frac{\mathrm{d}\phi}{\mathrm{d}z}\right)\frac{\mathrm{d}E}{\mathrm{d}z} + \left[\left(K - \frac{\mathrm{d}\phi}{\mathrm{d}z}\right)^2 + j\frac{\mathrm{d}^2\phi}{\mathrm{d}z^2} - \left(\frac{n_0\omega}{c}\right)^2\right]E = \mu_0\omega^2 P$$

This equation can be simplified considerably if one assumes that E and $\mathrm{d}\phi/\mathrm{d}z$ vary little in a wavelength, so that terms containing $\mathrm{d}^2E/\mathrm{d}z^2$, $\mathrm{d}^2\phi/\mathrm{d}z^2$ and $(\mathrm{d}E/\mathrm{d}z) \times (\mathrm{d}\phi/\mathrm{d}z)$ may be neglected. This slowly varying envelope approximation gives

$$-2jK\frac{\mathrm{d}E}{\mathrm{d}z} + K^2E - 2KE\frac{\mathrm{d}\phi}{\mathrm{d}z} - \left(\frac{n_0\omega}{c}\right)^2 E = \mu_0\omega^2 P$$

Then, i.e.,

$$\frac{\mathrm{d}E}{\mathrm{d}z} - jE\frac{\mathrm{d}\phi}{\mathrm{d}z} = \frac{j\mu_0\omega^2}{2K}P = \frac{j\omega}{2n_0\varepsilon_0 c}P = \frac{jK}{2}\chi E$$

Equating the real and imaginary parts, the *self-consistency* condition mentioned in Sect. 2.2.3 is

$$\frac{\mathrm{d}E(z)}{\mathrm{d}z} = -\frac{K}{2}\Im\mathrm{m}[\chi(z)]E(z) = -\frac{\omega}{2\varepsilon_0 n_0 c}\Im\mathrm{m}[P(z)]$$
$$\frac{\mathrm{d}\phi(z)}{\mathrm{d}z} = -\frac{K}{2}\Re\mathrm{e}[\chi(z)] = -\frac{\omega}{2\varepsilon_0 n_0 c E(z)}\Re\mathrm{e}[P(z)]$$

Therefore, the material gain g in intensity that satisfies $\mathrm{d}E/\mathrm{d}z = 1/2gE$ with units of inverse length, can be defined as

$$g = -K\Im\mathrm{m}[\chi(z)] = -\frac{\omega n_0}{c}\Im\mathrm{m}\,[\chi] = -\frac{\omega}{\varepsilon_0 n_0 c}\Im\mathrm{m}\,[P(z)]$$

To find the carrier-induced refractive index δn, note that the wavenumber of the laser field given by

$$K - \frac{\mathrm{d}\phi}{\mathrm{d}z} = (n_0 + \delta n)K_0$$

where K_0 is the wavenumber in vacuum. Since $K = n_0K_0$ and $\mathrm{d}\phi/\mathrm{d}z = -K_0\delta n$, hence

$$\delta n = \frac{n_0}{2}\Re\mathrm{e}\,[\chi] = \frac{1}{2\varepsilon_0 n_0 E(z)}\Re\mathrm{e}\,[P(z)]$$

Deriving from these two equations, the α_H-factor that describes the coupling between the material gain and the refractive index is expressed as follows [20]:

$$\alpha_H = \frac{\mathrm{d}\Re\mathrm{e}\left[\chi(N)\right]/\mathrm{d}N}{\mathrm{d}\Im\mathrm{m}\left[\chi(N)\right]/\mathrm{d}N} = -2\frac{\omega}{c}\frac{\mathrm{d}\delta n/\mathrm{d}N}{\mathrm{d}g/\mathrm{d}N} = -\frac{4\pi}{\lambda}\frac{\mathrm{d}\delta n/\mathrm{d}N}{\mathrm{d}g/\mathrm{d}N} \tag{3.14}$$

with N the carrier density, λ the lasing wavelength, $\mathrm{d}\delta n/\mathrm{d}N$ the differential refractive index, and $a = \mathrm{d}g/\mathrm{d}N$ the differential gain. The net modal gain in the laser system is defined as the modal gain $\Gamma_p g$ after subtracting the total loss α. The material gain g is obtained by assuming a linear dependence on carrier density N that can be described as follows [5]:

$$g = a(N - N_0) \tag{3.15}$$

where N_0 is the carrier density at the transparency point where the gain completely compensates for the absorption.

The α_H-factor plays a crucial role in characterizing various dynamic properties of semiconductor lasers, including spectral linewidth, sensitivity to optical feedback, nonlinear dynamics under optical injection, and four-wave mixing generation. Figure 3.9 illustrates a comparison of material gain (g) and refractive index (n) as a function of the lasing photon energy for both QD and QW lasers. In QW lasers, the refractive index zero point is displaced from the gain peak, both at low and high carrier densities, resulting in a nonzero α_H-factor. Typically, QW lasers exhibit α_H-factors ranging from 2 to 5 [5]. Conversely, in QD lasers with symmetric gain behaviour, the differential refractive index change is precisely zero at the lasing photon energy, regardless of the carrier density. Hence, QD lasers with symmetric gain can achieve a zero α_H-factor [24]. Nevertheless, achieving a zero

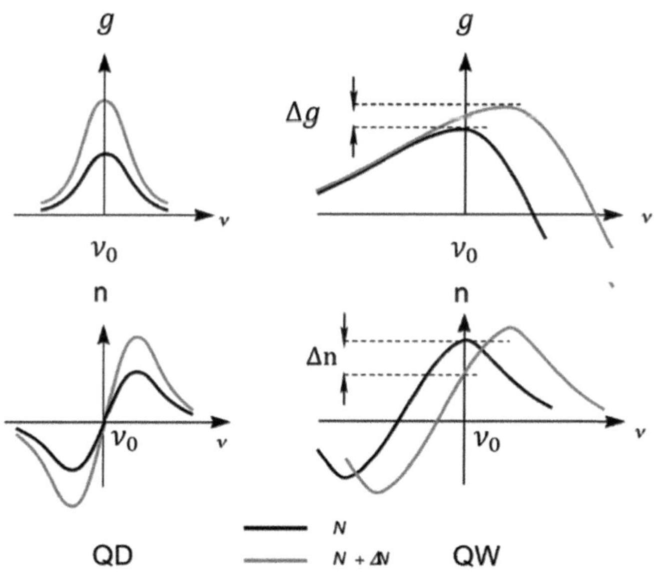

Fig. 3.9 Schematic of material gain g and refractive index n profiles variation with the carrier density changes for QD and QW. (*Source* On page 592 In Ref. [26])

α_H-factor has proven to be elusive in experimental observations, and the value of α_H in QD lasers can vary widely from nearly zero to more than 10. This unexpected behaviour is related to several mechanisms, including significant inhomogeneous broadening, the presence of off-resonant states (such as higher energy levels like ES), and the influence of free carrier plasma. To measure the α_H-factor, several techniques have been employed, which can be classified into three types [25]. The first type involves analyzing the optical spectrum, including the optical linewidth spectrum and the spectrum of amplified spontaneous emission (ASE). The second type relies on high-frequency modulation techniques, such as the FM/AM method and the fibre transfer function method. The third type focuses on external optical control methods, such as optical injection, optical feedback, and four-wave mixing techniques.

3.1.4 Correlation Between Intensity Noise and Frequency Noise

The frequency noise and intensity noise in semiconductor lasers are not solely connected to spontaneous emission noise events but also are influenced by the phase-amplitude coupling effect (α_H-factor), resulting in a strong correlation between them through carrier population dynamics within the active region [2]. Interestingly, this coupling is asymmetric, with amplitude fluctuations coupling into phase fluctuations rather than vice versa. This one-way coupling leads to excess phase fluctuations that amplify the fundamental linewidth, while the intensity noise remains unaffected [27]. As illustrated in Fig. 3.10a, spontaneous emission introduces a fluctuating field to the complex field of the laser mode, representing the addition of a single photon with a random phase θ_i. Then, the lasing field experiences both amplitude fluctuations ($\cos\theta_i$) and phase fluctuations ($\sin\theta_i$), acting as a "*detuned oscillator*". In Fig. 3.10b, an additional phase-amplitude coupling effect influences the phase fluctuations with a term of $-\alpha_H \cos\theta_i$ and the total phase diffusion is given by $\Delta\phi_i = \frac{1}{\sqrt{P}}(\sin\theta_i - \alpha_H \cos\theta_i)$. The amplitude fluctuations are usually attenuated due to the gain saturation of the amplifying medium, while the phase fluctuations persist since the mechanism driving laser oscillation does not impose a specific phase on the generated field. As a result, as successive spontaneous emission events occur, the phase of the laser field undergoes a random walk [1].

In this context, a partial "image" of the intensity noise lies in the frequency noise, which could be recovered from the phase fluctuations and used to reduce the amplitude fluctuations far below their intrinsic level [28, 29]. When the intensity noise is maximally reduced, the two types of fluctuations become decorrelated [27]. In the literature, M. A. Newkirk and Kerry J. Vahala demonstrated a substantial reduction of intensity noise (14.5 dB) in a DFB laser at low bias through amplitude-phase decorrelation. The reduction ratio can potentially reach $(1 + \alpha_H^2)$ [30]. Two methods were presented in their work. The first approach utilizes a frequency-dependent transmission passive element external to the laser cavity, while the second approach

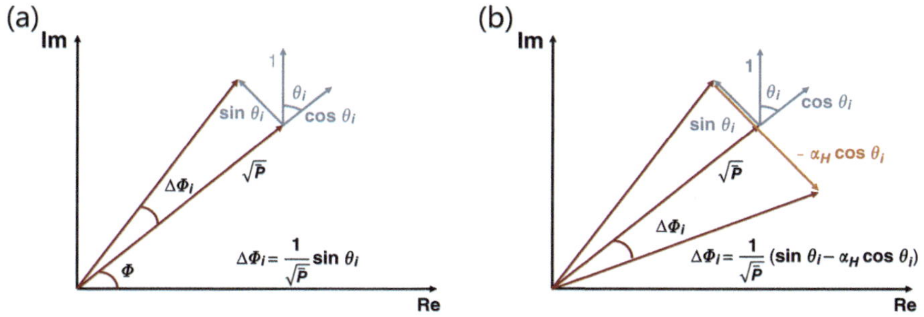

Fig. 3.10 Schematic illustrating the intensity fluctuation and phase diffusion ($\Delta\phi_i$) of the lasing field with **a** direct action of the spontaneous emission events only; **b** Joint action of both spontaneous emission events and phase-amplitude coupling. The angle θ_i is randomly determined. The amplitude and phase of the lasing field are represented as $\sqrt{P}$ and ϕ, respectively. The associated phase changes for each case are indicated in the diagram. (*Source* In Ref. [8])

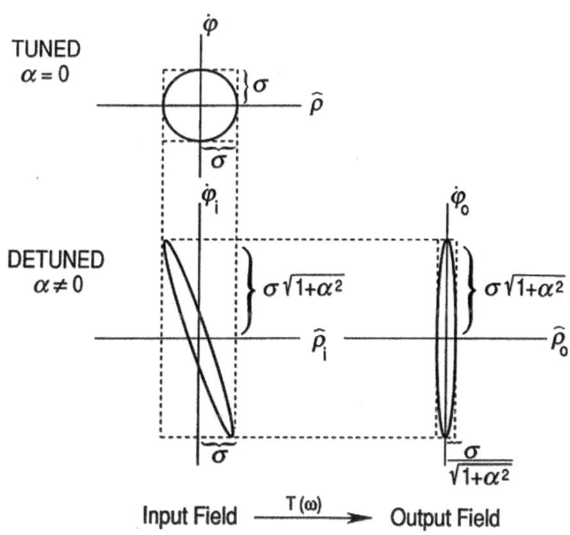

Fig. 3.11 Contours of constant probability density for tuned ($\alpha_H = 0$) and detuned ($\alpha_H \neq 0$) operation. For the detuned case, the tilted ellipse of the input field is rotated by $T(\omega)$ (i.e., Michelson interferometer) to decorrelate the variables and reduce the variance of $\hat{\rho}_0(t)$ by $1/(1+\alpha_H^2)$. (*Source* In Ref. [27])

employs a dispersive intracavity loss, sharing conceptual similarities with the detuned loading [31]. Another proposed scheme by Philippe Grangier involves the use of a Michelson interferometer to achieve phase-to-amplitude conversion [32]. When the laser operates in a tuned condition, corresponding to $\alpha_H = 0$, the contours of constant probability density appear as circles, as demonstrated in Fig. 3.11. However, in the detuned case, these contours take the shape of ellipses that are tilted concerning the coordinate axes. In this situation, the variance of the instantaneous frequency fluctuation is amplified by a factor of $(1+\alpha_H^2)$, while the variance of the amplitude fluctuation remains unchanged. This tilted ellipse represents the state of the input field, where the degree of tilt, and corresponding thinness of the ellipse signify the correlation between the fluctuations.

Furthermore, it has been established theoretically and experimentally that the correlation between intensity noise, frequency noise, and junction voltage noise persists even in the quantum regime [33]. The underlying principle is straightforward to comprehend. Fluctuations in the carrier population are influenced by spontaneous emission, stimulated emission, and absorption processes. When the carrier system is only coupled to a single lasing mode, and the output coupling rate significantly exceeds the internal loss rate, a perfect and negative correlation emerges between the junction-voltage fluctuation and the photon-number fluctuation in the output field. This negative correlation arises because a decrease of 1 in the number of carriers corresponds to a simultaneous increase of 1 in the photon number of the output field. Additionally, carrier fluctuations are coupled to laser frequency fluctuations through the gain-refractive-index coupling. A decrease in the number of carriers leads to a decrease in the lasing frequency, which corresponds to a positive value of α_H. Remarkably, even when the intensity noise is reduced below the standard quantum limit, the correlation between these noise sources remains important [28, 34]. By employing either of these signals in feedback or feedforward schemes, it becomes feasible to substantially reduce the intensity noise.

3.2 Theoretical Investigation of the Classical Noise

To capture the classical noise characteristics of quantum-confined lasers, it is essential to incorporate the fundamental and critical aspects of these devices. These characteristics can be described at a microscopic level, which often involves complex calculations but contributes to a deeper understanding of the underlying physics. In contrast, for practical device engineering purposes, it is beneficial to employ simplified models that incorporate phenomenological concepts and empirical parameters. This approach facilitates the analysis and optimization of device performance.

For instance, calculating carrier scattering rates rigorously is a complicated task, as it necessitates a sophisticated many-body quantum theory that accounts for interband collision processes [21, 23]. To address this, a semi-empirical model proposed [35, 36] introduces a formula that incorporates carrier-dependent capture and relaxation times, offering a practical and computationally efficient approach for modelling these processes:

$$\tau_i = \frac{1}{A_i + C_i N}$$

In the provided expression, the index i represents the capture or relaxation process, A_i denotes the phonon-assisted scattering rate, and C_i represents the coefficient that determines the Auger-assisted scattering by carriers in the different energy levels. Although this formulation yields favourable agreement with time-resolved photoluminescence (TRPL) experiments [37], it is worth noting that the parameters A_i and C_i can exhibit significant variations among different temperatures and devices, thereby limiting the generalizability of this expression [38]. Nevertheless, it is reasonable to make the assumption that once the laser is pumped above the threshold, the carrier reservoirs contain a high density of carriers that remain relatively constant with changes in the bias current. This justifies the approximation of treating the carrier scattering time as a constant value in the rate equation model, facilitating simplifications in the analysis.

3.2.1 Quantum Dot Lasers

The three-level rate equations model employed in this study is based on the electronic structure of quantum dots (QDs) depicted in Fig. 3.12. This numerical model assumes that the active region comprises a single ensemble of QDs. As a result, the model does not consider the dispersion in dot size, which is responsible for the inhomogeneous broadening of the gain profile. The QD density of states consists of two energy levels: a two-fold degenerate ground state (GS) and a four-fold degenerate excited state (ES).

As discussed before in Sect. 2.2.1, it is implicitly assumed that the QDs are always in a neutral state, and the electrons and holes are treated as electron-hole (eh) pairs rather than separately [38]. Therefore, the system exclusively consists of excitonic energy states. Initially, external injected carriers directly fill the two-dimensional carrier reservoir (RS), neglecting the carrier dynamics in the three-dimensional separate confinement heterostructure (barrier) layers. Some of these carriers are subsequently captured into the excited state (ES) within a capture time τ_{ES}^{RS}, while others undergo spontaneous recombination with a spontaneous emission time τ_{spon}^{RS}. Once in the ES,

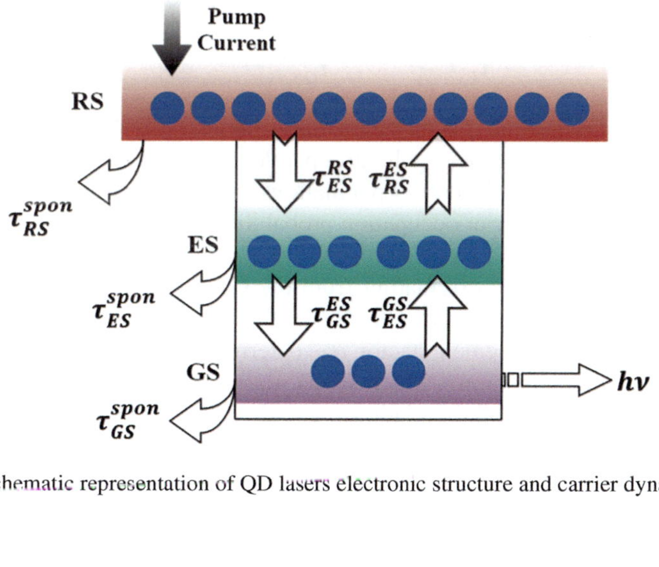

Fig. 3.12 Schematic representation of QD lasers electronic structure and carrier dynamics

carriers can relax to the ground state (GS) within a relaxation time τ_{GS}^{ES} or recombine spontaneously. On the other hand, carriers can also thermally reemit from the ES back to the RS, governed by the Fermi distribution following the detailed balance principle, with an escape time τ_{RS}^{ES} that characterizes quasi-thermal equilibrium in the absence of external excitation [39]:

$$\tau_{RS}^{ES} = \tau_{ES}^{RS} \frac{\sigma_{GS}}{\sigma_{ES}} e^{(\mathfrak{E}_{ES}-\mathfrak{E}_{GS})/(k_B T)}$$

$$\tau_{ES}^{GS} = \tau_{GS}^{ES} \frac{\sigma_{ES} N_B}{D_{RS}} e^{(\mathfrak{E}_{RS}-\mathfrak{E}_{ES})/(k_B T)}$$

where $\sigma_{ES} = 4, \sigma_{GS} = 2$ are the degeneracy of the considered confined states, $D_{RS} = \frac{m_{RS}^*}{\pi \hbar^2} k_B T$ is the effective density of states in the RS, N_B is the total QD density, $\mathfrak{E}_{RS,ES,GS}$ are the energy levels in different states. Similar dynamic behaviour is observed for the carrier population in the GS with respect to the ES. The model considers stimulated emission from the GS once the threshold is reached, while stimulated emission from the ES is not taken into account. Following the sketch of Fig. 3.12, the four coupled rate equations on carrier and photon densities are described:

$$\frac{\mathrm{d}N_{RS}}{\mathrm{d}t} = \frac{\eta I}{q} + \frac{N_{ES}}{\tau_{RS}^{ES}} - \frac{N_{RS}}{\tau_{ES}^{RS}}(1-\rho_{ES}) - \frac{N_{RS}}{\tau_{RS}^{spon}} + F_{RS} \tag{3.16}$$

$$\frac{\mathrm{d}N_{ES}}{\mathrm{d}t} = \left(\frac{N_{RS}}{\tau_{ES}^{RS}} + \frac{N_{GS}}{\tau_{ES}^{GS}}\right)(1-\rho_{ES}) - \frac{N_{ES}}{\tau_{RS}^{ES}} - \frac{N_{ES}}{\tau_{GS}^{ES}}(1-\rho_{GS}) - \frac{N_{ES}}{\tau_{ES}^{spon}} + F_{ES} \tag{3.17}$$

$$\frac{\mathrm{d}N_{GS}}{\mathrm{d}t} = \frac{N_{ES}}{\tau_{GS}^{ES}}(1-\rho_{GS}) - \frac{N_{GS}}{\tau_{ES}^{GS}}(1-\rho_{ES}) - \Gamma_p v_g g_{GS} S_{GS} - \frac{N_{GS}}{\tau_{GS}^{spon}} + F_{GS} \tag{3.18}$$

$$\frac{\mathrm{d}S_{GS}}{\mathrm{d}t} = \left(\Gamma_p v_g g_{GS} - \frac{1}{\tau_p}\right) S_{GS} + \beta_{sp} \frac{N_{GS}}{\tau_{GS}^{spon}} + F_S \tag{3.19}$$

where I is the bias current, q is the elementary charge, Γ_p is the optical confinement factor, τ_p the photon lifetime, v_g the group velocity, β_{sp} the spontaneous emission factor, and $N_{RS,ES,GS}$ are the carrier populations in the RS, ES, and GS, respectively, and $\rho_{GS,ES,RS}$ are the carrier occupation probabilities in the GS, ES, and RS, which are given by $\rho_{GS} = N_{GS}/2N_B$, $\rho_{ES} = N_{ES}/4N_B$, $\rho_{RS} = N_{RS}/D_{RS}$, respectively. Only stimulated emission originating from the GS level is considered hence S_{GS} accounts for the photon number in the GS level.

The linewidth enhancement factor α_H determines the fundamental characteristics of semiconductor lasers. Macroscopic derivations by Mariangela Gioannini employed the Kramers-Krönig relationship to describe the index variation resulting from the carrier population in localized states of the quantum dots [40]. In 2012, Benjamin Lingnau provided microscopic insights into the behaviour of QD lasers, highlighting the desynchronization between the real and imaginary parts of the optical susceptibility. This desynchronization arises from the resonant state's contribution to the gain change, while the off-resonant states predominantly influence the refractive index [41]. As discussed in Sect. 3.1.3, the active medium polarisation drives the electric field $E(t)$. In Ref. [23], Lingnau introduced the slowly varying inter-band microscopic polarization $p(t)$, which is characterized by the single decay time constant T_D, representing the lifetime of the microscopic polarization. The expressions for $p(t)$ are given by:

$$\frac{\partial}{\partial t}p_X(t) = -j(\omega_X - \omega_0)p_X(t) - j\frac{\mu_X E(t)}{2\hbar}(2\rho_X - 1) - \frac{1}{T_D}p_X(t)$$

where $\omega_X = \mathfrak{E}_X/\hbar$ is the lasing frequency in different energy states $X = \text{GS, ES, RS}$, ω_0 is the lasing frequency in the cold cavity, μ_X are the corresponding dipole transition matrix elements. To simplify the quantum-dot laser model further, it is possible to eliminate the dynamics of the microscopic inter-band polarization. This is justified by the relatively fast dephasing time T_D of the polarization, typically on the order of approximately 100 fs at room temperature [42]. Considering that charge-carrier scattering times are typically a few picoseconds and the photon lifetime is several picoseconds, this assumption is often valid. Consequently, all other dynamic variables can be treated as slowly varying, leading to a quasi-static relation for the microscopic polarization amplitudes: $\frac{\mathrm{d}}{\mathrm{d}t}p_X(t) = 0$. The adiabatically eliminated microscopic polarization under the quasi-static assumption is yielded:

$$p_X(t) = -j\frac{\mu_X T_D}{2\hbar}(2\rho_X - 1)\frac{1 - j(\omega_X - \omega_0)T_D}{1 + (\omega_X - \omega_0)^2 T_D^2}E(t)$$

Defining the macroscopic polarization amplitude of the active medium:

$$P(t) = \frac{2N_B}{V_B}\sum_{GS,ES}\sigma_{GS,ES}\mu^*_{GS,ES}p_X(t)$$

with N_B the total QD number, V_B the QD active region volume, and the factor of two accounts for spin degeneracy of the charge carrier states. Additionally, the RS macroscopic polarization amplitude has the same form, with the QW region volume V_{RS}:

$$P(t) = \frac{2D_{RS}}{V_{RS}}\mu^*_{RS}p_{RS}(t)$$

Following the relation between the slowly varying electric field envelope $E(t)$ and the macroscopic polarization amplitude $P(t)$ in Sect. 3.1.3, on can get:

$$\frac{\partial}{\partial t}E(t) = \frac{j\omega_0\Gamma_p}{2\varepsilon_0\varepsilon_{bg}}P(t) = \tilde{G}(\omega_0, t)E(t)$$

with ε_{bg} and ε_0 are the background and vacuum permittivity, respectively. It is worth noting that this analysis does not consider inhomogeneous broadening resulting from dot size fluctuations [43]. Therefore, the complex gain $\tilde{G}(\omega_0, t)$ can be obtained as:

$$\begin{aligned}\tilde{G}(\omega_0, t) &= \frac{j\omega_0\Gamma_p}{2\varepsilon_0\varepsilon_{bg}} \sum_{X=GS,ES,RS} \frac{P_X(t)}{E(t)} \\ &= \frac{\omega_0\Gamma_p}{\varepsilon_0\varepsilon_{bg}} \frac{N_B}{V_B} \left[\sigma_{GS}\tilde{g}_{GS}(2\rho_{GS} - 1) + \sigma_{ES}\tilde{g}_{ES}(2\rho_{ES} - 1)\right] \\ &+ \frac{\omega_0\Gamma_p}{\varepsilon_0\varepsilon_{bg}} \frac{D_{RS}}{V_{RS}} \tilde{g}_{RS}(2\rho_{RS} - 1)\end{aligned}$$

where $\tilde{g}_X$ is the complex gain coefficient in $X = GS, ES, RS$,

$$\tilde{g}_X = \frac{|\mu_X|^2\, T_D}{2\hbar} \frac{1 - j(\omega_X - \omega_0)T_D}{1 + (\omega_X - \omega_0)^2 T_D^2}$$

This equation gives rise to a Lorentzian-shaped gain spectrum for each optical transition, characterized by a full width at half maximum (FWHM) of $2\hbar T_D^{-1}$. The imaginary part of the gain vanishes directly at the transition frequency, but it becomes significant at $\hbar\omega \pm T_D^{-1}$ and gradually decays at higher detunings, as depicted in Fig. 3.13. Consequently, when the optical field frequency is significantly detuned from the reservoir transitions, the imaginary part of the gain coefficient dominates. Therefore, by assuming that the ES and RS lasing frequency is detuned sufficiently

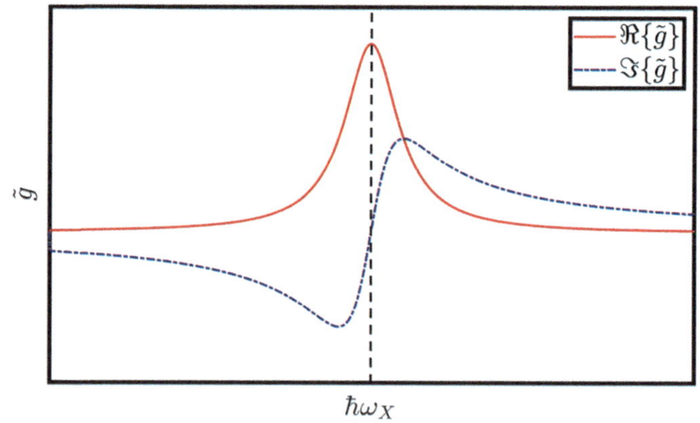

Fig. 3.13 Lorentzian gain profile of a single optical transition. The real part of the QD gain coefficient $\tilde{g}$ exhibits a Lorentzian-shaped profile centred around the frequency ω, while the imaginary part reaches its maximum at $|\omega - \omega_X| = T_D^{-1}$ and becomes negligible for resonant transitions ($\omega = \omega_X$). It is noteworthy that the decay of the imaginary part is slower compared to the real part as the optical frequencies move further away from the transition frequency

far from the reservoir transitions, one can neglect the real part of the ES and RS reservoir gain coefficients:

$$\Re\mathrm{e}\{\tilde{g}_{ES,RS}\} \approx 0$$

The resulting complex gain can then be written as

$$\begin{aligned}
\tilde{G}(\omega_0, t) &= \frac{N_B}{V_B}(2\rho_{GS} - 1)\frac{2\omega_0\Gamma_p}{\varepsilon_0\varepsilon_{bg}}\tilde{g}_{GS} + \frac{N_B}{V_B}(2\rho_{ES} - 1)\frac{4\omega_0\Gamma_p}{\varepsilon_0\varepsilon_{bg}}\tilde{g}_{ES} \\
&+ \frac{D_{RS}}{V_{RS}}(2\rho_{RS} - 1)\frac{\omega_0\Gamma_p}{\varepsilon_0\varepsilon_{bg}}\tilde{g}_{RS} \\
&= \frac{N_B}{V_B}(2\rho_{GS} - 1)\frac{2\left|\mu_{GS}\right|^2 T_D\omega_0\Gamma_p}{2\hbar\varepsilon_0\varepsilon_{bg}}\frac{1 - j(\omega_{GS} - \omega_0)T_D}{1 + (\omega_{GS} - \omega_0)^2 T_D^2} \\
&+ \frac{N_B}{V_B}(2\rho_{ES} - 1)\frac{4\left|\mu_{ES}\right|^2 T_D\omega_0\Gamma_p}{2\hbar\varepsilon_0\varepsilon_{bg}}\frac{1 - j(\omega_{ES} - \omega_0)T_D}{1 + (\omega_{ES} - \omega_0)^2 T_D^2} \\
&+ \frac{D_{RS}}{V_{RS}}(2\rho_{RS} - 1)\frac{\left|\mu_{RS}\right|^2 T_D\omega_0\Gamma_p}{2\hbar\varepsilon_0\varepsilon_{bg}}\frac{1 - j(\omega_{RS} - \omega_0)T_D}{1 + (\omega_{RS} - \omega_0)^2 T_D^2}
\end{aligned}$$

The optical susceptibility $\chi(t)$ can be derived from the complex gain expression through the relationship $\chi(t) = (2\varepsilon_{bg}\tilde{G}(\omega_0, t)/(j\omega\Gamma_p)$. The real part of the gain $\tilde{G}(\omega_0, t)$ is related to the laser gain while the imaginary part gives the instantaneous frequency shift of the electric field. The three terms on the right hand give contributions of the GS, the ES and the RS, respectively. Introducing the differential gain (a_X):

$$\begin{aligned}
a_{GS} &= \frac{2\left|\mu_{GS}\right|^2 T_D\omega_{GS}}{2\hbar v_g\varepsilon_0\varepsilon_{bg}} \\
a_{ES} &= \frac{4\left|\mu_{ES}\right|^2 T_D\omega_{ES}}{2\hbar v_g\varepsilon_0\varepsilon_{bg}} \\
a_{RS} &= \frac{\left|\mu_{RS}\right|^2 T_D\omega_{RS}}{2\hbar v_g\varepsilon_0\varepsilon_{bg}}
\end{aligned}$$

Then the $g_{RS,ES,GS}$ are the material gain of each state expressed as [36]:

$$\begin{aligned}
g_{GS} &= \frac{a_{GS}}{1 + \xi S_{GS}}\frac{N_B}{V_B}(2\rho_{GS} - 1) \\
g_{ES} &= a_{ES}\frac{N_B}{V_B}(2\rho_{ES} - 1) \\
g_{RS} &= a_{RS}\frac{D_{RS}}{V_{RS}}(2\rho_{RS} - 1)
\end{aligned}$$

where ξ is the gain compression factor. Due to the rapid decrease of the real part of the complex gain away from resonance, the field gain is primarily influenced by the resonant state. When considering lasing emission in resonance with the GS transition, denoted by $\omega_0 = \omega_{GS}$, one obtains the following expression:

$$\Re e\{\tilde{G}(\omega_0)\} \approx \Gamma_p v_g g_{GS}$$

In contrast, the imaginary part of the complex gain exhibits a slow decay for off-resonant frequencies (refer to Fig. 3.13). As a consequence, the off-resonant states can have a significant impact on the refractive index change, despite their negligible contribution to the gain of the GS lasing. The carrier populations in the off-resonant ES and RS induce frequency shifts in the laser field, given by:

$$\Delta\omega_N^{ES} = \frac{1}{2}\Gamma_p v_g g_{ES} \alpha_H^{ES}$$
$$\Delta\omega_N^{RS} = \frac{1}{2}\Gamma_p v_g g_{RS} \alpha_H^{RS}$$

with $\alpha_{ES,RS}$ the coefficients linked to the ES and RS contributions to the α_H-factor, respectively and defined as:

$$\alpha_H^{ES,RS} = \frac{\omega_{GS}}{\omega_{ES,RS}} \frac{(\omega_{ES,RS} - \omega_{GS})T_D}{1 + (\omega_{ES,RS} - \omega_{GS})^2 T_D^2}$$

While operating the laser at the gain peak with a symmetric gain distribution, the resonant GS does not contribute to the refractive index change. However, various factors, such as asymmetric QD size dispersion, can introduce a finite linewidth enhancement factor α_H in the resonant state, denoted as α_H^{GS}. The resulting frequency shift relative to the cold cavity can be expressed as:

$$\Delta\omega_N^{GS} = \frac{1}{2}\Gamma_p v_g g_{GS} \alpha_H^{GS}$$

Therefore, to extract the frequency noise (FN), the differential equation governing the phase dynamics of the electric field is also presented:

$$\frac{d\phi}{dt} = \Delta\omega_N^{GS} + \Delta\omega_N^{ES} + \Delta\omega_N^{RS} = \frac{1}{2}\Gamma_p v_g (g_{GS}\alpha_H^{GS} + g_{ES}\alpha_H^{ES} + g_{RS}\alpha_H^{RS}) + F_\phi \tag{3.20}$$

The relative intensity noise (RIN) modelling incorporates the inclusion of Langevin noise sources, which characterize both the carrier noises and spontaneous emission noises, as discussed in Sects. 2.2.1 and 3.1.1. These noise sources denoted as $F_{RS,ES,GS}$, F_S, and F_ϕ, introduce fluctuations to the laser deviating it from its steady-state condition. Due to their white noise nature, the expectation values of all Langevin noise terms are zero. Moreover, the correlation strength between two Langevin noise sources is given by $\langle F_i(t)F_j(t')\rangle = 2D_{ij}\delta(t - t')$, where the indexes

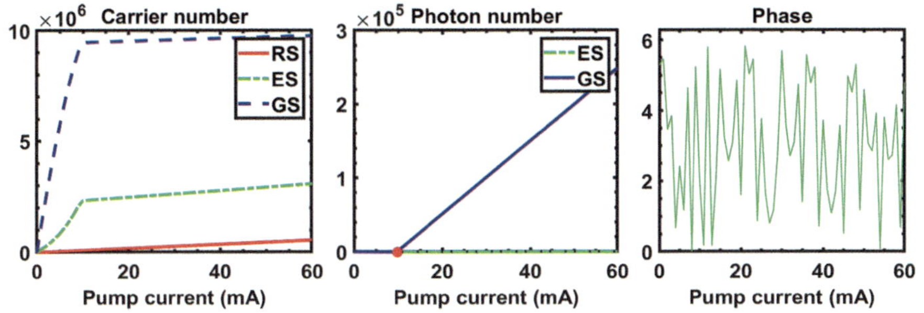

Fig. 3.14 Photon density, carrier density and phase versus injected pump current

i and j correspond to RS, ES, GS, S, and ϕ, and D_{ij} represents the diffusion coefficient between the two noise sources, which are delta-correlated. For additional details, please refer to Appendix B.

The properties of the laser are investigated using the proposed method. Unless otherwise specified, all material and optical parameters used in the analysis are listed in Table A.1. To validate the semi-analytical approach, the steady-state properties of the system are initially studied by numerically solving the five rate Eqs. (3.16)–(3.20). The results, shown in Fig. 3.14, demonstrate that both the GS and ES carrier populations increase with the injected current. Furthermore, for an injected current exceeding 9.7 mA (i.e., threshold current), the ES and GS populations become clamped, resulting in the emergence of GS lasing emission while the RS carrier population continues to increase.

Then, the simplified RIN expression is given as follows [44]:

$$\begin{aligned}\mathrm{RIN} =& \frac{|\delta S_{GS}(\omega)|^2}{S_{GS}^2}\\ =& \frac{1}{S_{GS}^2|\Delta|^2}\Big\{\langle F_{\mathrm{RS}}F_{\mathrm{RS}}^*\rangle|M_{\mathrm{RS-S}}|^2+\langle F_{\mathrm{ES}}F_{\mathrm{ES}}^*\rangle|M_{\mathrm{ES-S}}|^2+\langle F_{\mathrm{GS}}F_{\mathrm{GS}}^*\rangle|M_{\mathrm{GS-S}}|^2\\ &+\langle F_S F_S^*\rangle|M_{\mathrm{S-S}}|^2+\left\langle F_\phi F_\phi^*\right\rangle\left|M_{\phi-S}\right|^2-2\langle F_{\mathrm{RS}}F_{\mathrm{ES}}^*\rangle\Re\mathrm{e}\left\{M_{\mathrm{RS-S}}M_{\mathrm{ES-S}}^*\right\}\\ &-2\langle F_{\mathrm{ES}}F_{\mathrm{GS}}^*\rangle\Re\mathrm{e}\left\{M_{\mathrm{ES-S}}M_{\mathrm{GS-S}}^*\right\}-2\langle F_{\mathrm{GS}}F_S^*\rangle\Re\mathrm{e}\left\{M_{\mathrm{GS-S}}M_{\mathrm{S-S}}^*\right\}\Big\}\end{aligned} \tag{3.21}$$

The comprehensive derivation of (3.21) is available in the first Section of Appendix B.

Figure 3.15 shows the calculated RIN of the QD laser for various bias currents. Below 1.0 GHz, the RIN remains nearly constant, while at the resonance frequency of the QD laser, a prominent peak is observed. Beyond the resonance frequency, the RIN gradually decreases at higher frequencies. With increasing bias current, the RIN decreases across the entire spectral range due to a larger damping factor, and the peak of the RIN shifts to higher frequencies while exhibiting a reduced peak amplitude. Once again, the FN spectrum can be further simplified to:

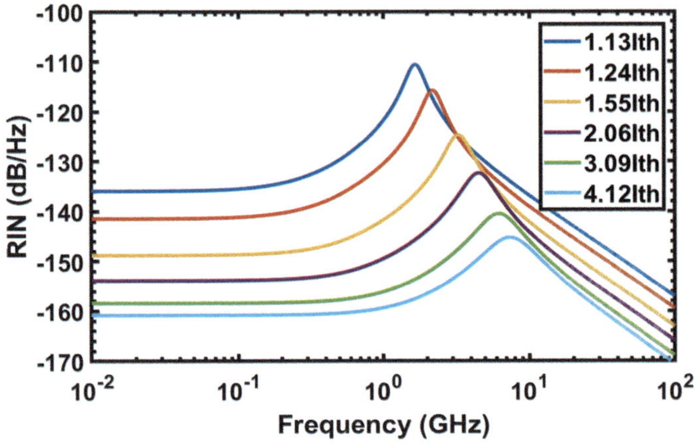

Fig. 3.15 Calculated RIN at several injected currents from (3.21)

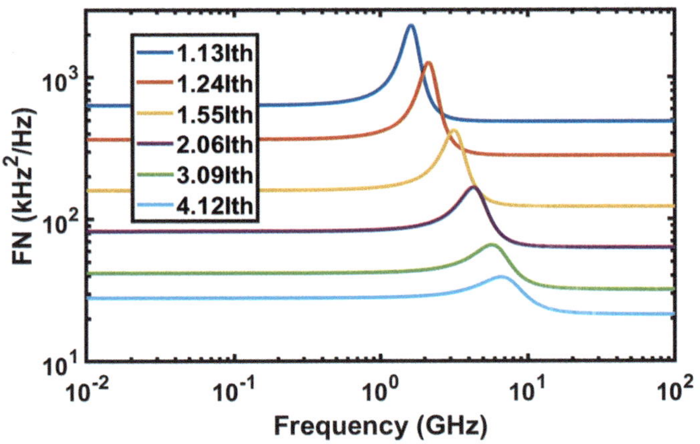

Fig. 3.16 Calculated FN at several injected currents from (3.22)

$$
\begin{aligned}
\mathrm{FN} &= |\delta\omega_S(\omega)|^2 \\
&= \frac{\omega^2}{|\Delta|^2}\Big\{\langle F_{\mathrm{RS}} F_{\mathrm{RS}}^*\rangle |M_{\mathrm{RS}-\phi}|^2 + \langle F_{\mathrm{ES}} F_{\mathrm{ES}}^*\rangle |M_{\mathrm{ES}-\phi}|^2 + \langle F_{\mathrm{GS}} F_{\mathrm{GS}}^*\rangle |M_{\mathrm{GS}-\phi}|^2 \\
&\quad + \langle F_S F_S^*\rangle |M_{S-\phi}|^2 + \left\langle F_\phi F_\phi^*\right\rangle |M_{\phi-\phi}|^2 - 2\langle F_{\mathrm{RS}} F_{\mathrm{ES}}^*\rangle \mathrm{Re}\left\{M_{\mathrm{RS}-\phi} M_{\mathrm{ES}-\phi}^*\right\} \\
&\quad -2\langle F_{\mathrm{ES}} F_{\mathrm{GS}}^*\rangle \mathrm{Re}\left\{M_{\mathrm{ES}-\phi} M_{\mathrm{GS}-\phi}^*\right\} - 2\langle F_{\mathrm{GS}} F_S^*\rangle \mathrm{Re}\left\{M_{\mathrm{GS}-\phi} M_{S-\phi}^*\right\}\Big\}
\end{aligned}
\tag{3.22}
$$

The comprehensive derivation of (3.22) is available in the first Section of Appendix B.

Figure 3.16 illustrates the FN spectra of the QD laser at different pump currents. The FN exhibits a prominent peak centred around the relaxation resonance frequency f_R. As the pump current increases, the peak amplitude decreases due to an increase in the damping factor. Beyond the resonance peak, the FN gradually decreases to a

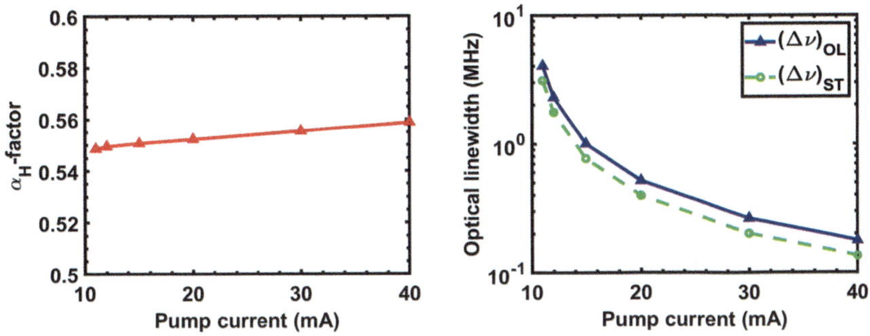

Fig. 3.17 Extracted α_H-factor and optical linewidth at several injected currents from (3.13) and (3.22)

constant level, which is solely determined by spontaneous emission. Therefore, the full width at half maximum (FWHM) of the Schawlow-Townes linewidth can be expressed as $(\Delta\nu)_{ST} = 2\pi \text{FN}|_{f \gg f_R}$. On the other hand, the FN spectrum below the resonance frequency is influenced not only by spontaneous emission but also by carrier fluctuations through the phase-amplitude coupling effect. Thus, the total optical linewidth is given by $(\Delta\nu)_{OL} = 2\pi \text{FN}|_{f \ll f_R}$. Moreover, the linewidth enhancement factor α_H can be also calculated by the relationship in (3.13). On the left side of Fig. 3.17, the linewidth enhancement factor α_H exhibits a slight increase with higher pump currents, mostly attributed to the contribution of the ES state rather than the RS state [45]. This is in agreement with the observation in Ref. [46]. On the right side of Fig. 3.17, both the total optical linewidth $(\Delta\nu)_{OL}$ (depicted as blue triangles) and the Schawlow-Townes linewidth $(\Delta\nu)_{ST}$ (represented by green circles) are plotted as a function of the pump current. Consistent with expectations, both linewidths demonstrate a decreasing trend as the pump current is increased due to the enhanced output power. For instance, at a pump current of 40 mA, which is nearly four times the threshold current (I_{th}), the optical linewidth measures 178 kHz. While this value is appropriate compared to recent measurements on QD lasers [8], it can be further reduced by achieving higher output power and implementing more refined design optimization techniques.

3.2.2 *Quantum Dot Lasers on Silicon*

The monolithic integration of QD lasers on silicon [47] presents an optimal solution for reducing fabrication complexity and cost, while simultaneously improving the integration density of silicon PICs. However, challenges arise from the mismatch in lattice constants and thermal expansion coefficients, resulting in the formation of threading dislocations (TDs) that act as non-radiative recombination centres through the Shockley–Read–Hall (SRH) process and degrade laser performance as shown in

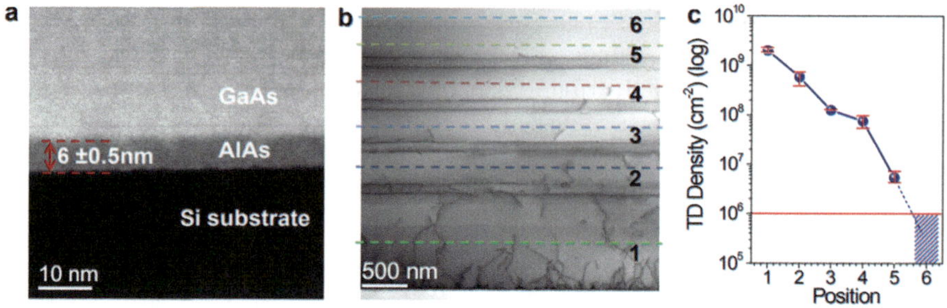

Fig. 3.18 Epitaxial growth and structural characterization of QD lasers on Si substrate. **a** High-angle annular dark field scanning transmission electron microscopy (TEM) image of the interface between the 6 nm AlAs nucleation layer and a silicon substrate. **b** Bright-field scanning TEM image of dislocation filter layers. **c** Dislocation density measured at different positions indicated in (**b**). (*Source* In Ref. [51])

Fig. 3.18. For instance, when compared to silicon, GaAs and InP exhibit lattice mismatches of 4.1% and 8.1%, respectively, along with thermal expansion coefficient mismatches of 120.4% and 76.9%, respectively [48]. Hence, the growth of either compound on a silicon substrate leads to threading or misfit dislocation densities of 10^8–10^{10}cm^{-2}. Despite recent advancements [49, 50], the achievable density of TDs still falls within the range of 10^5–10^6cm^{-2}, which remains approximately two orders of magnitude higher than that of native substrates [51]. At such elevated densities, quantum well (QW) lasers exhibit poor performance or may even fail to lase entirely. In contrast, substituting quantum dot active regions in place of quantum wells can offer an additional means to mitigate the detrimental impact of residual dislocations on laser performance [8]. The efficient capture and three-dimensional confinement of injected carriers by individual quantum dots result in reduced nonradiative recombination occurring at defects or dislocations [52, 53]. Consequently, the influence of dislocations that may still exist in the active layer is greatly diluted by the total number of independent dots, as they operate independently of each other, as illustrated in Fig. 3.19. Apart from this feature, QD lasers on silicon share several similarities with ordinary QD laser diodes.

In order to address these limitations, the development of a suitable strategy is crucial, and modelling serves as a valuable platform for achieving a comprehensive understanding of the device's operation. In this regard, I employ numerical analysis to investigate the properties of QD lasers on silicon, using an enhanced rate-equation model that incorporates the nonradiative recombination induced by the Shockley-Read-Hall (SRH) process. By integrating the SRH process into the model, a more comprehensive exploration of the device's behaviour becomes possible, leading to the development of effective solutions.

The proposed three-level rate equations model builds upon the native III–V QD lasers discussed in the previous section, with the addition of τ_{SRH} representing the nonradiative recombination lifetime attributed to defects. It is assumed that this nonradiative recombination lifetime remains the same for all three carrier states,

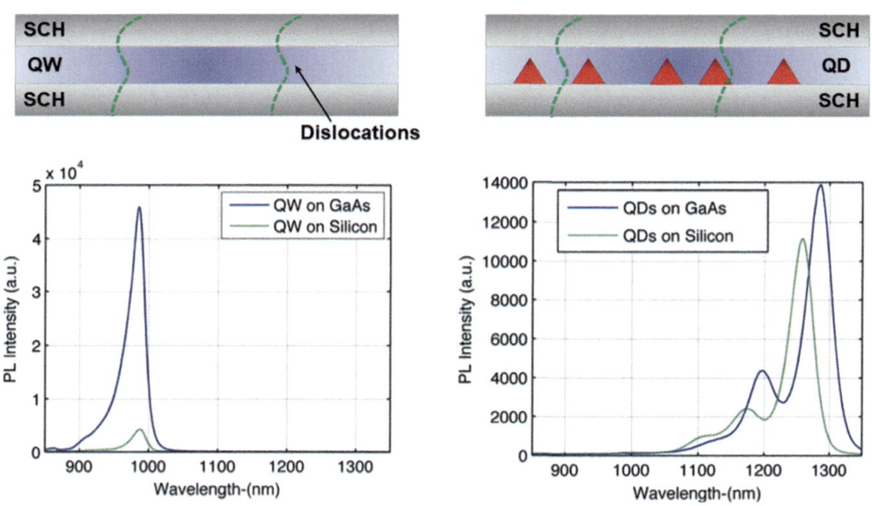

Fig. 3.19 Room-temperature photoluminescence comparison of single InAs quantum dot layer and single quantum well grown on GaAs versus silicon substrates. (*Source* In Ref. [53])

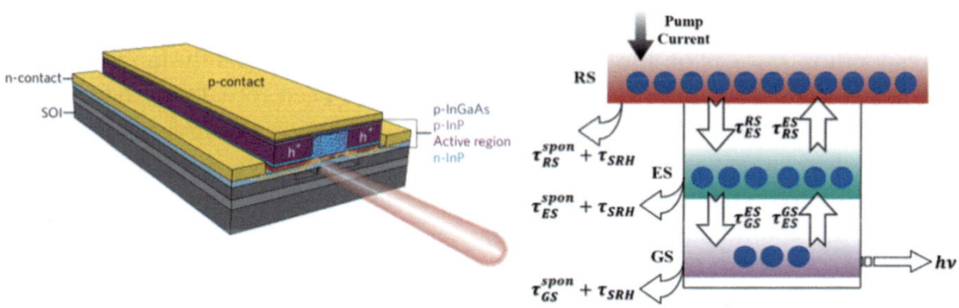

Fig. 3.20 Left panel: Schematic of a QD laser on silicon, showing recombination scheme of injected carriers in III–V epitaxial layers on top of a Si waveguide. Right panel: Schematic representation of the electronic structure and carrier dynamics of QD lasers on silicon. (*Source* In Refs. [48, 54])

as depicted in Fig. 3.20. Then the coupled rate equations for the carrier numbers (N_{RS}, N_{ES}, N_{GS}), the photon number (S_{GS}), and the phase (ϕ) of the electric field are given by [54]:

$$\frac{dN_{RS}}{dt} = \frac{\eta I}{q} + \frac{N_{ES}}{\tau_{RS}^{ES}} - \frac{N_{RS}}{\tau_{ES}^{RS}}(1-\rho_{ES}) - \frac{N_{RS}}{\tau_{RS}^{spon}} - \frac{N_{RS}}{\tau_{SRH}} + F_{RS} \tag{3.23}$$

$$\frac{dN_{ES}}{dt} = \left(\frac{N_{RS}}{\tau_{ES}^{RS}} + \frac{N_{GS}}{\tau_{ES}^{GS}})(1-\rho_{ES}\right) - \frac{N_{ES}}{\tau_{RS}^{ES}} - \frac{N_{ES}}{\tau_{GS}^{ES}}(1-\rho_{GS}) - \frac{N_{ES}}{\tau_{ES}^{spon}} - \frac{N_{ES}}{\tau_{SRH}} + F_{ES} \tag{3.24}$$

$$\frac{dN_{GS}}{dt} = \frac{N_{ES}}{\tau_{GS}^{ES}}(1-\rho_{GS}) - \frac{N_{GS}}{\tau_{ES}^{GS}}(1-\rho_{ES}) - \Gamma_p v_g g_{GS} S_{GS} - \frac{N_{GS}}{\tau_{GS}^{spon}} - \frac{N_{GS}}{\tau_{SRH}} + F_{GS} \tag{3.25}$$

$$\frac{dS_{GS}}{dt} = \left(\Gamma_p v_g g_{GS} - \frac{1}{\tau_p}\right) S_{GS} + \beta_{sp} \frac{N_{GS}}{\tau_{GS}^{spon}} + F_S \tag{3.26}$$

$$\frac{d\phi}{dt} = \frac{1}{2} \Gamma_p v_g (g_{GS} \alpha_H^{GS} + g_{ES} \alpha_H^{ES} + g_{RS} \alpha_H^{RS}) + F_\phi \tag{3.27}$$

Using the same parameter definitions as in the previous section, unless explicitly mentioned, all material and optical parameters employed in the analysis are provided in Table A.2. It's important to highlight that, in reality, the SRH contribution predominantly affects the RS within the wetting layer, with minimal impact on the ES or GS. However, in this study, we make the simplifying assumption of an equal contribution across all three states. The relationship between the nonradiative recombination lifetime and the defect density can be expressed as [55]:

$$\frac{1}{\tau_{SRH}} = \frac{1}{\tau_{SRH}^0} + \frac{D\pi^3\kappa}{4} \tag{3.28}$$

With $\tau_{SRH}^0 = 1877$ ns representing the lifetime of dislocation-free GaAs-based QD lasers, $D = 10\,\mathrm{cm^2 s^{-1}}$ denoting the average diffusion coefficient, and κ representing the TDs density value, the epitaxial defect density in QD lasers on native substrates typically falls within the range of 10^3–$10^4\,\mathrm{cm^{-2}}$ or even lower. Consequently, the corresponding τ_{SRH} is on the order of 10 ns, which is significantly longer than the spontaneous emission lifetime (approximately 1.0 ns). Therefore, for such QD lasers, the inclusion of this additional nonradiative recombination term in our model can be safely disregarded without sacrificing generality. However, a different scenario arises in Si-based QD lasers, where the defect density ranges from 10^6-$10^8\,\mathrm{cm^{-2}}$. In this case, the nonradiative carrier lifetime can be as short as 0.1 ns, surpassing the spontaneous emission lifetime (0.5 ns), as illustrated in Fig. 3.21.

Through a small signal analysis, assuming a sinusoidal current modulation $\delta I = Ie^{j\omega t}$ with modulation frequency ω, this change will induce the variations of carriers $\delta N_{RS,ES,GS}$ photon number δS_{GS}, and phase $\delta\phi$. Therefore, one can linearize the rate Eqs. (3.23)–(3.27), and yield:

$$\begin{bmatrix} \gamma_{11} + j\omega & -\gamma_{12} & 0 & 0 & 0 \\ -\gamma_{21} & \gamma_{22} + j\omega & -\gamma_{23} & 0 & 0 \\ 0 & -\gamma_{32} & \gamma_{33} + j\omega & -\gamma_{34} & 0 \\ 0 & 0 & -\gamma_{43} & \gamma_{44} + j\omega & 0 \\ -\gamma_{51} & -\gamma_{52} & -\gamma_{53} & -\gamma_{54} & j\omega \end{bmatrix} \begin{bmatrix} \delta N_{RS} \\ \delta N_{ES} \\ \delta N_{GS} \\ \delta S_{GS} \\ \delta\phi \end{bmatrix} = \frac{\eta \delta I}{q} \begin{bmatrix} 1 \\ 0 \\ 0 \\ 0 \\ 0 \end{bmatrix} \tag{3.29}$$

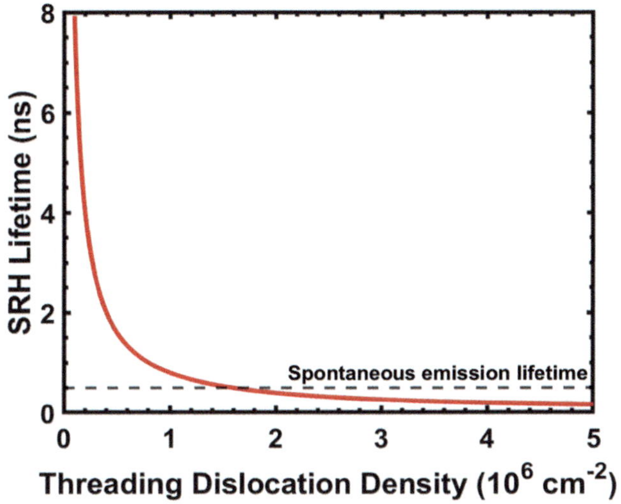

Fig. 3.21 Calculated SRH recombination lifetime obtained with an approximation method in (3.28) as a function of TD density. (*Source* In Ref. [54])

with

$$\gamma_{11} = \frac{1-\rho_{ES}}{\tau_{ES}^{RS}} + \frac{1}{\tau_{RS}^{spon}} + \frac{1}{\tau_{SRH}}$$

$$\gamma_{12} = \frac{1}{\tau_{RS}^{ES}} + \frac{1}{4N_B}\frac{N_{RS}}{\tau_{ES}^{RS}}$$

$$\gamma_{21} = \frac{1-\rho_{ES}}{\tau_{ES}^{RS}}$$

$$\gamma_{22} = \frac{1-\rho_{GS}}{\tau_{GS}^{ES}} + \frac{1}{\tau_{RS}^{ES}} + \frac{1}{\tau_{ES}^{spon}} + \frac{1}{4N_B}\left(\frac{N_{RS}}{\tau_{ES}^{RS}} + \frac{N_{GS}}{\tau_{ES}^{GS}}\right) + \frac{1}{\tau_{SRH}}$$

$$\gamma_{23} = \frac{1-\rho_{ES}}{\tau_{ES}^{GS}} + \frac{1}{2N_B}\frac{N_{ES}}{\tau_{GS}^{ES}}$$

$$\gamma_{32} = \frac{1-\rho_{GS}}{\tau_{GS}^{ES}} + \frac{1}{4N_B}\frac{N_{GS}}{\tau_{ES}^{GS}}$$

$$\gamma_{33} = \frac{1-\rho_{ES}}{\tau_{ES}^{GS}} + \frac{1}{2N_B}\frac{N_{ES}}{\tau_{GS}^{ES}} + \frac{1}{\tau_{GS}^{spon}} + \Gamma_p v_g a S_{GS} + \frac{1}{\tau_{SRH}}$$

$$\gamma_{34} = -\Gamma_p v_g g_{GS} + \Gamma_p v_g a_p S_{GS}$$

$$\gamma_{43} = \Gamma_p v_g a S_{GS} + \frac{\beta_{sp}}{\tau_{GS}^{spon}}$$

$$\gamma_{44} = \frac{1}{\tau_p} - \Gamma_p v_g g_{GS} + \Gamma_p v_g a_p S_{GS}$$

$$\gamma_{51} = \Gamma_p v_g a_{RS} \alpha_H^{RS}$$

$$\gamma_{52} = \frac{1}{4}\Gamma_p v_g a_{ES} \alpha_H^{ES}$$

$$\gamma_{53} = \frac{1}{2}\Gamma_p v_g a \alpha_H^{GS}$$

$$\gamma_{54} = -\frac{1}{2}\Gamma_p v_g a_p \alpha_H^{GS}$$

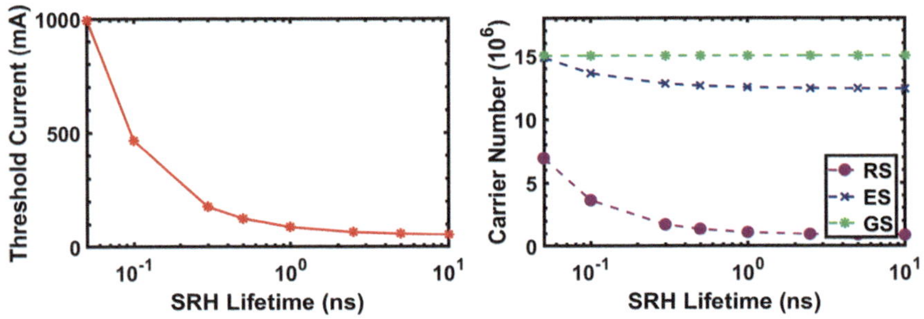

Fig. 3.22 Nonradiative recombination effects on the threshold current and the carrier numbers in GS, ES, and RS at 2×10^5 photon number

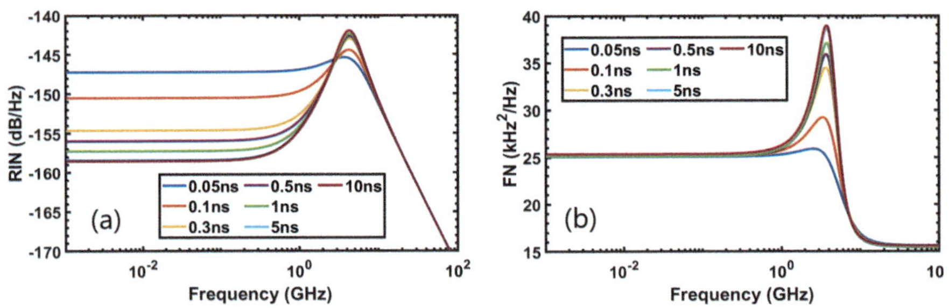

Fig. 3.23 Non-radiative recombination effects on the RIN spectrum and the FN spectrum

To ensure clarity, the subsequent calculations do not replicate earlier ones but incorporate τ_{SRH} into the Langevin diffusion coefficients $2D_{ii}$, similar to the inclusion of γ_{ii}. The simulations presented in this section primarily investigate the impact of non-radiative recombination or defect density on the laser dynamics. A fixed emitted photon number of 2×10^5 is maintained throughout all the simulations. As depicted in Fig. 3.22, it can be observed that the presence of fast nonradiative recombination processes or high defect densities leads to an increase in the threshold current, consistent with experimental observations. Furthermore, the carrier populations in the ES and RS also exhibit an increase. Conversely, the carrier population in the GS remains unchanged due to the gain clamping effect.

Figure 3.23a illustrates that a shorter nonradiative recombination lifetime results in an elevated level of RIN at low frequencies. Specifically, the low-frequency RIN increases from -158 dB/Hz for $\tau_{SRH} = 10$ ns to -150 dB/Hz for $\tau_{SRH} = 0.1$ ns. This increase in RIN can be attributed to the reduced carrier lifetime caused by high-density defects. Additionally, Fig. 3.23b also demonstrates that nonradiative recombination has minimal impact on the Schawlow-Townes linewidth and the total spectral linewidth of Qdot lasers. However, it significantly suppresses the FN resonance peak. For example, the amplitude of the FN peak decreases from 39 kHz^2/Hz at $\tau_{SRH} = 10$ ns to 29 kHz^2/Hz at $\tau_{SRH} = 0.1$ ns.

As discussed in the previous section, the constant α_H-factor of the QD laser on silicon can be calculated from the FN spectrum by (3.13):

$$\alpha_H^{FN} = \sqrt{\frac{(\Delta\nu)_{OL}}{(\Delta\nu)_{ST}} - 1} \tag{3.30}$$

Several techniques have been proposed for measuring the α_H-factor in recent decades. In this study, I employ the widely recognized amplified spontaneous emission (ASE) method for below-threshold analysis and the well-established "FM/AM" technique for above-threshold analysis [20]. For semiconductor lasers operating below the threshold, the ASE method relies on directly measuring the optical spectra of ASE in the laser cavity. By incrementally adjusting the pump current (ΔI), the change in gain can be extracted by the ASE method and the corresponding wavelength variation can be recorded using an optical spectrum analyzer. Subsequently, the below-threshold α_H^{ASE}-factor can be derived using the following approach [54]:

$$\alpha_H^{ASE}(\omega) = \frac{2}{\Gamma_p v_g}\frac{[\Delta\omega_N^{Total}(N)]}{\delta g_{GS}(N)} = \alpha_H^{GS} + \frac{1}{2}\alpha_H^{ES}\frac{a_{ES}\delta N_{ES}}{a\delta N_{GS}} + 2\alpha_H^{RS}\frac{a_{RS}\delta N_{RS}}{a\delta N_{GS}} \tag{3.31}$$

It should be emphasized that only the carrier contribution (δN) is included in the above equation, while the photon contribution (δS) and phase contribution $\delta\phi$ are excluded. Actually, the ratios $\frac{\delta N_{RS}}{\delta N_{GS}}$ and $\frac{\delta N_{ES}}{\delta N_{GS}}$ include implicitly the τ_{SRH} if using Cramer's rule considering Eq. (3.29):

$$\frac{\delta N_{ES}}{\delta N_{GS}} = \begin{vmatrix} \gamma_{11}+j\omega & 1 & 0 & 0 & 0 \\ -\gamma_{21} & 0 & -\gamma_{23} & 0 & 0 \\ 0 & 0 & \gamma_{33}+j\omega & -\gamma_{34} & 0 \\ 0 & 0 & -\gamma_{43} & \gamma_{44}+j\omega & 0 \\ -\gamma_{51} & 0 & -\gamma_{53} & -\gamma_{54} & j\omega \end{vmatrix} \Big/ \begin{vmatrix} \gamma_{11}+j\omega & -\gamma_{12} & 1 & 0 & 0 \\ -\gamma_{21} & \gamma_{22}+j\omega & 0 & 0 & 0 \\ 0 & \gamma_{32} & 0 & -\gamma_{34} & 0 \\ 0 & 0 & 0 & \gamma_{44}+j\omega & 0 \\ -\gamma_{51} & -\gamma_{52} & 0 & -\gamma_{54} & j\omega \end{vmatrix}$$

$$\frac{\delta N_{RS}}{\delta N_{GS}} = \begin{vmatrix} 1 & -\gamma_{12} & 0 & 0 & 0 \\ 0 & \gamma_{22}+j\omega & -\gamma_{23} & 0 & 0 \\ 0 & -\gamma_{32} & \gamma_{33}+j\omega & -\gamma_{34} & 0 \\ 0 & 0 & -\gamma_{43} & \gamma_{44}+j\omega & 0 \\ 0 & -\gamma_{52} & -\gamma_{53} & -\gamma_{54} & j\omega \end{vmatrix} \Big/ \begin{vmatrix} \gamma_{11}+j\omega & -\gamma_{12} & 1 & 0 & 0 \\ -\gamma_{21} & \gamma_{22}+j\omega & 0 & 0 & 0 \\ 0 & -\gamma_{32} & 0 & -\gamma_{34} & 0 \\ 0 & 0 & 0 & \gamma_{44}+j\omega & 0 \\ -\gamma_{51} & -\gamma_{52} & 0 & -\gamma_{54} & j\omega \end{vmatrix}$$

The carrier density variations in the three states under small signal modulation are illustrated in Fig. 3.24. For frequencies below 1 GHz, the carrier density variations in all states remain nearly constant. Consequently, it is possible to normalize all these variations with respect to the value δN_{GS} at 0.01 GHz. The variations in the ES (δN_{ES}) and RS (δN_{RS}) populations are approximately 5 dB larger than that of the GS (δN_{GS}) population. The limited variation in the GS carrier population can be attributed to the gain clamping effect above the threshold. Resonances around 4 GHz

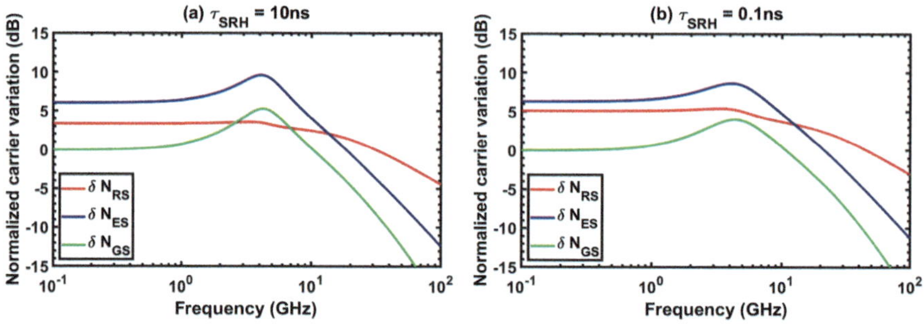

Fig. 3.24 Small-signal carrier density variations in the GS (green line), ES (blue line), and RS (red line) versus modulation frequency **a** when $\tau_{SRH} = 10$ ns and **b** when $\tau_{SRH} = 0.1$ ns. The carrier variation is normalized to the value δN_{GS} of 0.01 GHz

are observed in both δN_{GS} and δN_{ES}. Beyond the resonance frequency, δN_{GS} decays at a faster rate compared to δN_{ES} and δN_{RS}. These characteristics have a significant impact on the behaviour of the α_H^{ASE}-factor, as described in (3.31).

The "FM/AM" technique relies on the direct current modulation of the laser, which generates both the optical frequency (FM) and amplitude (AM) modulations [25]. With respect to the linearized rate equations, the ratio of the FM-to-AM index is derived as

$$\alpha_H^{FM/AM}(\omega) = \frac{2\beta(\omega)}{m(\omega)} = 2\frac{\delta\omega_S/\omega}{\delta S_{GS}/S_{GS}} = 2S_{GS}\frac{j\delta\phi}{\delta S_{GS}} \tag{3.32}$$

where the relation $\delta\omega_S = j\omega\delta\phi$ is used in the above derivation. In this approach, the laser's $\alpha_H^{FM/AM}$-factor is usually extracted through the formula $\alpha_H^{FM/AM} = \min\{2\beta(\omega)/m(\omega)\}$. Again, the ratio $\frac{\delta\phi}{\delta S_{GS}}$ can be found if using Cramer's rule

$$\frac{\delta\phi}{\delta S_{GS}} = \begin{bmatrix} \gamma_{11}+j\omega & -\gamma_{12} & 0 & 0 & 1 \\ -\gamma_{21} & \gamma_{22}+j\omega & -\gamma_{23} & 0 & 0 \\ 0 & -\gamma_{32} & \gamma_{33}+j\omega & -\gamma_{34} & 0 \\ 0 & 0 & -\gamma_{43} & \gamma_{44}+j\omega & 0 \\ -\gamma_{51} & -\gamma_{52} & -\gamma_{53} & -\gamma_{54} & 0 \end{bmatrix} \Bigg/ \begin{bmatrix} \gamma_{11}+j\omega & -\gamma_{12} & 0 & 1 & 0 \\ -\gamma_{21} & \gamma_{22}+j\omega & -\gamma_{23} & 0 & 0 \\ 0 & -\gamma_{32} & \gamma_{33}+j\omega & 0 & 0 \\ 0 & 0 & -\gamma_{43} & 0 & 0 \\ -\gamma_{51} & -\gamma_{52} & -\gamma_{53} & 0 & j\omega \end{bmatrix}$$

Figure 3.25 illustrates the comparison between $\alpha_H^{FM/AM}(\omega)$ and $\alpha_H^{ASE}(\omega)$ as a function of the modulation frequency. At low modulation frequencies, a substantial disparity is evident between these two parameters, signifying that a smaller τ_{SRH} leads to a lower FM-to-AM index ratio. As expected, $\alpha_H^{FM/AM}(\omega)$ exhibits large values due to gain compression and the considerable carrier variations in the ES and RS. Conversely, $\alpha_H^{ASE}(\omega)$ remains constant. However, there are no notable differences at high modulation frequencies. As the modulation frequency increases beyond several GHz, both parameter values decrease and converge to a *plateau*, representing the conventional α_H-factor. Recent experiments have even demonstrated the measurement of an ultralow α_H^{FN} factor below unity [56], revealing the immense potential for isolator-free silicon-based PICs.

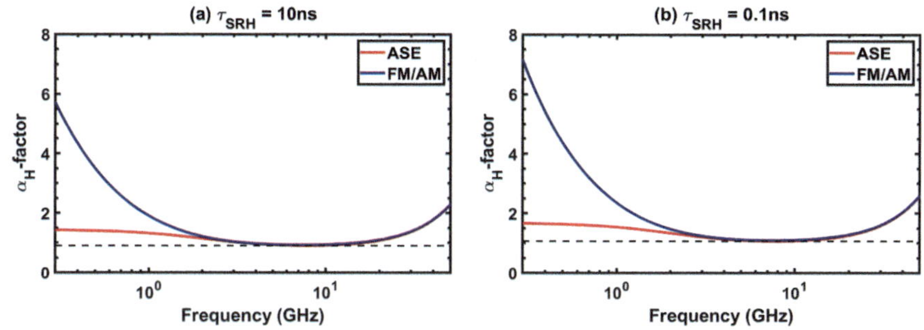

Fig. 3.25 Modulation-frequency dependence of the FM/AM method $\alpha_H^{FM/AM}(\omega)$ (blue line) and of the ASE method $\alpha_H^{ASE}(\omega)$ (red line) near the lasing threshold **a** when $\tau_{SRH} = 10\,\text{ns}$ and **b** when $\tau_{SRH} = 0.1\,\text{ns}$. The minimum level indicated by the black horizontal line gives the laser's conventional α_H-factor

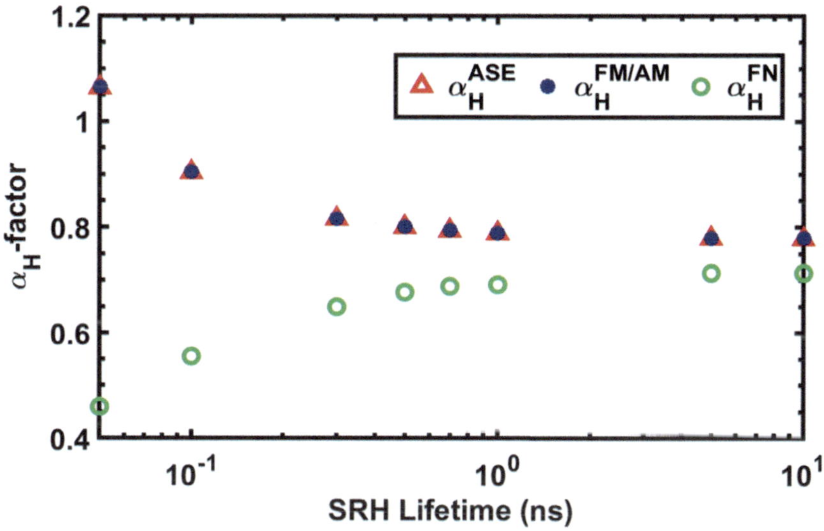

Fig. 3.26 Non-radiative recombination effects on the α_H-factor near the threshold. The green open circles are obtained by (3.30) (FN method), the red triangles are obtained by (3.31) (ASE method), and the blue closed circles are obtained by (3.32) (FM/AM method)

To ensure a fair and comprehensive comparison among all three α_H-factor measurements, it is essential to operate the QD lasers near the lasing threshold. For the subsequent simulation, the photon number is fixed at 3×10^3, which satisfies the condition ($I \approx I_{th}$) for different values of τ_{SRH}. The green circles in Fig. 3.26 clearly demonstrate that a reduction in the nonradiative recombination rate has a great impact on the α_H^{FN}-factor, which changes from 0.71 (for $\tau_{SRH} = 10\,\text{ns}$) to 0.46 (for $\tau_{SRH} = 0.05\,\text{ns}$). The red triangles and blue stars in Fig. 3.26 represent the ASE method and FM/AM method, respectively. Interestingly, it is worth noting that the smaller values of τ_{SRH} result in larger values of $\alpha_H^{ASE} =$ and $\alpha_H^{FM/AM}$, but their changing trends agree closely, as confirmed by experimental observations [36]. Furthermore, for larger values of τ_{SRH}, all three methods converge to an approximately same α_H-factor of about 0.75.

As previously mentioned, direct current modulation of a QD laser leads to simultaneous modulation of photon and carrier densities and the modulation of carrier density affects both the gain and refractive index. Consequently, the optical cavity length is modulated by the current, resulting in the resonant mode shifting in frequency. This frequency modulation, known as *chirp*, causes a broadening of the laser's modulated spectrum, which poses challenges for optical fibre communications. The chirping characteristic of semiconductor lasers can be quantified using the chirp-to-power ratio (CPR), defined as follows:

$$\mathrm{CPR}(\omega) = \frac{\delta\omega_S(\omega)}{\delta S_{GS}(\omega)} \tag{3.33}$$

Figure 3.27 shows the CPR for various τ_{SRH} lifetime. The CPR remains constant for modulation frequencies below 1 GHz but rises almost parabolically with the modulation frequency for higher modulation frequencies. It can be clearly shown that a lower τ_{SRH} lifetime effectively increases the CPR of QD lasers from 0.165 GHz/mW ($\tau_{SRH} = 10$ ns) to 0.206 GHz/mW ($\tau_{SRH} = 0.1$ ns).

The modulation transfer function for the QD laser on silicon can be easily extracted by only temporarily omitting the phase equation ϕ:

$$H_{QD}(\omega) = \frac{R_0}{\Delta} = \frac{R_0}{R_0 + j\omega R_1 - \omega^2 R_2 - j\omega^3 R_3 + \omega^4} \tag{3.34}$$

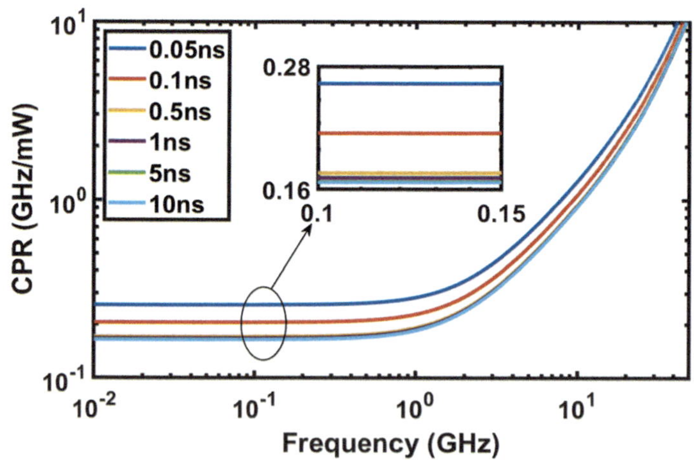

Fig. 3.27 Non-radiative recombination effects on the CPR

where Δ is the determinant of the matrix of 4×4 without (3.27) and four parameters that characterize $H_{QD}(\omega)$ are given by:

$$
\begin{aligned}
R_0 &= \omega_R^2\omega_{R0}^2 - \gamma_{23}\gamma_{44}(\gamma_{11}\gamma_{32} + \gamma_{12}\gamma_{31}) \\
R_1 &= \omega_R^2\Gamma_0 + \omega_{R0}^2\Gamma - \gamma_{23}\gamma_{32}(\gamma_{11} + \gamma_{44}) - \gamma_{12}\gamma_{23}\gamma_{31} \\
R_2 &= \omega_R^2 + \Gamma_0\Gamma + \omega_{R0}^2 - \gamma_{23}\gamma_{32} \\
R_3 &= \Gamma_0 + \Gamma
\end{aligned}
$$

where the relaxation resonance frequency ω_R and damping factor Γ are approximately defined as:

$$
\begin{aligned}
\omega_R^2 &= \gamma_{33}\gamma_{44} - \gamma_{34}\gamma_{43} \\
\Gamma &= \gamma_{33} + \gamma_{44}
\end{aligned}
$$

and with the same method, the two other parameters are:

$$
\begin{aligned}
\omega_{R0}^2 &= \gamma_{11}\gamma_{22} - \gamma_{12}\gamma_{21} \\
\Gamma_0 &= \gamma_{11} + \gamma_{22}
\end{aligned}
$$

By utilizing the already defined set of γ, the damping factor Γ can be rewritten as following:

$$
\Gamma = \frac{1-\rho_{ES}}{\tau_{ES}^{GS}} + \frac{1}{2N_B}\frac{N_{ES}}{\tau_{GS}^{ES}} + \frac{1}{\tau_{GS}^{spon}} + \frac{1}{\tau_{SRH}} + \Gamma_p v_g a S_{GS} + \frac{1}{\tau_p} - \Gamma_p v_g g_{GS} + \Gamma_p v_g a_p S_{GS} \tag{3.35}
$$

The modulation dynamics of the QD laser $H_{QD}(\omega)$ are depicted on the left of Fig. 3.28. It is evident that the smaller nonradiative recombination rate leads to the suppression of the resonance peak due to the shortened total carrier lifetime caused by nonradiative recombination. As a result, as shown on the right of Fig. 3.28, the damping factor is enhanced, and the 3-dB bandwidth shrinks. This finding aligns

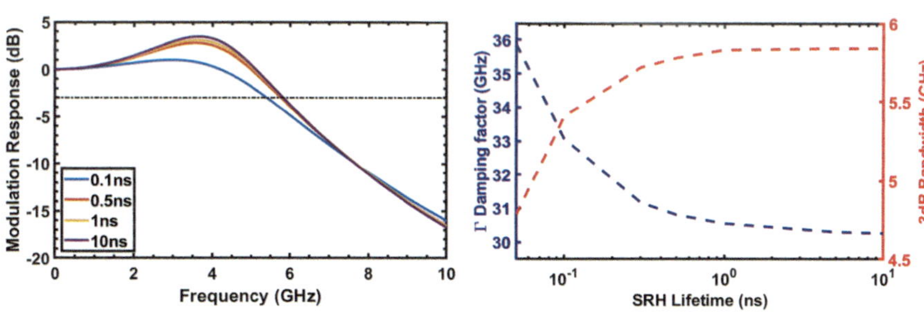

Fig. 3.28 Left panel: Nonradiative recombination effects on the QD laser's modulation dynamics. The black dashed line signifies the position of 3-dB modulation response power. Right panel: Nonradiative recombination effects on the QD laser's damping factor and 3-dB bandwidth

with a recent numerical study about the QD laser model in which the turn-on delay dynamics were simulated by varying the timescale ratio between carrier and photon lifetimes [57]. Based on this hypothesis, altering the SRH recombination time will inevitably modify the aforementioned ratio, subsequently affecting the damping behaviour of the solitary laser. This, in turn, may limit its practical suitability for direct modulation applications [58].

3.2.3 Interband Cascade Lasers

Interband cascade lasers (ICLs) have emerged as highly efficient mid-infrared (MIR) laser sources [59], catering to the growing demand for gas sensing applications. These lasers prove particularly valuable as they can target the fingerprint absorption lines of numerous molecules such as carbon monoxide, carbon dioxide, and methane, which are prevalent in the MIR range. Furthermore, the wavelength of 3–6 μm at the MWIR range coincides with the atmospheric transmission window, enabling the utilization of ICLs in free space optical (FSO) communication. Notably, ICLs employ a carrier cascading configuration similar to quantum cascade lasers (QCLs), but their laser emission is based on interband transitions rather than intersubband transitions. Several key improvements have been made to enhance ICL performance. One of these improvements is the adoption of a W-shape configuration in the active region, replacing the single type-II interface. This configuration involves sandwiching a hole quantum well (QW) between two electron QWs. The W-shape active region significantly increases the overlap of electron-hole wavefunctions, resulting in an improved optical gain. Another crucial enhancement is the concept of carrier rebalancing, which involves increasing the n-doping concentration in the electron injector to rebalance the electron and hole densities in the active region [60]. This approach greatly reduces the threshold power density of ICLs.

In the rate equation modelling of ICLs, all the cascading stages in the ICL are assumed identical, following the same assumption in the modelling of QCLs [62]. As shown in Fig. 3.29, each cascading stage in ICLs contains three parts: a hole injector, an electron injector, and a W-shape active region. Under the external electric field, holes and electrons flow into the active region via the tunnelling effect, where they recombine through spontaneous emission, Auger recombination, and stimulated emission due to population inversion. Then, in order to replenish the carriers removed by the field, equal numbers of additional electrons and holes must be regenerated constantly and transported to the next stage. Therefore, ICLs' rate equations for carrier N and photon S can be written as [61, 63]:

$$\begin{aligned}\frac{\mathrm{d}N}{\mathrm{d}t} &= \eta\frac{I}{q} - \Gamma_p v_g g S - \frac{N}{\tau_{sp}} - \frac{N}{\tau_{aug}} + F_N \\ \frac{\mathrm{d}S}{\mathrm{d}t} &= m\Gamma_p v_g g S - \frac{S}{\tau_p} + m\beta\frac{N}{\tau_{sp}} + F_S\end{aligned} \tag{3.36}$$

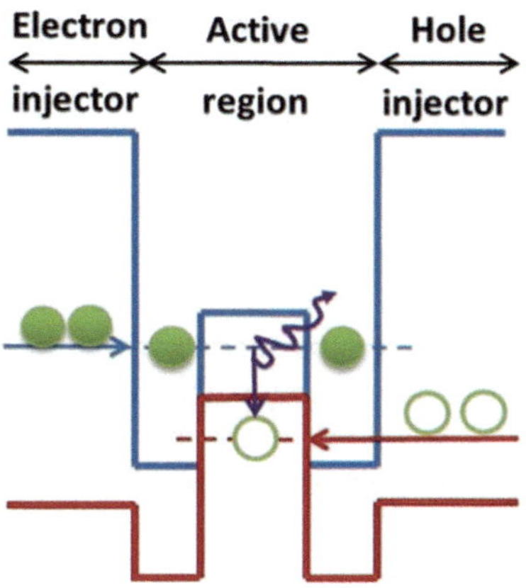

Fig. 3.29 Schematic band structure of one stage in ICLs. (*Source* In Ref. [61])

The pump current I, elementary charge q, current injection efficiency η, m number of cascading gain stage, τ_p is the photon lifetime, v_g the group velocity of light, and β is the spontaneous emission factor are related by the equation. The optical confinement factor per gain stage, Γ_p, which accounts for the decrease in value with increasing gain stage number and thickness of separate confinement layers, is assumed to be a constant in this work. In ICLs, the Auger recombination lifetime τ_{aug} is shorter than the spontaneous emission lifetime τ_{sp} and must be considered in the carrier dynamics. This is in contrast to standard QW lasers, where the Auger process is much slower than spontaneous emission and can be neglected in the rate equation [5]. The Auger lifetime is inversely proportional to the square of the carrier sheet density, given by $\tau_{aug} = 1/\gamma_3(N/A)^2$, where γ_3 is the 2-D Auger coefficient and A is the area of the active region [64]. However, for the sake of simplicity in discussing ICL dynamics, the Auger lifetime is treated as a constant value. This assumption is reasonable when the laser operates above the threshold because the carrier population is nearly clamped [61]. The material gain per stage is denoted by $g = a_0(N - N_0)/A$, with a_0 being the differential gain and N_0 being the transparent carrier number.

The properties of the laser are investigated using the proposed method. Unless otherwise specified, all material and optical parameters used in the analysis are listed in Table A.3. Figure 3.30 illustrates the carrier behaviour in ICLs. It can be observed that the clamped carrier number for $m = 10$ is slightly smaller compared to that for $m = 5$, while the photon number for the former is significantly higher. The threshold current for $m = 5$ is measured at 19.7 mA and decreases to 17.5 mA for $m = 10$. Remarkably, the threshold currents consistently decrease with an increasing stage number. The lower threshold current limit is primarily determined by the transparent carrier number N_0.

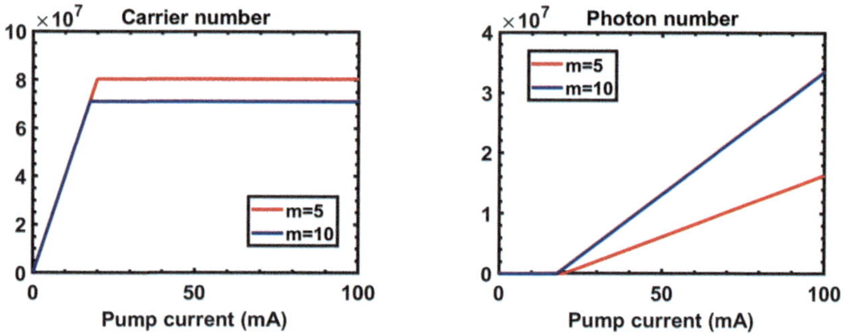

Fig. 3.30 Steady-state carrier number and photon number versus the pump current for m = 5 (red) and m = 10 (blue)

In order to investigate both the relative intensity noise (RIN) and the frequency noise (FN) through the Langevin approach, the phase dynamics of the electric field have to be taken into account:

$$\frac{\mathrm{d}\phi}{\mathrm{d}t} = \frac{\alpha_H}{2}\left(m\Gamma_p v_g g - \frac{1}{\tau_p}\right) + F_\phi \tag{3.37}$$

The diffusion coefficients among the carrier noise source F_N, the photon noise source F_S, and the phase noise source F_ϕ of ICLs are derived in the same manner as for QD lasers. For more detailed information, please refer to Appendix B. Therefore, the RIN expression is:

$$\begin{aligned}\mathrm{RIN} = \frac{|\delta S(\omega)|^2}{S^2} &= \frac{1}{S^2|\Delta|^2}\Big\{\langle F_N F_N^*\rangle |M_{\mathrm{N-S}}|^2 + \langle F_S F_S^*\rangle |M_{\mathrm{S-S}}|^2 + \left\langle F_\phi F_\phi^*\right\rangle |M_{\phi\mathrm{-S}}|^2 \\ &\quad -2\langle F_N F_S^*\rangle \Re\mathrm{e}\left\{M_{\mathrm{N-S}}M_{\mathrm{S-S}}^*\right\} - 2\left\langle F_N F_\phi^*\right\rangle \Re\mathrm{e}\left\{M_{\mathrm{N-S}}M_{\phi\mathrm{-S}}^*\right\} - 2\left\langle F_S F_\phi^*\right\rangle \Re\mathrm{e}\left\{M_{\mathrm{S-S}}M_{\phi\mathrm{-S}}^*\right\}\Big\} \\ &= \frac{1}{S^2|\Delta|^2}\left\{\langle F_N F_N^*\rangle |M_{\mathrm{N-S}}|^2 + \langle F_S F_S^*\rangle |M_{\mathrm{S-S}}|^2 - 2\langle F_N F_S^*\rangle \Re\mathrm{e}\left\{M_{\mathrm{N-S}}M_{\mathrm{S-S}}^*\right\}\right\}\end{aligned} \tag{3.38}$$

The comprehensive derivation of (3.38) is available in the second Section of Appendix B. Figure 3.31 illustrates the resonance frequency of the ICL, as recently reported in experiments [9]. It highlights that reducing the number of gain stages results in the suppression of RIN across the entire frequency range. The low-frequency RIN exhibits a nonlinear decline from −161 dB/Hz for m = 10 to −169 dB/Hz for m = 3. This behaviour can be attributed to the self-correlation of the photon noise source, $\langle F_S F_S^*\rangle$. Since the photon noise sources in each cascading stage are uncorrelated, the total photon noise of the ICL is the summation of all stages. Therefore, having more gain stages in an ICL results in a higher RIN. In the same spirit, the FN of the ICLs can be expressed as follows:

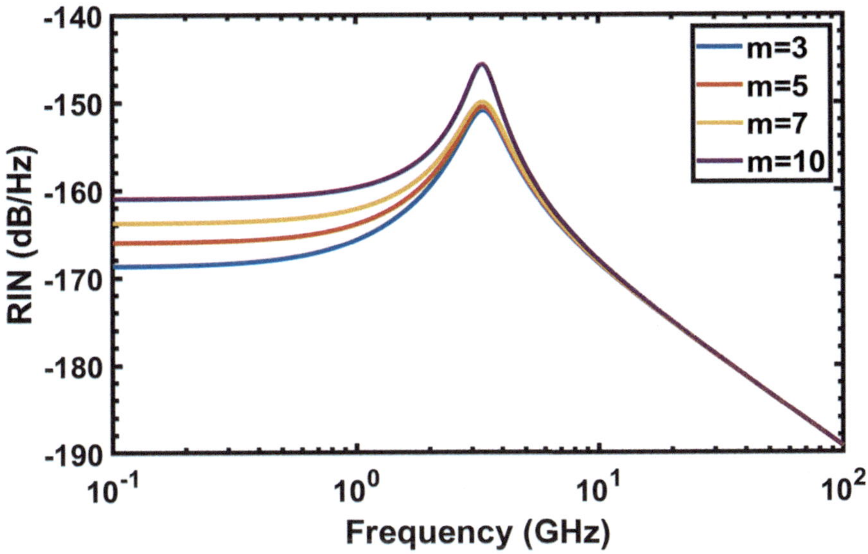

Fig. 3.31 Calculated RIN at different stage number m from (3.38)

$$\begin{aligned}
\mathrm{FN} &= \left| j\omega \frac{\Delta\phi}{\Delta} \right|^2 = \frac{\omega^2}{|\Delta|^2} \left| F_N M_{N-\phi} - F_S M_{S-\phi} + F_\phi M_{\phi-\phi} \right|^2 \\
&= \frac{\omega^2}{|\Delta|^2} \Big\{ \langle F_N F_N^* \rangle |M_{N-\phi}|^2 + \langle F_S F_S^* \rangle |M_{S-\phi}|^2 + \left\langle F_\phi F_\phi^* \right\rangle |M_{\phi-\phi}|^2 \\
&\quad -2 \langle F_N F_S^* \rangle \Re\mathrm{e} \left\{ M_{\mathrm{N}-\phi} M_{\mathrm{S}-\phi}^* \right\} - 2 \left\langle F_N F_\phi^* \right\rangle \Re\mathrm{e} \left\{ M_{\mathrm{N}-\phi} M_{\phi-\phi}^* \right\} - 2 \left\langle F_S F_\phi^* \right\rangle \Re\mathrm{e} \left\{ M_{\mathrm{S}-\phi} M_{\phi-\phi}^* \right\} \Big\} \\
&= \frac{\omega^2}{|\Delta|^2} \Big\{ \langle F_N F_N^* \rangle |M_{N-\phi}|^2 + \langle F_S F_S^* \rangle |M_{S-\phi}|^2 + \left\langle F_\phi F_\phi^* \right\rangle |M_{\phi-\phi}|^2 \\
&\quad -2 \langle F_N F_S^* \rangle \Re\mathrm{e} \left\{ M_{\mathrm{N}-\phi} M_{\mathrm{S}-\phi}^* \right\} \Big\}
\end{aligned} \tag{3.39}$$

The comprehensive derivation of (3.39) is available in the second Section of Appendix B.

Figure 3.32 demonstrates that increasing the number of cascading stages in the ICL only amplifies the amplitude of the resonant peak in the FN spectrum, without affecting the linewidth. In the context of free-space communications, direct modulation of ICLs presents an attractive scheme compared to external modulation, as it eliminates the need for an additional optical modulator. Applying Cramer's rule without the phase equation (3.37), the modulation transfer function of the ICLs, which represents the normalized IM response, can be expressed as follows:

$$H(\omega) \equiv \frac{\delta S(\omega)/\delta I}{\delta S(0)/\delta I} = \frac{\omega_R^2}{\omega_R^2 - \omega^2 + j\omega\gamma}$$

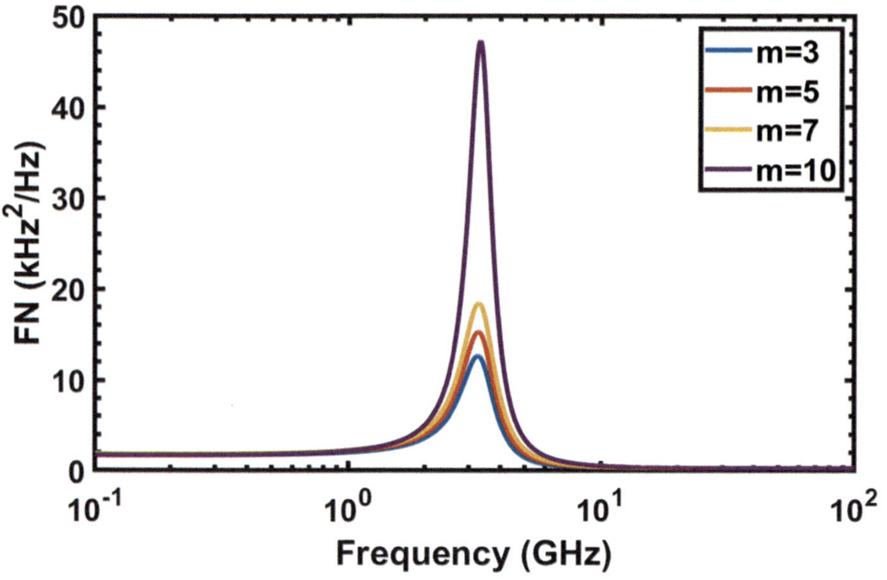

Fig. 3.32 Calculated FN at different stage number m from (3.39)

with the resonance frequency ω_R

$$\omega_R = \gamma_{12}\gamma_{21} + \gamma_{11}\gamma_{22} = G_0\,(N - N_0)\,(mG_0S + m\frac{\beta}{\tau_{sp}}) + (G_0S + \frac{1}{\tau_{sp}} + \frac{1}{\tau_{aug}})\left[\frac{1}{\tau_p} - mG_0\,(N - N_0)\right]$$

and the damping factor γ

$$\gamma = \gamma_{11} + \gamma_{22} = G_0S + \frac{1}{\tau_{sp}} + \frac{1}{\tau_{aug}} + \frac{1}{\tau_p} - mG_0\,(N - N_0)$$

Figure 3.33 illustrates the evident resonance observed in the normalized response of the ICL (m = 5), similar to that observed in QW lasers. However, QCLs, characterized by their ultrashort carrier lifetime, do not exhibit any resonance. The amplitude of the resonance peak decreases with higher pump currents due to the increased damping factor. Moreover, ICLs demonstrate a comparable modulation bandwidth to conventional laser diodes, as they share similar characteristics such as carrier lifetime and differential gain. In a recent proof-of-concept experiment, interband cascade lasers were employed for energy-efficient mid-infrared free-space communication, as reported in [10].

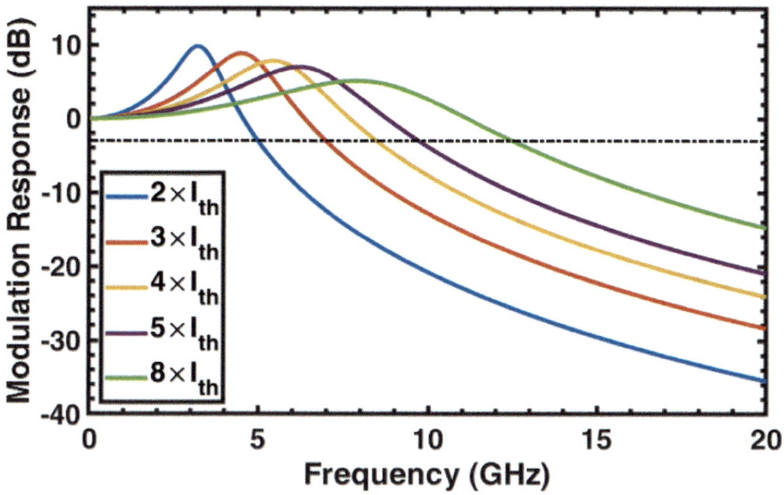

Fig. 3.33 IM response at various pump currents for m = 5. The dashed black line indicates the 3-dB bandwidth

3.3 Introduction to the Time-Delayed Feedback on Semiconductor Lasers

Semiconductor lasers subject to time-delayed feedback are of significant interest from a dynamical system perspective [65]. In semiconductor lasers, polarization relaxation usually occurs much faster than the relaxation of the optical field and carrier inversion. Therefore, laser dynamics can be described as a driven damped nonlinear oscillator, leading to a spiralling flow towards a steady state, referred to as relaxation oscillations. Various feedback configurations can be employed to overcome damped relaxation oscillations and generate chaotic behaviour. The introduction of feedback into the system leads to a delay, which in turn induces high dimensionality. This high dimensionality gives rise to a diverse range of phenomena, including limit cycle, multistability, bursting, intermittency, irregular intensity dropouts (low-frequency fluctuations, LFFs), and fully developed chaos. A comprehensive review summarizing the complex dynamics of delay-coupled lasers is available in Ref. [66]. These complex dynamics hold great promise for various applications, such as secure communication [67], LIDAR systems [68], parallel random number generation [69], speckle-free imaging [70], and high-speed photonic reservoir computing [71, 72].

Semiconductor lasers possess a remarkable property that renders them highly sensitive to external perturbations, particularly optical feedback. The physical processes involved in this sensitivity can be depicted schematically as shown in Fig. 3.34. The active region of the laser exhibits phase-amplitude coupling between the intracavity field and the returned light field, leading to fluctuations in both amplitude ($|\Delta E|$) and phase ($|\Delta\phi|$). Optical feedback is introduced into the laser cavity through the

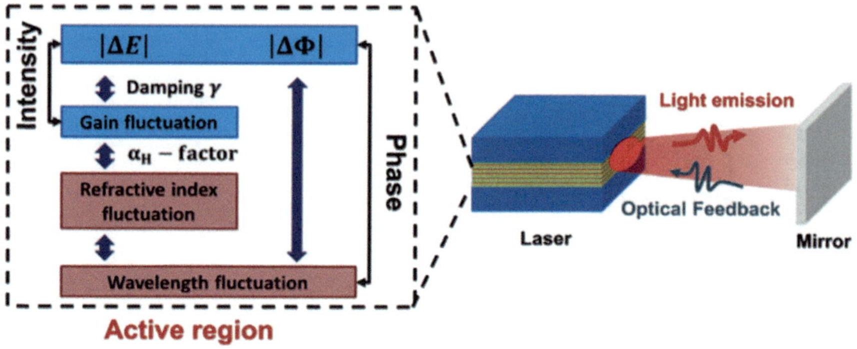

Fig. 3.34 Physical processes involved in a semiconductor laser with external optical feedback. (*Source* In Ref. [73])

output facet, causing a perturbation in the photon density. This perturbation induces fluctuations in the carrier density, subsequently affecting the optical gain. The resulting intensity fluctuation is modulated by the damping effect and is linked to the optical gain. Moreover, variations in the gain impact the refractive index through the α_H-factor, leading to wavelength shifts. Additionally, the phase fluctuation caused by the returned field is also associated with wavelength fluctuations. The interplay between the intensity and phase loops gives rise to highly complex dynamics in laser systems subjected to optical feedback.

Furthermore, optoelectronic oscillators (OEOs) [74] are autonomous, dissipative, and nonlinear systems that feature a closed feedback loop comprising two interconnected branches: an optical branch and an electronic branch. Energy is cyclically exchanged between the optical and electronic domains within this loop. Notably, the overall round-trip time in the loop is significant in relation to the other time constants within the system, as illustrated in Fig. 3.35. These oscillators have the

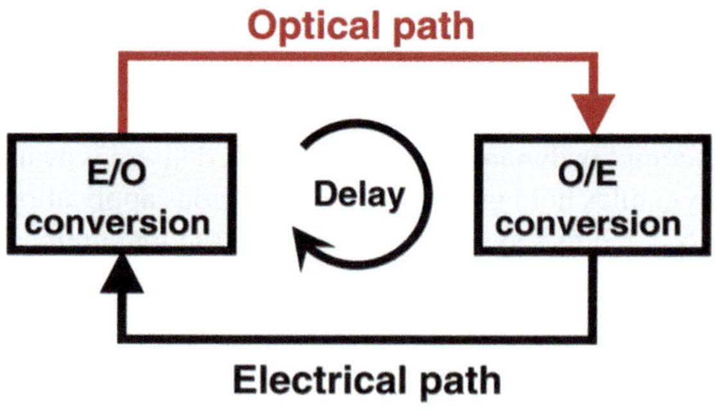

Fig. 3.35 Generic representation of an optoelectronic oscillator with time-delayed feedback. Energy flows alternatively in the optical and electrical forms, and conversion between both paths is performed by electrical-to-optical (E/O) and optical-to-electrical (O/E) converters. (*Source* In Ref. [74])

capability to simultaneously output signals in both the electrical range (approximately 0–100 GHz) and the optical range (approximately 50–500 THz). Thus, from an applied perspective, semiconductor lasers functioning as optoelectronic oscillators hold great promise for generating diverse dynamic behaviour.

In this section, I shall introduce two distinct types of time-delayed feedback: external optical feedback using the Lang–Kobayashi (LK) model and optoelectronic feedback using the Ikeda-like model.

3.3.1 *External Optical Feedback: Lang–Kobayashi Model*

The effects of external optical feedback (EOF) in semiconductor lasers have been a subject of study since their early stages of development. In fact, it is practically impossible to completely avoid EOF in experiments. Whenever an optical element, like a detector, a modulator or an optical lens, is positioned in front of a laser, it inevitably scatters a portion of the laser beam backwards. This retroreflected light has the potential to re-enter the laser cavity, thereby interfering with the pre-existing optical field inside, resulting in a perturbation of the laser output. In 1970, Broom et al. made the pioneering observation of intensity self-modulation at the relaxation oscillation frequency, which remained undamped due to an external cavity created by a distant mirror. In 1977, Risch et al. reported the self-pulsation of the light output of a continuous-wave driven semiconductor laser in an external cavity, exhibiting resonance at the frequency of the external cavity or at a lower frequency within the range of 3–30 MHz. These lower frequencies, falling below the intrinsic frequencies of both the laser and the cavity, were termed low-frequency fluctuations (LFFs).

The complex nature of the feedback from an external mirror prompted the development of models aiming to reproduce the static and dynamic properties of the system. In a landmark paper published in the early 1980s, Lang and Kobayashi [75] provided a significant contribution by investigating the effects of optical feedback on single-mode lasers. They described the temporal evolution of the complex optical field and carriers, taking into account the influence of optical feedback through the interference of the laser field with its own weak but coherent delayed field that had undergone a *single* propagation through the external cavity. The Lang–Kobayashi (LK) equations can be written as

$$\frac{\mathrm{d}\tilde{E}(t)}{\mathrm{d}t} = \frac{1+\alpha_H}{2}\left[G_N(N-N_0) - \frac{1}{\tau_p}\right]\tilde{E}(t) + \kappa\tilde{E}(t-\tau)e^{-i\omega_0\tau}$$
$$\frac{\mathrm{d}N(t)}{\mathrm{d}t} = \eta\frac{I}{q} - \frac{N}{\tau_c} - G_N(N-N_0)\left|\tilde{E}(t)\right|^2$$

with $\tilde{E}(t)$ the slowly varying complex amplitude of the optical electric field, $N(t)$ the carrier density, τ_c the carrier lifetime, τ_p the photon lifetime, ω_0 the angular optical frequency of the solitary laser, G_N the differential gain and α_H the linewidth enhancement factor. The optical feedback is taken into account via the feedback term, including κ as the feedback coefficient and τ as the delay time. The delay time τ is proportional to the external cavity length L with $\tau = 2L/v_g$, where v_g denotes the group velocity. Moreover, the feedback coefficient is mathematically defined as $\kappa = \frac{2C_l}{\tau_{in}}\sqrt{f_{ext}}$, where $\tau_{in} = \frac{2n_r L_{in}}{c}$ represents the internal round-trip time, $f_{ext} = \frac{P_f}{P_{out}}$ indicates the feedback strength, and P_f and P_{out} stand for the returned power and output power, respectively. The coupling coefficient C_l from the facet to the external cavity for the FP structure is expressed as [76]

$$C_l = \frac{1 - R}{2\sqrt{R}}$$

Assuming $\tilde{E}(t) = E_0(t)e^{j[\omega_0 t + \phi(t)]}$, then

$$\frac{\mathrm{d}E_0(t)}{\mathrm{d}t} = \frac{1}{2}\left[G_N(N - N_0) - \frac{1}{\tau_p}\right]E_0(t) + \kappa E_0(t - \tau)\cos(\Delta(t))$$
$$\frac{\mathrm{d}\phi(t)}{\mathrm{d}t} = \frac{\alpha_H}{2}\left[G_N(N - N_0) - \frac{1}{\tau_p}\right] - \kappa\frac{E_0(t - \tau)}{E_0(t)}\sin(\Delta(t))$$

where $\Delta(t) = \omega_0\tau + \phi(t) - \phi(t - \tau)$. This work laid the foundation for understanding the complex dynamics resulting from optical feedback in semiconductor lasers. The three coupled equations employed in this study reveal unstable oscillations and chaotic dynamics in their output powers, corresponding to the behaviour observed in three coupled equations of Lorenz systems. To accurately calculate and simulate these dynamics, numerical simulations often utilize the fourth-order Runge–Kutta algorithm, chosen for its superior performance in the calculations [67].

At the steady-state, where all time derivatives are equal to zero ($\frac{d}{dt} = 0$), the constant output power results in $E_0(t) = E_0(t - \tau) = E_s$ and $N(t) = N_s$. The steady-state phase ϕ_s can be expressed as $\phi_s = (\omega_s - \omega_0)t$, where ω_s represents the frequency at the steady-state. Under these conditions, one can derive the steady-state solutions as follows

$$E_s^2 = \frac{\eta I/q - N_s/\tau_c}{G_N(N_S - N_0)}$$
$$\Delta n = N_S - N_0 = -\frac{2\kappa}{G_N}\cos(\omega_s\tau)$$
$$\Delta\omega_s = \omega_s - \omega_0 = \kappa\left[\alpha_H\cos(\omega_s\tau) + \sin(\omega_s\tau)\right]$$

For a zero feedback coefficient ($\kappa = 0$), the above equations simplify to the solutions for the solitary laser. The last frequency equation can be rewritten as

$$\omega_0\tau = \omega_s\tau + C\sin[\omega_s\tau + \arctan(\alpha_H)] \tag{3.40}$$

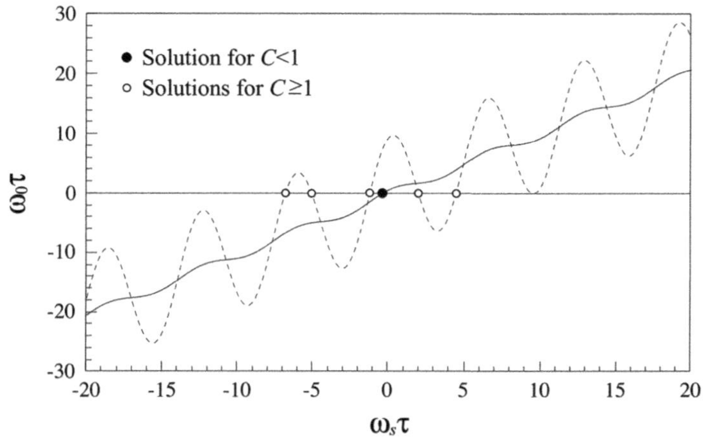

Fig. 3.36 Dependence of steady-state solutions for the phase on the parameter value C. Solid and dashed lines correspond to $C < 1$ and $C > 1$, respectively. The black circle denotes only one solution for $C < 1$ and the white circles represent multiple solutions for $C > 1$. (*Source* In Ref. [67])

where the C parameter is defined by $C = \kappa\tau\sqrt{1 + \alpha_H^2}$. Using this definition, the solutions of laser oscillation modes in the presence of optical feedback can be graphically determined as the intersections of the curves $y = \omega_0\tau$ and $y = \omega_s\tau + C \sin\left[\omega_s\tau + \arctan\left(\alpha_H\right)\right]$, as illustrated in Fig. 3.36. When $C < 1$ (indicating small optical feedback and a short external cavity), there exists only a single solution, and the laser exhibits stable oscillation. However, if $C > 1$, multiple possible modes for laser oscillations are generated, establishing a relationship between the internal laser modes and the excited external modes through (3.40). As a result, the laser demonstrates unstable operation [77].

When the C parameter significantly exceeds unity, a multitude of modes is excited in the laser output, leading to a state of true instability. To represent the many oscillation modes effectively, researchers often resort to a phase space plot of the oscillation frequency and the carrier density. Figure 3.37 illustrates such a representation in the parameter space, specifically in the $\Delta\omega_s$ versus Δn plane. This relationship is calculated by eliminating the sine and cosine functions and is expressed as

$$\left(\Delta\omega_s\tau - \frac{\alpha_H\tau}{2}G_N\Delta n\right)^2 + \left(\frac{\tau}{2}G_N\Delta n\right)^2 = (\kappa\tau)^2$$

In Fig. 3.37, the dashed sinusoidal curve represents the deviation from the steady state of the oscillation angular frequency ($\Delta\omega_s$), while the other sinusoidal curve represents the deviation of the carrier density (Δn). The crossing points of these two curves correspond to the possible oscillation states and are situated on the ellipsoid (depicted as a thick solid curve in the figure). The solutions found in the lower half

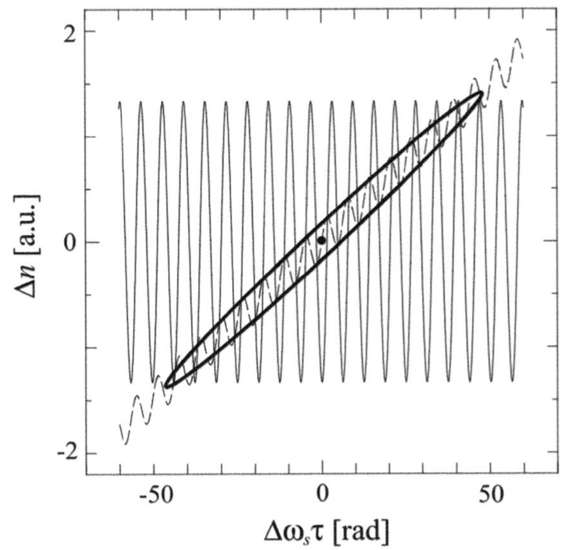

Fig. 3.37 Carrier density change (Δn) plotted against frequency change ($\Delta\omega_s$) for the potential steady states under external feedback. The intersections of the solid and dashed sinusoidal waves indicate the precise locations of the modes, which are arranged on an ellipsoid. The solitary oscillation mode is represented by a solid dot positioned at the centre (0, 0). (*Source* In Ref. [67])

of the ellipsoid correspond to stable oscillations (external modes) and are associated with constructive interference between the light inside the laser cavity and the feedback light. On the other hand, the solutions in the upper half indicate unstable oscillations (anti-modes) and are related to destructive interference conditions. During stable laser oscillation, the laser operates at one of the external modes, with the maximum gain mode being the most probable choice. However, when the laser oscillation becomes unstable due to external feedback, it exhibits mode hopping among the external modes and the anti-modes, leading to chaotic oscillations [67].

Additionally, it is also crucial to distinguish between two main cases that significantly influence the cavity length and, consequently, the feedback delays, giving rise to complex behaviour. These cases are determined based on the ratio of the delay time τ and the relaxation oscillation period T_{RO}, leading to the classification of the long delay regime ($\tau \gg T_{RO}$) and the short delay regime ($\tau \ll T_{RO}$). In each regime, the resulting dynamics exhibit differences and hold practical relevance in real applications [77]. The long delay regime is usually pertinent to most standard free-space and optical fibre-based configurations. In contrast, the short delay regime becomes particularly relevant for integrated structures, although in some cases, the feedback loop can still be quite long.

For long delays and moderate feedback levels: a few hundred to thousands of intersection points would exhibit when $C \gg 1$ as shown in Fig. 3.36. In this regime, a chaotic oscillation known as low-frequency fluctuations (LFFs) emerges due to intermittent chaos resulting from saddle-node instability. A distinctive characteristic of LFFs is the sudden power dropout followed by a gradual power recovery. These fluctuations occur irregularly in time, depending on the system parameters, and their frequency typically ranges from MHz to a hundred MHz [78]. As the frequency of

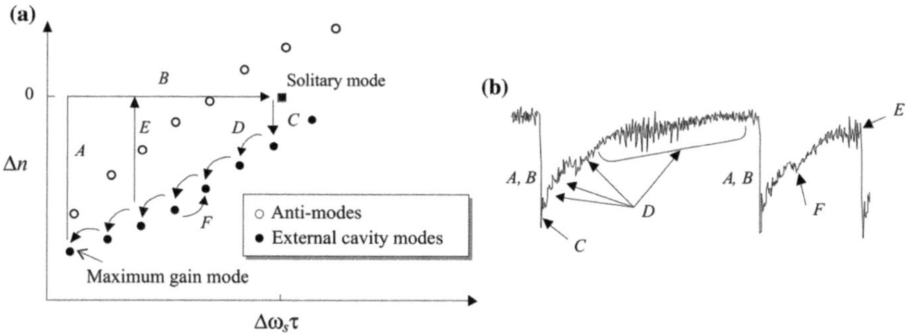

Fig. 3.38 **a** External- and anti-modes in the phase space. **b** Corresponding LFFs waveform to (**a**). (*Source* In Ref. [67])

LFFs is significantly lower than the ordinary chaotic fluctuations associated with relaxation oscillations, they are aptly named LFFs. When observing the waveforms of LFFs using a fast digital oscilloscope, they appear to be continuous signals. However, further investigation reveals that LFFs possess rapid time structures within the waveform, consisting of a series of fast pulses on the order of picoseconds. These ultrafast pulses are responsible for the complex and irregular behaviour observed in LFFs, contributing to their chaotic nature [78].

Figure 3.38 illustrates the lower half of the distribution of modes in the phase space, similar to the representation in Fig. 3.37. This configuration corresponds to a scenario characterized by a long external cavity length and a high level of external feedback. Consequently, the modes are densely distributed in the phase space.[2] At first, it is assumed that the laser initially oscillates around the maximum gain mode, and the state of laser oscillation fluctuates near this mode with a small amplitude. However, in reality, the laser's behaviour is not confined to a fixed and stable mode. Eventually, the state of the laser reaches a point very close to the corresponding anti-mode, and at this juncture, the laser becomes trapped in the anti-mode. During this transition, the phase of the laser remains unchanged, but the carrier density abruptly jumps to the value of the solitary oscillation mode. This sudden jump in carrier density causes an increase in the phase, leading to a sudden power dropout in the laser output. At this state, the laser's output power becomes nearly equivalent to that of free-running oscillation. This phenomenon corresponds to the sudden power dropout characteristic of LFFs. After the power dropout, the laser becomes trapped by one of the external modes, which is close to the solitary oscillation mode. The laser then proceeds to traverse around successive external modes, eventually returning toward the maximum gain mode. This process represents the power recovery phase of LFFs. The above process is subsequently repeated, and the occurrence of LFFs is not periodic but rather irregular. This irregularity arises because the laser may

[2] For the case of short external cavity length with low external feedback, the mode distribution will be sparse.

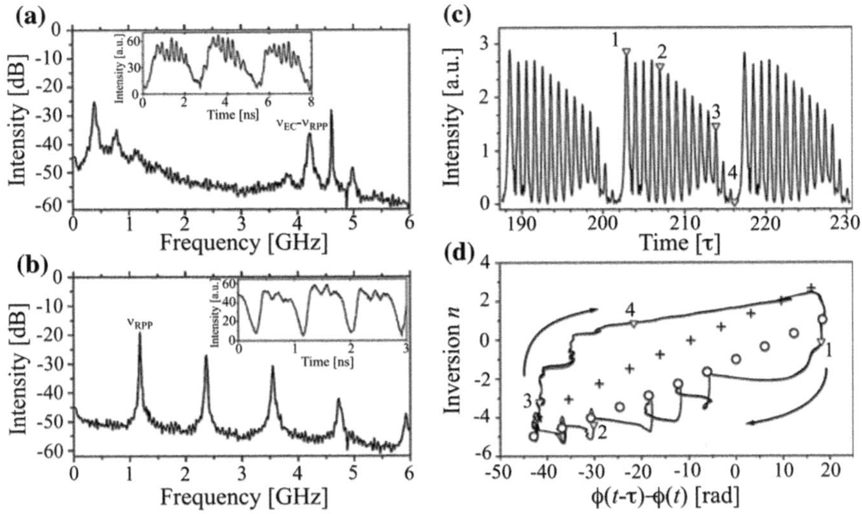

Fig. 3.39 **a** and **b** Experimentally observed RF spectra and time series for different configurations. **c** Numerical calculation of RPP and **d** trajectory plotted in the phase space. (*Source* In Ref. [79])

become trapped in the associated anti-mode, leading to a power dropout even before reaching the maximum gain mode.

For short delays and strong feedback levels: For a short external cavity ($C < 1$), the stability of a semiconductor laser is significantly enhanced, as previously mentioned. In this regime, the laser tends to oscillate in the LFFs regime rather than exhibiting irregular, fast chaotic oscillations. Additionally, the laser becomes highly sensitive to the optical phase due to the strong coupling between the internal and external cavities. Heil et al. conducted an experimental and theoretical investigation into the dynamics of a semiconductor laser with a short external cavity [79]. A distinctive characteristic of the laser's behaviour was the observation of a regular pulse package (RPP), resembling LFFs oscillation, with a frequency of several hundreds of MHz. Each pulse package comprised around 10 pulses, and the period of each pulse corresponded to the length of the external cavity (~10 MHz or less).

In Fig. 3.39a, the laser exhibits a RPP characterized by LFFs frequency of 390 MHz. Each pulse package displays a high degree of regularity, consisting of a sequence of uniformly spaced light pulses that occur at the round trip frequency. Additionally, the spectrum reveals a peak corresponding to the external cavity of 4.5 GHz, which corresponds to a short external cavity length of 3.3 cm. The inset depicts the direct waveform observed using a digital oscilloscope. In Fig. 3.39b, the LFFs frequency is 1.195 GHz with an external cavity length of 1.1 cm, which is notably fast when compared to the case of long external cavity feedback. In Fig. 3.39c, it presents the simulation results for RPP oscillations based on the LK equations. In Fig. 3.39d, external modes are indicated by circles, while anti-modes are marked by crosses. The temporal evolution follows a clockwise direction, and the numbers in the figure provide a one-to-one correspondence of time series in Fig. 3.39c. The LFFs itinerary for one cycle in the phase space is comparable to that of a long external cavity. One remarkable characteristic of this trajectory is that it consistently visits the same external cavity modes, resulting in the laser exhibiting regular pulses. The phenomenon of RPP in QD lasers was initially theoretically predicted in 2010 [80], and its observation was subsequently reported in Refs. [81, 82].

3.3.2 *Optoelectronic Feedback: Ikeda-Like Model*

In addition to optical feedback, optoelectronic feedback schemes have gained intense interest in the study of complex dynamics. These schemes take advantage of the direct modulation capability of the pump current in semiconductor lasers with GHz bandwidths. In this context, the laser's intensity is detected either directly or at a specific point (e.g., after a Mach-Zehnder Modulator), and the amplified electrical signal is subsequently fed back to the pump current of the semiconductor laser. This configuration enables the design and exploration of various types of feedback nonlinearities. By manipulating the electrical signal fed back into the pump current, one can achieve and investigate different feedback effects. A comprehensive review of this topic can be found in [74].

The inception of optoelectronic oscillators (OEOs) can be traced back to the late 1960s when the early OEOs were designed to deliberately suppress oscillations. In the 1970s, researchers started incorporating lithium niobate ($LiNbO_3$) crystals into Fabry-Perot (FP) resonators and $LiNbO_3$ waveguide modulators as the nonlinear part of the feedback loop. However, the main limitation was the requirement for relatively high voltages (around 50–100 V) to control the nonlinear transfer function. This obstacle was overcome with the introduction of integrated $LiNbO_3$ Mach-Zehnder interferometers, which allowed for full controllability of nonlinearity using voltages of just a few volts. In recent times, even systems with direct detection of the laser's output and delayed electronic feedback to the pump demonstrate a rich variety of complex dynamics. In the 2000s, J. M. Liu et al. observed deterministic chaos with positive and negative feedback [83]. Currently, broadband OEOs are distinguished by their electrical bandwidth, which spans at least an octave, often extending over more than a decade in frequency. These systems exhibit a remarkable interplay of high gain, nonlinearity, broad bandwidth, and time delay, enabling the observation of a wide array of complex and high-dimensional dynamical behaviour. Among them are phenomena like multistability, excitability, chaotic breathers, pulse packages, chimaera states, and fully developed chaos [74, 84, 85].

In optical feedback applied to semiconductor lasers, phase sensitivity assumes a critical role in laser dynamics [80]. However, in optoelectronic feedback schemes, there is no need to account for the phase effect. This is due to the phase information being effectively eliminated during the feedback process through photodetection, which, in turn, reduces the complexity of modelling such architectures. In a groundbreaking paper published in 1979, Ikeda [86] put forth an equation (now bearing his

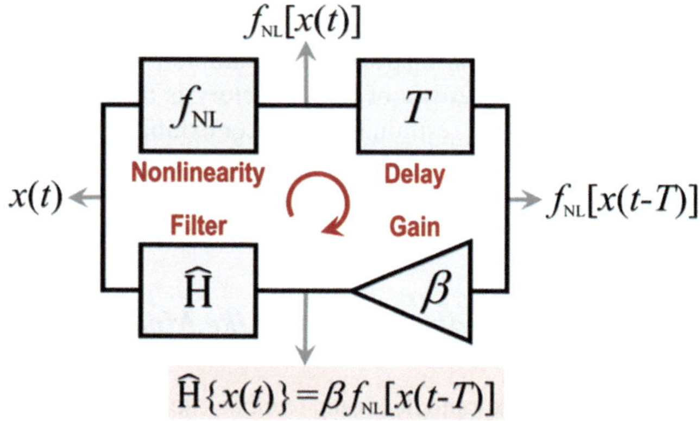

Fig. 3.40 Time-domain block diagram for an Ikeda-like model. The variable $x(t)$ circulates in the clockwise direction and is subjected to the four main elements of the loop. (*Source* In Ref. [74])

name) to elucidate the complex dynamics of a laser field confined in a four-mirror ring optical cavity, wherein the laser light propagated freely, except for a small segment incorporating a nonlinear element. This model stands as a milestone in optoelectronic oscillator research and can be simplified with four indispensable elements [74]: the linear gain β, the nonlinear function f_{NL}, the spectral linear filtering function $H(i\omega)$, and the time delay T. The closure condition of the feedback loop is

$$\hat{H}\{x(t)\} = \beta f_{NL}[x(t-T)] \tag{3.41}$$

In this context, $\hat{H}\{x(t)\}$ represents a linear integrodifferential operator, which can alternatively be seen as an input-output system following the relationship $\hat{H}\{x_{out}\} = x_{in}$. According to its definition, (3.41) is referred to as an Ikeda-like model. The interplay among the four elements of the OEO block diagram will determine the complex dynamical behaviour, as depicted schematically in Fig. 3.40.

The two most popular implementations of Ikeda-like OEOs are depicted in Fig. 3.41. The first method involves employing an integrated Mach-Zehnder modulator (MZM) as the nonlinear element, as seen in Fig. 3.41a, where the variable x corresponds to the RF voltage V_{rf} at the input of the modulator. The second approach entails using the laser itself for electrical-to-optical conversion, achieved via its pump current electrode, as demonstrated in Fig. 3.41b. The prime distinction between these OEOs arises from the utilization of different nonlinearity functions f_{NL}. For instance, the MZM employs a sinusoidal-like f_{NL} [84]:

$$f_{NL} = \cos^2[x(t-T)+\phi]$$

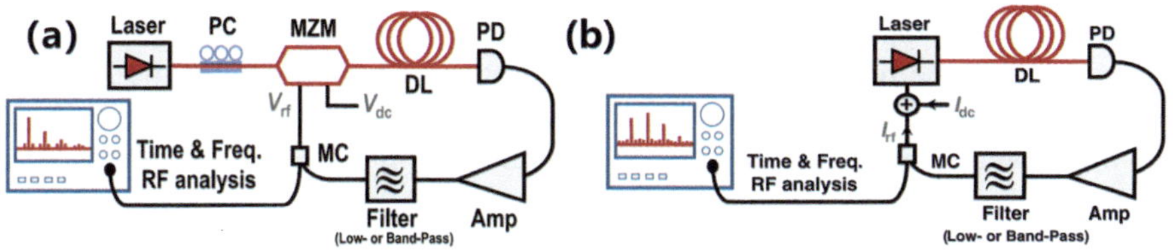

Fig. 3.41 Experimental setup for the most common implementations of Ikeda-like OEOs. **a** The external Mach-Zehnder modulator (MZM) induces a sinusoidal nonlinearity function f_{NL} in the feedback loop. **b** The semiconductor laser itself with optoelectronic feedback. (*Source* In Ref. [74])

In the case of semiconductor lasers, the nonlinearity functions f_{NL} take the form of a piecewise function from the light-current (L-I) curve. Detailed explanations are in Sect. 3.4.3.

3.4 Investigation of the Time-Delayed Feedback on Quantum-Confined Lasers

3.4.1 Optical Feedback Insensitivity in Quantum Dot Lasers on Silicon

The integration of optical functions on a microelectronic chip opens up a plethora of innovative possibilities and holds the potential to enhance the performance of photonic integrated circuits (PICs) [87]. In particular, silicon photonics stands out by profiting from the well-established complementary-metal-oxide-semiconductor (CMOS) industry, enabling the development of low-cost, compact, and high-density PICs [88]. However, it is important to be mindful that incorporating additional on-chip components can introduce potential sources for reflections to any on-chip laser. These reflections may arise from various sources, including coherent optical feedback caused by vertical grating couplers and multiple passive and active interfaces/transitions between the III–V material and silicon. In addition, incoherent feedback originating from amplified spontaneous emission (ASE) noises generated by active building blocks like semiconductor optical amplifiers (SOAs) or active waveguides can also impact the laser's behaviour. As a result, the laser's stability could be compromised, leading to the emergence of undesirable nonlinear dynamics, such as periodic or chaotic oscillations. Careful consideration and design are necessary to mitigate these effects and ensure the reliable operation of PICs in practical applications, particularly in optical fibre communication systems. Meeting industry standards, IEEE 802.3ah specification suggests that the maximum allowable amount of relative intensity noise for a 1 Gbps link is under −12 dB of reflection at −113 dB/Hz, and the reflection tolerance of an optical interconnect should be higher than −26 dB [89].

One common approach to address this issue is to insert an optical isolator that serves to block the feedback light from reaching the active region of the laser and prevents unwanted reflections. However, it is important to recognize that this solution may come at the cost of poorer performance and higher expenses. For instance, integrated isolators are categorized into three groups: those relying on nonlinear effects [90], those based on electro-optic effect [91], and those using magneto-optic effects [92]. Nevertheless, to date, there has been no demonstration of an integrated isolator that offers both strong isolation and negligible insertion loss. This highlights the importance of developing feedback-insensitive lasers. In this context, developing reflection-resistant optical transmitters remains a major objective that has the potential to revolutionize the core technology of the physical layer. In this section, I will introduce quantum dot (QD) lasers directly grown on silicon, which exhibit remarkable insensitivity to optical feedback.

Similar to (3.23)–(3.27) in Sect. 3.2.2, the Lang–Kobayashi (LK) feedback terms are added in this model [54]. Langevin forces are omitted from this model as the intention is to differentiate between laser noise and deterministic chaos.

$$\frac{dN_{RS}}{dt} = \frac{\eta I}{q} + \frac{N_{ES}}{\tau_{RS}^{ES}} - \frac{N_{RS}}{\tau_{ES}^{RS}}(1-\rho_{ES}) - \frac{N_{RS}}{\tau_{RS}^{spon}} - \frac{N_{RS}}{\tau_{SRH}} \tag{3.42}$$

$$\frac{dN_{ES}}{dt} = \left(\frac{N_{RS}}{\tau_{ES}^{RS}} + \frac{N_{GS}}{\tau_{ES}^{GS}}\right)(1-\rho_{ES}) - \frac{N_{ES}}{\tau_{RS}^{ES}} - \frac{N_{ES}}{\tau_{GS}^{ES}}(1-\rho_{GS}) - \frac{N_{ES}}{\tau_{ES}^{spon}} - \frac{N_{ES}}{\tau_{SRH}} \tag{3.43}$$

$$\frac{dN_{GS}}{dt} = \frac{N_{ES}}{\tau_{GS}^{ES}}(1-\rho_{GS}) - \frac{N_{GS}}{\tau_{ES}^{GS}}(1-\rho_{ES}) - \Gamma_p v_g g_{GS} S_{GS} - \frac{N_{GS}}{\tau_{GS}^{spon}} - \frac{N_{GS}}{\tau_{SRH}} \tag{3.44}$$

$$\frac{\mathrm{d}S_{GS}}{\mathrm{d}t} = (\Gamma_p v_g g_{GS} - \frac{1}{\tau_p})S_{GS} + \beta_{sp}\frac{N_{GS}}{\tau_{GS}^{spon}} + \frac{k_c}{\tau_{in}}\sqrt{S_{GS}(t-\tau)S_{GS}(t)}\cos(\Delta) \tag{3.45}$$

$$\frac{\mathrm{d}\phi}{\mathrm{d}t} = \frac{1}{2}\Gamma_p v_g (g_{GS}\alpha_H^{GS} + g_{ES}\alpha_H^{ES} + g_{RS}\alpha_H^{RS}) - \frac{k_c}{\tau_{in}}\sqrt{\frac{S_{GS}(t-\tau)}{S_{GS}(t)}}\sin(\Delta) \tag{3.46}$$

with τ being the external cavity photon round-trip time and k_c is the optical feedback term written as $k_c = \frac{1-R}{\sqrt{R}} f_{ext}$, where R is the mirror reflectivity and f_{ext} is the feedback strength. The values of other parameters can be found in Table A.2. The phase difference Δ is defined as $\Delta = \psi_0 + \phi(t) - \phi(t-\tau)$ according to the LK model, with ψ_0 being the initial phase shift. To solve these equations numerically, the fourth-order Runge–Kutta algorithm is employed, utilizing the initial values obtained from the steady-state solution. Throughout the simulation process, special attention is given to using a time step of 0.5 ps and a time span of 2 μs. For simplicity, the initial phase is set to zero, and the pump current is maintained at $2\times I_{th}$.

In PICs, the short cavity regime is prevalent because of the presence of numerous integrated optical components, resulting in reflection distances of only a few centimetres. In this scenario, the relaxation frequency ν_{RO} is smaller than the external cavity frequency ν_{ext}, leading to a substantial reduction in the number of possible degrees of freedom. Figure 3.42 illustrates the bifurcation diagrams, time series, and phase portraits extracted from (3.42) to (3.46). An external cavity length of 4 cm is taken. For each value of the feedback strength f_{ext}, a normalized output electrical intensity is plotted at local maxima and minima in the time series. The impact of the SRH lifetime on the bifurcation diagram is evident in Fig. 3.42, showing that a decrease in τ_{SRH} from (a) to (d) results in a less complex bifurcation diagram. At $\tau_{SRH} = 0.1$ ns, the periodic region nearly disappears, leaving stable operation. The first Hopf bifurcation[3] is closely tied to τ_{SRH}, as obvious from Fig. 3.42a–d, where decreasing τ_{SRH} from 10 to 0.1 ns shifts the onset of the first Hopf bifurcation to larger values of f_{ext}, expanding the stable operation area. This demonstrates that the SRH recombination influences the evolution of reflection sensitivity, as lower τ_{SRH} increases the resistance against optical feedback. This is consistent with epitaxial QD lasers on silicon, which show improved resistance to feedback compared to their counterparts grown on native substrates [93]. For comparison, Fig. 3.42 (e1) displays the corresponding bifurcation diagram of a quantum well (QW) laser. In this case, the major difference is the α_H-factor, which is about 2 near the threshold, and the SRH recombination lifetime is neglected. As a result, the QW laser exhibits a typical bifurcation scenario with a cascade leading to chaotic operation as f_{ext} increases [94].

In recent experiments, the epitaxial QD laser on silicon demonstrated excellent stability in the short cavity regime, exhibiting possible periodic oscillations without chaotic operation [82], which aligns with these simulation results. This inherent property can be attributed to the small α_H-factor, the large damping, and the absence of higher energy states involved in the lasing emission process. Figure 3.43 depicts the RF spectrum (left column), time series (middle column), and phase portraits (right column) of the device operating at 160 mA ($4.4\times I_{th}$) with different feedback strengths and external cavity lengths. In Fig. 3.43a, the feedback strength is 68.2% (-1.66 dB), and the external cavity length is 4 cm (3.75 GHz, corresponding ratio of $\nu_{RO}/\nu_{ext} = 0.5$). The time series in Fig. 3.43a-ii exhibits a series of regular pulses oscillating at ν_{ext}, which are modulated by a slower envelope with a repetition frequency ν_{RPP} of 55 MHz, clearly depicted in the inset of Fig. 3.43a-i. Moreover, the inset of Fig. 3.43a (a-ii) highlights the fast oscillations at ν_{ext} within one envelope. Extending the external cavity length to 6.5 cm ($\nu_{RO}/\nu_{ext} \approx 0.9$) while ensuring the same feedback strength of 68.2%, the device enters into the period-one state in Fig. 3.43b. Further increasing the ratio of $\nu_{RO}/\nu_{ext} \approx 2$ by extending the external cavity length to 15 cm, the device enters the quasiperiodic state, as shown in Fig. 3.43c-i. Both the ν_{ext} and the self-pulsation frequency ν_{sp} shown in Fig. 3.43c-i contribute to shaping the waveform in this state. When the feedback strength is decreased to 33% (-4.8 dB) while keeping the external cavity length fixed at 15 cm,

[3] A Hopf bifurcation is a critical point in which, as a parameter changes, a system's stability undergoes a switch, leading to the emergence of a periodic solution.

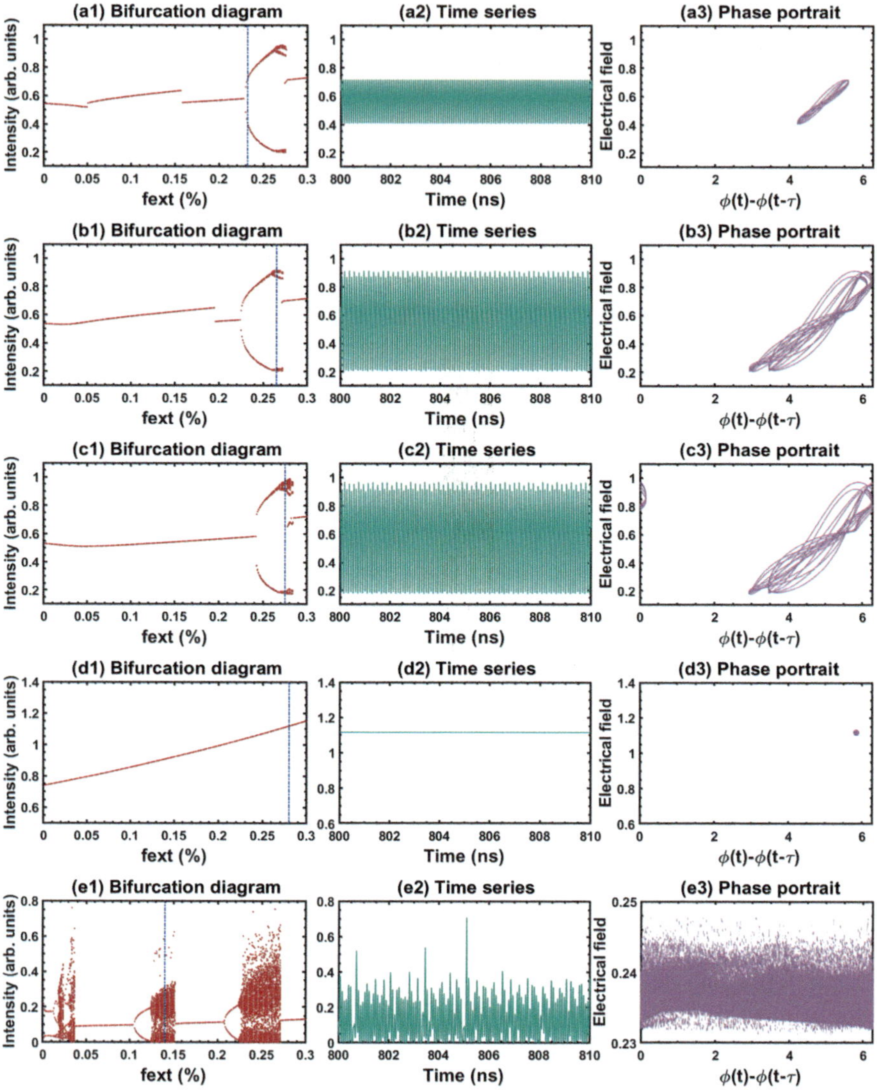

Fig. 3.42 Numerically computed bifurcation diagram, time series and phase portrait in the column (1), (2), (3) respectively for different values of τ_{SRH} in the short cavity regime (4 cm): τ_{SRH} = 10 ns and $f_{ext} = 0.232$ (row a), $\tau_{SRH} = 5$ ns and $f_{ext} = 0.265$ (row b), $\tau_{SRH} = 1$ ns and $f_{ext} = 0.275$ (row c), $\tau_{SRH} = 0.1$ ns and $f_{ext} = 0.28$ (row d) and the corresponding QW laser and $f_{ext} = 0.14$ (row e). The blue vertical dashed lines in (1) mark the exact f_{ext} value taken in (2) and (3)

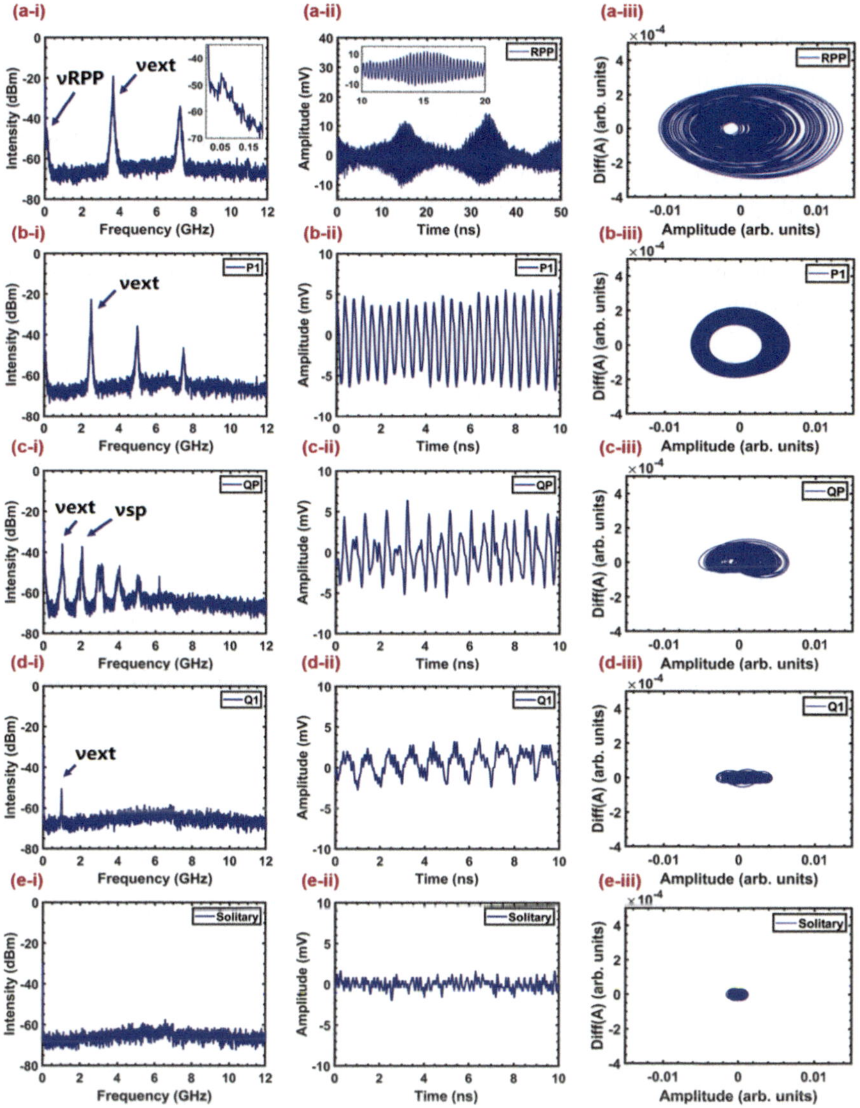

Fig. 3.43 RF spectra (left), time series (middle), and phase portraits (right) of the QD laser with (feedback strength, external cavity length) = **a** (68.2%, 4 cm), **b** (68.2%, 6.5 cm), **c** (68.2%, 15 cm), **d** (33%, 15 cm), and **e** solitary, when it was biased to 160 mA. ν_{ext}, frequency of the external cavity; ν_{sp}, self-pulsation frequency (*Source* In Ref. [82])

the device becomes more stable and enters the quasi period-one state, as shown in Fig. 3.43d. For reference, the spectra of the solitary state are shown in Fig. 3.43e.

In addition to analyzing the dynamics of epitaxial QD lasers on silicon in the short cavity regime, one also investigates the long cavity regime, which is equally important as there may be situations in PICs where the external cavity length is longer than expected. In this regime, the number of possible degrees of freedom drastically increases, leading to different dynamical behaviour compared to the short cavity regime. Figure 3.44 displays the bifurcation diagrams, time series, and phase portraits for a fixed external cavity length of 30 cm and different values of the nonradiative recombination lifetime τ_{SRH}. In the first column of Fig. 3.44a–c, the route to chaos in the bifurcation diagram is observed, and chaotic bubbles disappear, which is also evident from the time series in the second column. Additionally, the increase in τ_{SRH} has an impact on the first Hopf bifurcation point, slightly shifting it to higher values of f_{ext}, consistent with the results obtained in the short cavity regime. It is noteworthy that in Fig. 3.44d, only stable operation is observed at all times, which is in line with recent experimental feedback results in epitaxial QD lasers where no chaotic operation was observed [93]. Finally, when comparing the results to the dynamics of a QW laser with a larger α_H-factor and negligible SRH recombination, multiple chaotic bubbles occur as depicted in Fig. 3.44e, and a period-doubling bifurcation cascade leads to the outbreak of chaos [94], highlighting the much more complex dynamics in this case.

Concerning the experiments, Fig. 3.45 presents the spectral evolution of both QD and QW lasers as the feedback strength increases at $3\times I_{th}$ with a much longer external cavity length of 7 m [93]. The first row illustrates the mapping in the optical domain, while the second row shows the RF domain. In Fig. 3.45a, the QD laser on silicon demonstrates remarkable stability against optical feedback at all feedback strength levels. Only a slight red-shift of the modal wavelength is observed, and the RF response in Fig. 3.45c does not exhibit any signs of nonlinear oscillations. This implies that the laser operates in a chaos-free state within the range of feedback levels considered in the experiment, even at $6\times I_{th}$. On the other hand, the QW laser remains undisturbed until the feedback strength reaches 0.3% (−25 dB), which corresponds to the critical level associated with the onset of the undamping of the relaxation oscillations. Beyond this level, the laser starts to experience the route to chaos through coherence collapse [95], with a significant broadening of the optical spectrum (Fig. 3.45b) and intense chaotic oscillations in the frequency domain (Fig. 3.45d).

In conclusion, the rate equation model for the QD lasers on silicon presented in Sect. 3.2.2 has been extended by incorporating the LK equations to account for delayed optical feedback. Through simulations, QD lasers on silicon exhibit a reduction in the chaotic region and a shift in the first Hopf bifurcation point to higher feedback values across different external cavity regimes. These simulation results are consistent with experimental findings, showcasing the validity of the model. The insights gained from these simulations are crucial for designing feedback-resistant lasers for future high-performance PICs without the need for optical isolators.

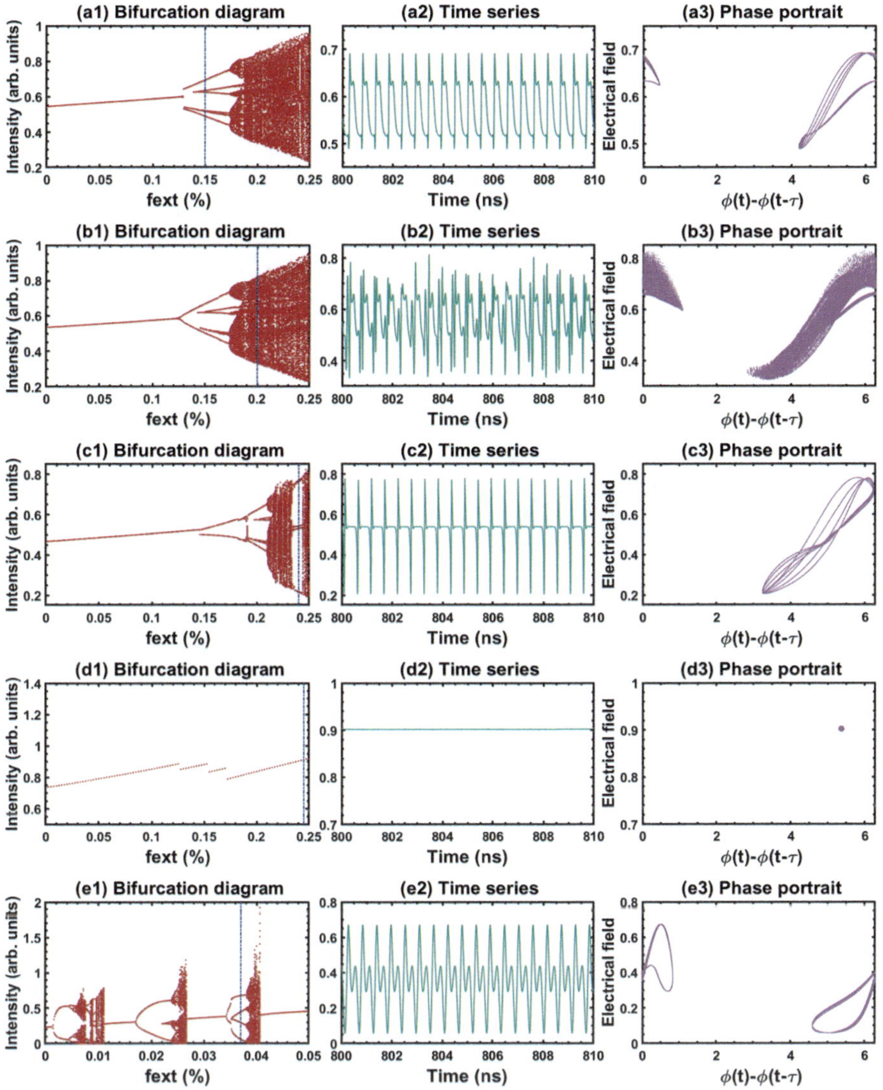

Fig. 3.44 Numerically computed bifurcation diagram, time series and phase portrait in the column (1), (2), (3) respectively for different values of τ_{SRH} in the long cavity regime (30 cm): $\tau_{SRH} = 10$ ns and $f_{ext} = 0.15$ (row a), $\tau_{SRH} = 5$ ns and $f_{ext} = 0.20$ (row b), $\tau_{SRH} = 1$ ns and $f_{ext} = 0.24$ (row c), $\tau_{SRH} = 0.1$ ns and $f_{ext} = 0.245$ (row d) and the corresponding QW laser and $f_{ext} = 0.037$ (row e). Blue vertical dashed lines in (1) mark the exact f_{ext} value taken in (2) and (3)

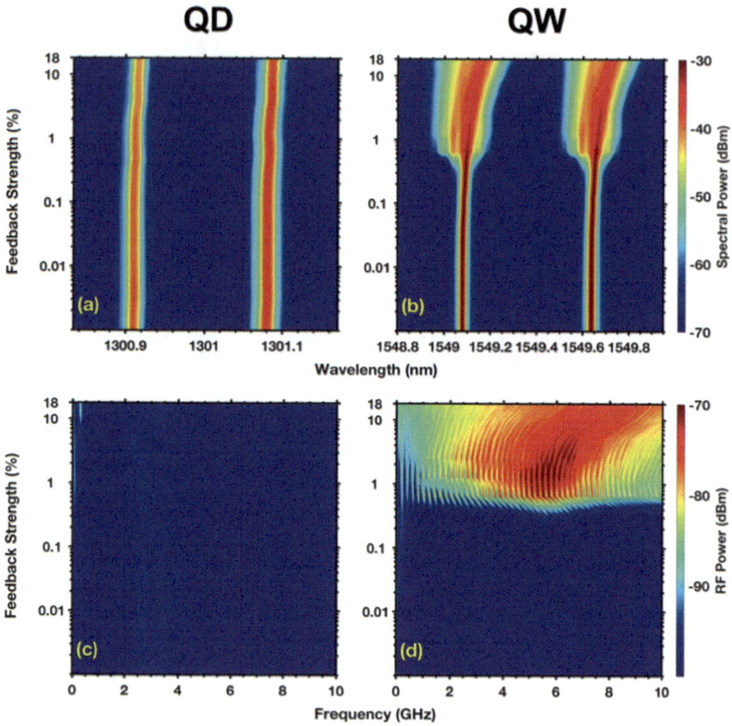

Fig. 3.45 **a** Optical and **c** RF spectral mappings of the QD laser on silicon as a function of the feedback strength at $3\times I_{th}$; **b** Optical and **d** RF spectral mappings of the QW laser as a function of the feedback strength at $3\times I_{th}$. The vertical axis is on a logarithmic scale. (*Source* In Ref. [93])

3.4.2 Broadband Chaos Generation in Interband Cascade Lasers

Chaos is a widespread phenomenon observed in numerous nonlinear dynamical systems. It is characterized by its unpredictable and random-looking behaviour, along with extreme sensitivity to initial conditions. In chaotic systems, even tiny differences in the starting conditions can lead to vastly divergent outcomes over time, making long-term predictions challenging or even impossible. Semiconductor lasers have become popular testbeds for studying chaos [65], however, most reported chaos in semiconductor lasers occurs in the near-infrared regime, specifically within the O-band and C-band communication windows of optical fibres. On the other hand, mid-infrared chaos holds great potential for applications in chaos-based free-space optical (FSO) communication [96]. This is due to the favourable high-transmission windows (3–5 and 8–12 μm) of the Earth's atmosphere.

In the 1980s, mid-infrared chaos was demonstrated in gas lasers, such as CO_2 lasers at 10.6 μm and He-Xe lasers at 3.5 μm. However, at that time, the chaos bandwidth was limited to the MHz range. Since the demonstration of chaos generation in

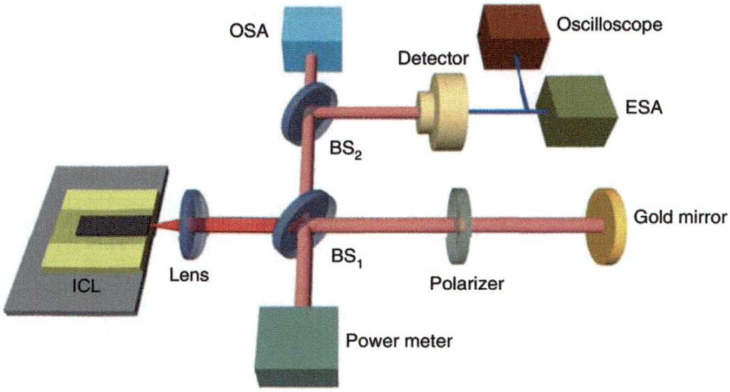

Fig. 3.46 Experimental setup for the chaos generation in an ICL subject to external optical feedback. OSA (optical spectrum analyzer), ESA (electrical spectrum analyzer), BS (beam splitter). (*Source* In Ref. [98])

quantum cascade lasers (QCLs), a wide variety of mid-infrared nonlinear dynamics has been experimentally observed [97]. Despite the intersubband technology of QCLs, most of the observed nonlinear dynamics still also have a limited bandwidth, resulting in a restricted transmission rate for chaos-based communications. Additionally, QCLs mostly produce low-frequency fluctuations rather than chaos [96]. In order to circumvent this limitation, in this section, I shall discuss the exciting discovery of fully developed chaos generated from mid-infrared interband cascade lasers (ICLs). The mid-infrared chaotic signal achieved exhibits an impressive bandwidth that reaches the GHz range, showing a significant breakthrough in chaos generation using ICLs.

Many methods have been demonstrated as effective means to generate nonlinear dynamics in semiconductor lasers, such as electro-optical feedback, optical injection, and direct modulation [65]. Since experiments with mid-infrared semiconductor lasers have mostly concentrated on external optical feedback [97], this configuration will be the focus of the following investigation. The multimode ICL being studied comprises seven cascading gain stages, consisting of W-shaped InAs/GaInSb type-II quantum wells. To induce chaos, the ICL is perturbed by external optical feedback. As depicted in Fig. 3.46, the optical feedback is achieved using a gold mirror, and the feedback strength can be controlled by rotating the polarizer. The ICL exhibits a lasing threshold of $I_{th} = 78$ mA and saturates at 140 mA. At a pump current of 85 mA, the laser emits multiple longitudinal modes around 3.4 μm. Figure 3.47a illustrates the evolution of the time traces as the feedback strength increases. The ICL exhibits continuous-wave output for weak feedback levels, up to a feedback ratio of about -16 dB (example of -21.3 dB). However, for feedback ratios ranging from -16 to -10 dB (example of -12.7 dB), weak oscillations emerge in the time trace. This is because the optical feedback reduces the damping of the relaxation oscillation (RO) of the ICL. When the feedback ratios are increased to -10 to -8.0 dB (example of -9.4 dB), the ICL experiences destabilization and exhibits

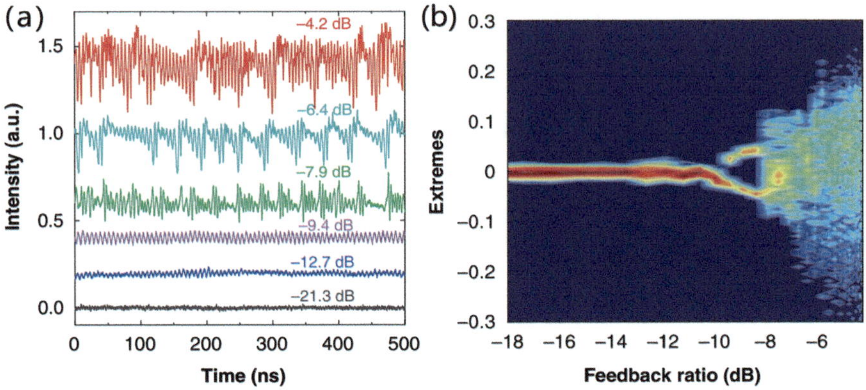

Fig. 3.47 Evolution of time traces towards chaos at 85 mA. **a** Time traces and **b** Bifurcation diagram of the electrical power extremes. (*Source* In Ref. [98])

period-one (P1) oscillation characterized by a single period in the time trace. The underlying mechanism for P1 oscillations stems from the undamped RO through Hopf bifurcation due to the optical feedback. The frequency of the P1 oscillation is found to agree with the frequency of the RO, further supporting this phenomenon. When the feedback strength surpasses a critical level, indicating the onset of chaos, the ICL begins to exhibit chaotic oscillations. The critical feedback level for the ICL is found to be approximately −8.0 dB, which is higher than that of common quantum well lasers by more than 20 dB [99]. For feedback ratios of −7.9, −6.4, and −4.2 dB, the ICL exhibits typical irregular pulse oscillations. The bifurcation diagram in Fig. 3.47b depicts the power extremes (both maxima and minima) extracted from the time series. It clearly illustrates that the Hopf bifurcation point occurs around −10 dB, representing the P1 oscillations with a single maximum and a single minimum. Beyond the critical feedback level of approximately −8.0 dB, the laser enters a chaotic state, producing multiple extremes.

The ICL demonstrates the capability to produce chaos not only at near-threshold pump currents but also at high ones. This contrasts with QCLs, where chaotic low-frequency fluctuations are mainly observed at near-threshold currents [96]. When the ICL is pumped at 105 mA, both the bifurcation diagram in Fig. 3.48a and the RF spectra map in Fig. 3.48b illustrate that the ICL does not display any periodic oscillations. Instead, the ICL produces LFFs for feedback ratios ranging from −14 to −8.0 dB (example of −11.9 dB) before transitioning to a regime of fully developed chaos. The LFFs, as shown in Fig. 3.48c, are characterized by irregular power jump-ups with gradual power increase and drastic power decrease. This behaviour stands in contrast to the typical LFFs observed in common laser diodes [78], which are characterized by random power dropouts with sudden power decreases and gradual power recovery. As the feedback level is further increased to the range from −8.0 to −6.0 dB (example of −7.1 dB), both the LFFs and the chaos coexist in the time trace, and the bandwidth of the power spectrum in Fig. 3.48d continues to broaden.

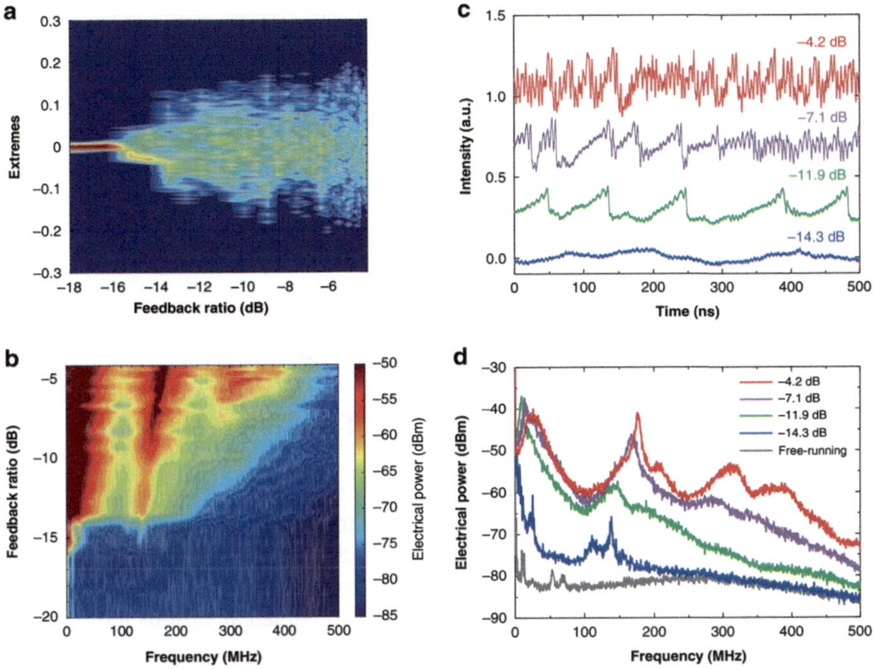

Fig. 3.48 Chaos at a high pump current of 105 mA. **a** Bifurcation diagram and **b** RF spectra map towards chaos. Examples of **c** time traces and **d** RF spectra for several feedback ratios. (*Source* In Ref. [98])

Eventually, the dynamics evolve into fully developed chaos when the feedback level surpasses −6.0 dB. In general, this onset of fully developed chaos occurs at a higher feedback level compared to the low pump current case.

The statistical characteristics of LFFs in semiconductor lasers are characterized by a probability distribution of the time interval between power dropouts or jump-ups. This distribution has a zero probability (dead zone) at short time intervals, followed by a rise to the maximum probability at a certain interval, and then an exponential decay for longer intervals [100]. In the case of the ICL pumped at a high current of 105 mA, LFFs are observed for feedback ratios in the range of −14 to −8.0 dB. As an example, Fig. 3.49 displays the statistical distribution of the LFFs at a feedback ratio of −11.9 dB. In this case, the LFFs indeed exhibit a dead zone for short time intervals up to 40 ns. Then, the probability climbs up to the peak value at a time interval of 120 ns. The distribution probability has a long tail for large time intervals, and it declines with the typical exponential trend. The mean time interval of the power dropout events is approximately 124 ns (dashed line), and it is found that the mean time interval decreases with increasing feedback strength.

In a related experiment, another multimode ICL operating at 4.1 μm was used. Unlike previous experiments, external optical feedback was not provided by a feedback mirror with a few tens of centimetres of the external cavity. Instead, in this case,

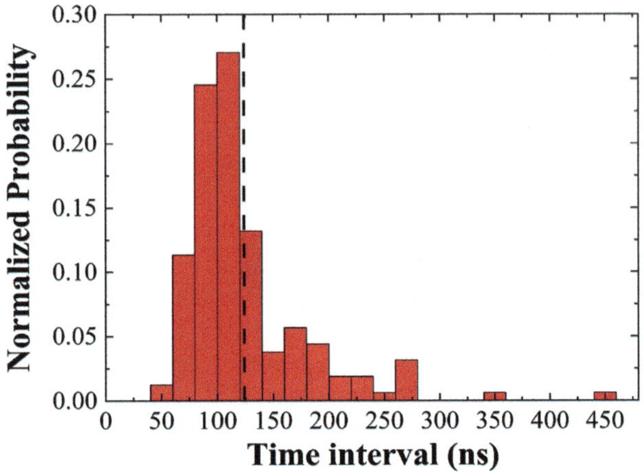

Fig. 3.49 Statistical distribution of the time interval between the power dropouts of the LFFs at the feedback ratio of −11.9 dB. The dashed line indicates the average time interval. (*Source* In Ref. [98])

the laser exhibited complex dynamics under solitary operation. This was made possible by a combination of reflection close to the back facet and the additional beating between numerous modes of the FP optical spectrum. Figure 3.50 displays the spectrograms, time traces, and RF spectra of the signal obtained at various temperatures while varying the laser bias current from 75 mA (slightly above threshold) to 180 mA (2.5 times the threshold). At 15 °C, as shown in Fig. 3.50a, the ICL remains stable for low bias currents and starts to exhibit destabilization above 110 mA. Then, up to the maximum bias current, the nonlinear dynamics produce a broad spectrum that is limited by the bandwidth of the HgCdTe detector. The combination of the broadest and flattest spectrum is found around 150 mA. However, for this temperature of operation, it is challenging to clearly identify the RO frequency (f_R) that could correspond to the orange area spanning around 0.5–0.6 GHz between 150 and 180 mA. At a higher temperature of 17.5 °C, as shown in Fig. 3.50b, the overall behaviour remains similar. However, now the RO frequency can be clearly observed from 140 mA onwards. This RO frequency (f_R) increases with the bias current, which is consistent with the findings from another ICL optimized for high-frequency modulation [9]. At an even higher temperature of 20 °C, as shown in Fig. 3.50c, the RF spectrum is no longer broad but mainly displays the evolution of the RO frequency (f_R) with temperature. Then, at around 150 mA, a period-doubling phenomenon can be observed in the RF spectrum. This phenomenon exhibits four main frequency components: one around $\frac{1}{2} \times f_R$, another around f_R, one around $\frac{3}{2} \times f_R$, and finally one around $2 \times f_R$. Interestingly, there is a discontinuous pattern around f_R and $2 \times f_R$, which locally broadens the frequency spectrum around these two values. At the highest temperature (22.5 °C) in Fig. 3.50d, the main frequency component observed is f_R, and once again, the complexity of the signal is reduced as the temperature increases. Additionally, the frequency doubling effect between 1.6 and 1.8 GHz can be seen, but the signal is relatively weak due to the frequency cut-off of the detector at higher frequencies.

The nonlinear dynamics observed in an ICL under weak external optical feedback can be analyzed using the Lang–Kobayashi model based on a rate equation

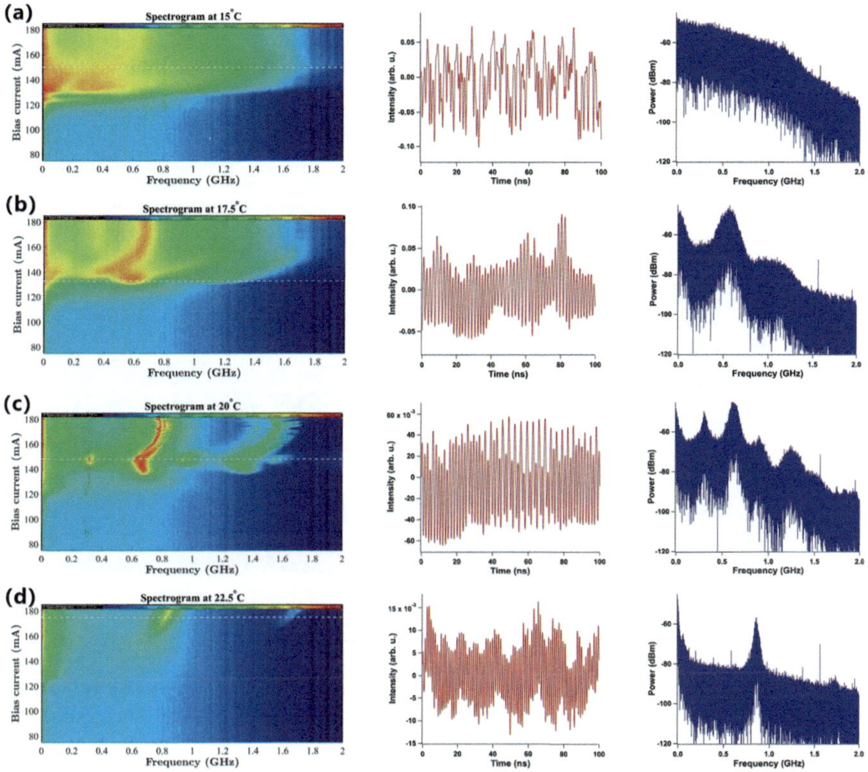

Fig. 3.50 Experimental characteristics of the optical chaos of ICL at different temperatures. The first row is the spectrogram, the second one is the temporal time trace, and the third one is the RF spectra. **a** 15 °C with a dashed white line highlighting the bias current of interest (150 mA), **b** 17.5 °C with a dashed white line highlighting the bias current of interest (133 mA), **c** 20 °C with a dashed white line highlighting the bias current of interest (148 mA), **d** 22.5 °C with a dashed white line highlighting the bias current of interest (175 mA)

approach [75]. This model is extensively utilized in numerical simulations to study various types of semiconductor lasers under feedback, including quantum dot lasers [54] and quantum cascade lasers [97]. For ICLs [98, 101], the modified rate equations governing the carrier density (N), the photon density (S), and the field phase (ϕ) are as follows:

$$\begin{aligned}
\frac{dN(t)}{dt} &= \eta\frac{I}{q} - \Gamma_p v_g g S(t) - \frac{N(t)}{\tau_{sp}} - \frac{N(t)}{\tau_{aug}} \\
\frac{dS(t)}{dt} &= [m\Gamma_p v_g g - \frac{1}{\tau_p}]S(t) + m\beta\frac{N(t)}{\tau_{sp}} + 2\kappa\sqrt{S(t)S(t-\tau)}\cos\Delta(t) \\
\frac{d\phi(t)}{dt} &= \frac{\alpha_H}{2}[m\Gamma_p v_g g - \frac{1}{\tau_p}] - \kappa\sqrt{\frac{S(t-\tau)}{S(t)}}\sin\Delta(t)
\end{aligned} \tag{3.47}$$

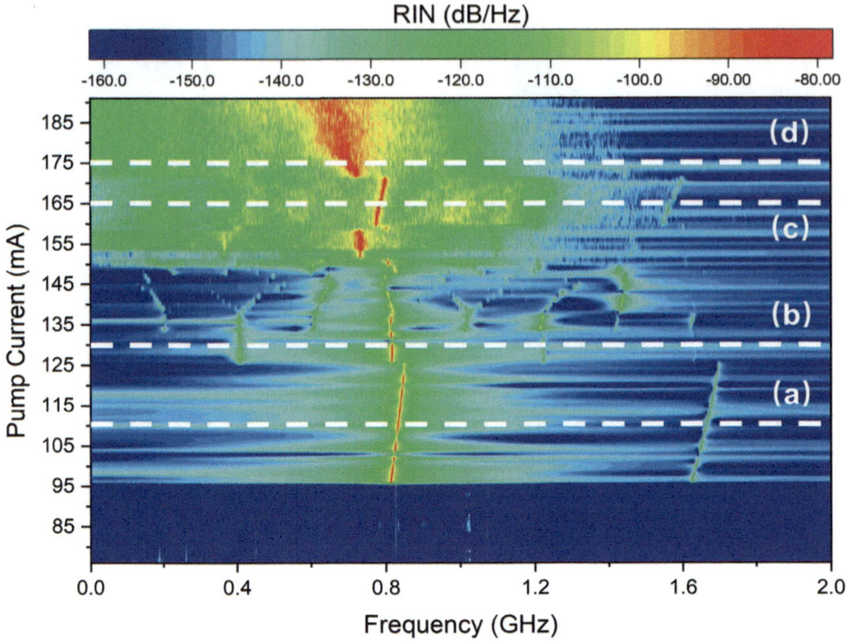

Fig. 3.51 Simulated spectrogram when varying the bias current in a range similar to the experimental scheme. The horizontal dashed line corresponds to the four bias current conditions in Fig. 3.52

where the parameter definitions can be found in Sect. 3.2.3, and their values are in Table A.4. $\Delta(t)$ is given by $\Delta(t) = \Psi_0 + \phi(t) - \phi(t - \tau)$ with Ψ_0 as the initial phase shift. The optical feedback term κ is defined as in the previous section. Simulations use an external cavity length of 4 cm to simulate the short-cavity feedback regime. The feedback strength is set at a constant value of 0.3% to match experimental observations. The initial phase shift Ψ_0 is set to $1.21 \times \pi$ to maximize dynamic behaviour during the simulations.

Figure 3.51 illustrates the simulated spectral evolution of the ICL as the pump bias current gradually increases. The simulation reveals a typical period-doubling route to chaos through Hopf bifurcations, leading to strong line broadening at a high bias current. Near the threshold, at 75 mA, the signal remains stable. As the bias current increases to around $1.5 \times I_{th}$, two main frequency components emerge, corresponding to the RO frequency f_R and its first harmonic at $2 \times f_R$. Further increasing the current to $2 \times I_{th}$ leads to more discrete frequency peaks, but chaos patterns are not observed yet. However, when the bias current reaches the maximum experimental value of $2.5 \times I_{th}$, a much more complex RF spectrum emerges, featuring strong relaxation frequency broadening followed by a dense area with remarkable nonlinear dynamics. The numerical simulations include a 0.8 GHz low-pass filter to account for the limited bandwidth of the HgCdTe detector. Although experimentally, temperature

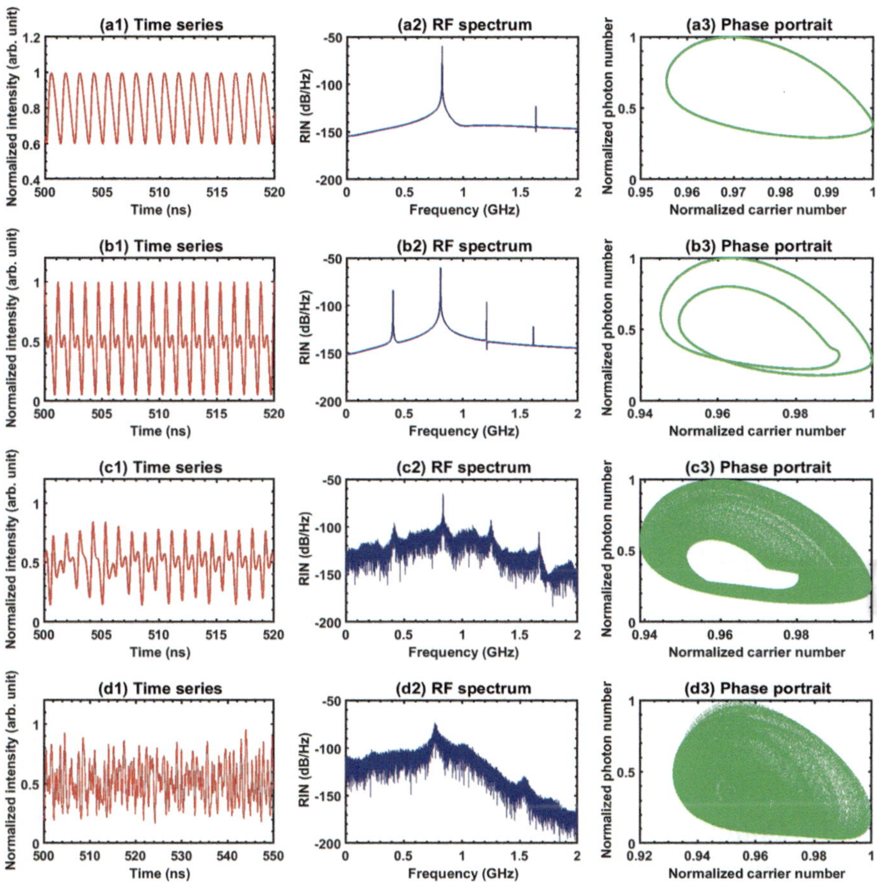

Fig. 3.52 Simulated characteristics of the dynamics for four bias currents compatible with the experimental scheme; **a** 110 mA, **b** 130 mA, **c** 165 mA, **d** 175 mA

adjustments between 15 and 22.5 °C were required to observe different nonlinear phenomena, the numerical simulations show these dynamics for a fixed temperature value.

To provide detailed insights into the dynamics exhibited by the simulations, four specific pump currents are indicated by white dashed lines in Fig. 3.51. Figure 3.52 presents the corresponding results, including time traces, RF spectra obtained through the Fourier transform, and phase portraits with normalized carrier and photon numbers. In the left column of Fig. 3.52, the time traces illustrate the ICL's output behaviour as the bias current increases. At low bias currents, stable operation is observed, while periodic oscillations are seen at intermediate bias currents, and at high bias currents, the output shows an irregular signature indicative of chaos. The RF spectra in the centre column confirm the presence of oscillations at the relaxation frequency and the period doubling. For instance, Fig. 3.52b exhibits a period-two time trace with four main frequency components in the RF spectrum located at $\frac{1}{2} \times f_R$, f_R, $\frac{3}{2} \times f_R$, and $2 \times f_R$, consistent with the experimental data in Fig. 3.50c. However, this period-doubling behaviour may not be suitable for practical applications due to its low complexity. On the other hand, at the maximum bias current (Fig. 3.52d), chaos is observed in both the time trace and the broad RF spectrum. The right column of Fig. 3.52 displays the associated phase portraits, revealing limit cycle solutions and chaotic attractors. Overall, these numerical results closely align with the experimental observations, where relaxation oscillations and period doubling have been identified. Moreover, the simulations demonstrate that chaos can be observed at high bias currents, which offer the advantage of producing intense optical power and are compatible with potential applications.

In conclusion, this section presented the findings of broad and flat chaos extending beyond 1 GHz in mid-infrared wavelength, achieved using an ICL. The numerical simulations, conducted with the Lang–Kobayashi model in the short-cavity regime, confirmed that ICLs exhibit a period-doubling route to chaos, and also demonstrated the presence of LFFs and deterministic chaos far from the bias threshold. These discoveries pave the way for future room-temperature mid-infrared applications leveraging high-speed ICLs. Notably, the use of ICLs in secured chaos transmission can be expected to expand as demonstrated in a recent proof-of-concept experiment [96]. However, this technique also necessitates the development of sensitive, fast detectors in the mid-infrared domain. In particular, the next generation of detectors, based on frequency-conversion technology or unipolar quantum optoelectronics, holds the potential to achieve bandwidths of dozens of GHz.

3.4.3 Optoelectronic Feedback in Quantum Dot Lasers on Silicon

Silicon photonics is a rapidly evolving field that aims to merge photonics and electronics onto a silicon platform. In parallel, optoelectronic oscillators (OEOs) represent nonlinear dynamical systems defined by a feedback loop that connects an optical and an electrical branch. The appeal of these systems lies in their inherent ability to bridge signals across optical and radio-frequency (RF) domains. This fundamental characteristic has paved the path for applications in diverse areas, including microwave generation, optical chaos cryptography, photonic neuromorphic computing, and sensing. Within this array of applications, the concept of reservoir computing systems has garnered considerable attention and relevance in the last decade. For example, the nonlinear dynamics inherent in semiconductor lasers are harnessed to mimic the behaviour of nonlinear nodes within the reservoir layer [71] or within the Ising machine [72].

In certain studies, quantum well (QW) lasers have been employed to construct an AC-coupled optoelectronic feedback (OEF) loop. However, this approach has only resulted in limited dynamics, showing periodic resonances or chaotic states [83]. In order to broaden the spectrum of nonlinear behaviour within the OEF loop, researchers have incorporated additional nonlinear elements. This has been achieved by processing the light emitted from the QW laser through a modulator, such as a phase modulator (PM) or a Mach-Zehnder Modulator (MZM). Through this configuration, a more diverse array of dynamics is introduced, including the emergence of novel dynamical states known as "breathers," characterized by a mixture of slow and fast behaviour [84]. Nonetheless, this strategy presents a challenge to the feasibility of silicon-based integration, as a consequence of the increased complexity associated with the integration of an electro-optical modulator. Given factors such as cost-effectiveness, process simplicity, and the ability to maintain stability, there is an urgent need to identify an OEF loop configuration that tactically minimizes the number of components while preserving the underlying potential for integration. At this moment, QD lasers emerge as alluring contenders for OEOs, particularly when compared to their QW counterparts. Despite the well-established resilience of QD lasers against optical feedback, the utilization of an AC-coupled OEF loop with a QD laser serves as a versatile approach to exploring the inherent nonlinearity of QD lasers. The approach being suggested involves using the light-current (L-I) transfer function of the emitting laser, which possesses an effective "elbow" piecewise linear nonlinearity [102]. Correspondingly, at the receiving end of the optical fibre delay line, the photodiode can display an opposing nature of nonlinearity because of saturation effects. As a result, the combined transfer function incorporates the nonlinearities of both the laser and the photodiode. The primary goal of this section is to illustrate that this composite system can be intentionally guided into a diverse range of dynamic behaviour inherent to QD lasers.

Figure 3.53 presents the time traces of the QD laser with increasing feedback strength. The QD laser displays various waveforms, transitioning from (a) stable operation, (b) 2 ns quasi-square wave due to the high cut-off frequency in the loop, (c) 250 ns square wave that precisely corresponds to the delay time, and eventually to (d) a mixed state characterized by slow and fast oscillations. The slow period of 20 μs corresponds to the wave packet shape, while the 250 ns fast period is associated with the square wave dynamics within the wave packet, as highlighted in (e). To conduct a thorough analysis, a comparative study was carried out with a 1310 nm QW FP laser. Employing an identical setup, the QW laser failed to manifest the fast oscillations observed in the QD laser, even at the highest feedback strength. By varying the OEF strength with electronic attenuators, distinct regimes were observed, ranging from steady-state to sine wave and even quasi-triangular wave patterns, as illustrated in Fig. 3.54a–b. To study the possibility of more complex dynamics emerging from the QW laser under high feedback strengths, an additional electronic amplifier (30 dB) was added to the loop. Due to constraints in the availability of electronic attenuators with a broader range, the adjustments were limited to a 10 dB attenuation following a total 60 dB RF signal amplification achieved through the use of two electrical amplifiers. With this modified setup, the QW laser showed a mixed waveform of slow and fast oscillations under high feedback strengths in (c)-(e). As the feedback strength increased, the slow period underwent gradual changes, meanwhile, the fast

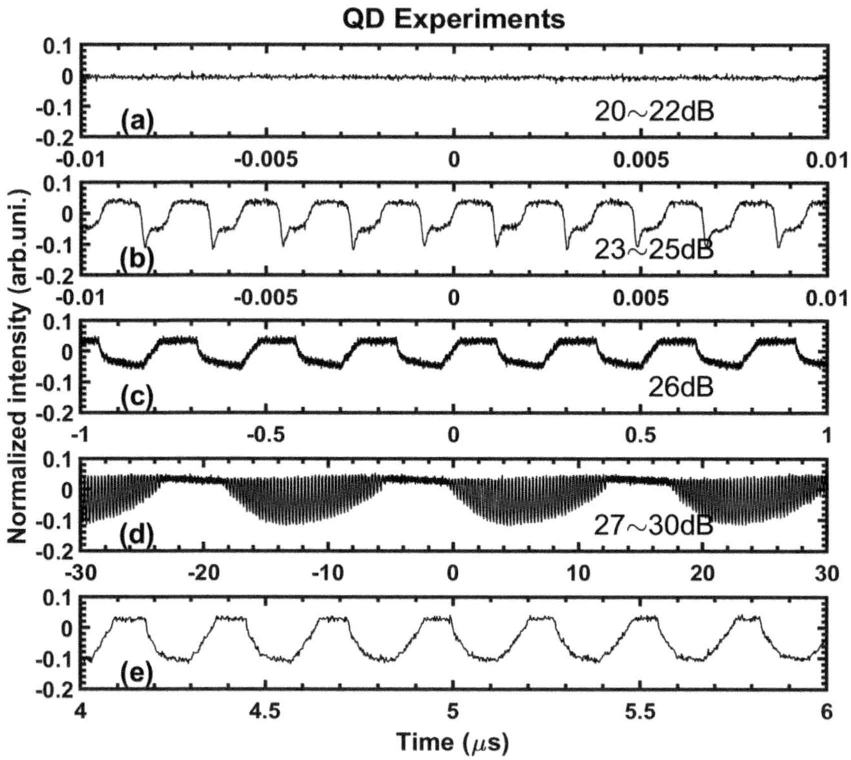

Fig. 3.53 Experimental time traces of the QD laser with OEF at different feedback strengths. Partial zoom-in views of (**d**) are presented in (**e**)

period remained a square wave with a consistent period. This observation aligned with the behaviour of the QD laser under high feedback strength with one electronic amplifier.

The experimental setup can be modelled, as depicted in Fig. 3.55, using the Ikeda-like model [86, 102]. The characteristics of this OEF loop are governed by the combined bandpass filtering effect, resulting from the aggregate bandwidths of the RF amplifier, the photodiode, the RF attenuator, and the bias tee. By utilizing the different spectral separation between the low cutoff frequency, $f_L = 50\,\text{kHz}$, and the high cutoff frequency, $f_H = 0.5\,\text{GHz}$, it is feasible to model this bandpass filter as a sequence of two first-order linear filters, comprising a high-pass filter and a low-pass filter. The loop delay time is $T = 250\,\text{ns}$. The relationship between the input voltage $V_{\text{in}}(t)$ and the output voltage $V_{\text{out}}(t)$ of a cascaded bandpass filter is given by the equation:

$$\hat{H}(V_{\text{out}}(t)) \equiv \left[1 + \frac{f_L}{f_H}\right] V_{\text{out}}(t) + \frac{1}{2\pi f_H}\frac{\mathrm{d}V_{\text{out}}(t)}{\mathrm{d}t} + 2\pi f_L \int_{t_0}^{t} V_{\text{out}}(s)ds = V_{\text{in}}(t) \tag{3.48}$$

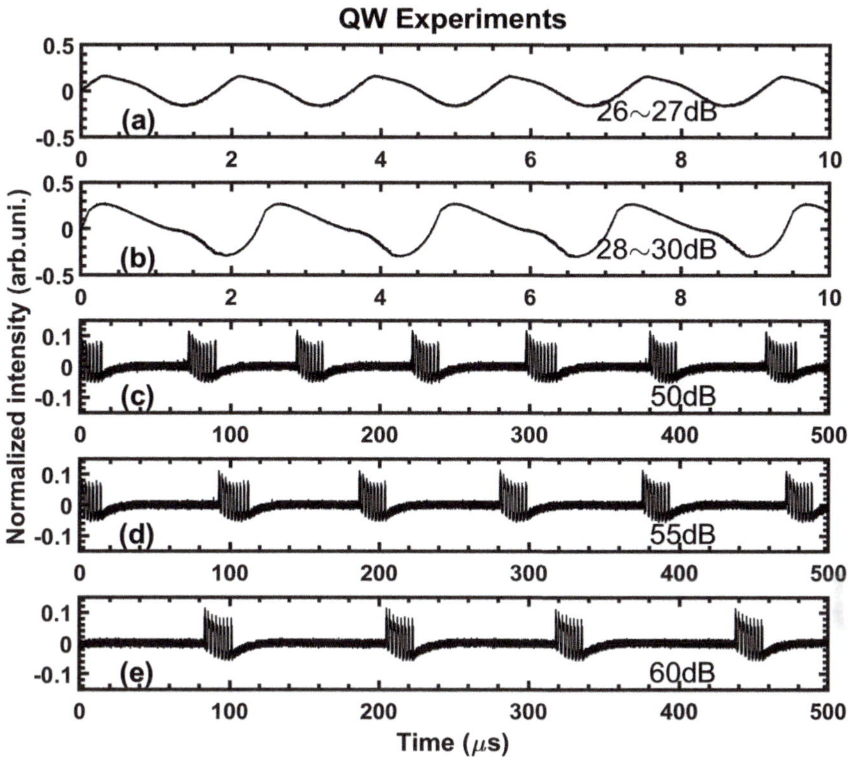

Fig. 3.54 Experimental time traces of the QW laser with OEF at different feedback strengths. **c–e** Time traces after the inclusion of another electrical amplifier (30 dB)

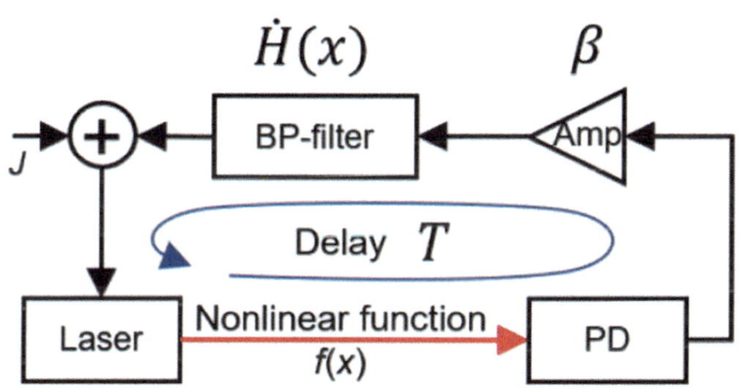

Fig. 3.55 Theoretical modelling of an OEF loop. A block diagram depicting the time domain representation of the Ikeda-like OEF loop. The variable $x(t)$ undergoes counterclockwise circulation and interacts with the four key components within the loop

The optical power P at the output of the optical fibre transforms into an electrical signal through the photodiode, governed by a time-delayed relationship denoted as $S \times P(t-T)$. Here, S represents the power-dependent responsivity of the photodiode. The photodiode's output voltage corresponds to the voltage $V_{\mathrm{in}}(t)$ applied at the input of the cascaded bandpass filter. Moreover, the connection between $V_{\mathrm{RF}}(t)$ and the output voltage can be expressed as $V_{\mathrm{RF}}(t) = \eta V_{\mathrm{out}}(t)$, where η serves as a conversion factor including all linear gains and losses (both electrical and optical) within the feedback loop. Consequently, it can be demonstrated that the RF voltage $V_{\mathrm{RF}}(t)$ obeys the following integrodifferential delay equation:

$$\left[1+\frac{f_L}{f_H}\right] V_{\mathrm{RF}}(t) + \frac{1}{2\pi f_H}\frac{\mathrm{d}V_{\mathrm{RF}}(t)}{\mathrm{d}t} + 2\pi f_L \int_{t_0}^{t} V_{\mathrm{RF}}(s)ds = \eta f_{NL}\left[V_{\mathrm{RF}}(t-T)\right] \tag{3.49}$$

The nonlinearity transfer function of the feedback loop, $f_{NL}\left[V_{\mathrm{RF}}(t-T)\right] = S \times P(t-T)$, is directly linked to the RF voltage. Considering $f_L \ll f_H$, this function can be recast into a dimensionless form as follows:

$$x(t) + \tau\frac{\mathrm{d}x(t)}{\mathrm{d}t} + \frac{1}{\theta}\int_{t_0}^{t} x(s)ds = \eta f_{NL}[x(t-T))] \tag{3.50}$$

Here, the system variable is defined as $x(t) = V_{\mathrm{RF}}(t)/V_{\mathrm{REF}}$, where $V_{\mathrm{REF}} = 1\,\mathrm{V}$ serves as a convenient reference voltage, ensuring that $|x(t)| = |V_{\mathrm{RF}}|$. The dimensionless feedback loop linear gain is η. Additionally, several time parameters are introduced: $\tau = 1/(2\pi f_H)$, and $\theta = 1/(2\pi f_L)$, which correspond to the high-pass and low-pass cutoff frequencies, respectively.

In many architectures of OEF systems, the conversion between electrical and optical signals often employs a phase or intensity modulator with a sinusoidal transfer function (MZM). However, an alternative approach involves using the laser itself as an electrical-to-optical converter [102]. Building upon this concept, a modelling strategy can be developed that capitalizes on the inherent "elbow" nonlinearity present in the light-current transfer function inherent inside different types of lasers. This nonlinearity stems from factors like the laser threshold, carrier leakage, gain compression effects, and photodiode saturation, contributing to a piecewise characteristic. To determine the nonlinearity transfer functions of both QD and QW lasers, a setup is employed, consisting of distinct laser diodes while maintaining the same photodiode in an open-loop configuration. In this setup, an input voltage V is gradually applied to the laser diode, and as this voltage increases, both the applied voltage and the resulting output voltage V_{PD} at the photodiode are recorded. Given this context, the piecewise transfer functions are formulated for the QD laser (f_{QD}) and the QW laser (f_{QW}) as follows:

$$f_{QD}(x) = \begin{cases} 0, & x \leq -0.14 \\ 1.05x + 0.147, & -0.14 < x \leq 0.47 \\ 1.45x - 0.041, & 0.47 < x \leq 1.15 \\ 1.624, & 1.15 < x \leq 1.5 \\ -1.62x + 4.054, & x > 1.5 \end{cases} \tag{3.51}$$

$$f_{QW}(x) = \begin{cases} 0, & x \leq -0.11 \\ 2.44x + 0.2684, & -0.11 < x \leq 1.31 \\ -2.2x + 6.3468, & 1.31 < x \leq 1.61 \\ 2.80, & x > 1.61 \end{cases} \tag{3.52}$$

The coefficients of the piecewise transfer functions are computed by linear interpolation based on the available experimental data points. It should be mentioned that the variable x should not be confused with the input voltage (V) applied to the laser diode, as recorded in previous measurements. Instead, x signifies the RF voltage V_{RF}, derived from the difference between the input voltage (V) and the polarization voltage (V_{bias}) of the laser when operating at twice the threshold current during OEF operation. For the QD laser, the value of V_{bias} is 1.30 V, while for the QW laser, it is 1.19 V. Additionally, the function $f(x)$ corresponds to the detection voltage (V_{PD}) of the photodiode. Once all the simulation parameters are ready, (3.50) can be reformulated into an integrodifferential delay equation:

$$\dot{y} = x$$

$$\tau\dot{x} = -x - \frac{1}{\theta}y + \eta f_{NL}[x_T]$$

where $y = \int_{t_0}^{t} x(s)ds, x_T \equiv x(t-T)$. $f_{NL}[x_T]$ is represented in detail by (3.51) and (3.52). The time step is set to 10 ps.

A comparison of dynamic behaviour between QD and QW lasers can be undertaken by changing the linear gain (η), which simultaneously acts as the feedback strength. The bifurcation diagrams of the QD and QW lasers are shown in Fig. 3.56 a and b, respectively. The QD laser has a high cut-off frequency of $f_H = 500\,\text{MHz}$ and a low cut-off frequency of $f_L = 50\,\text{kHz}$. The bifurcation diagram of the QD laser unveils three operational regions: a steady state, a period-one region, and a multi-period region. Moreover, the QW laser has a high cut-off frequency of $f_H = 1\,\text{MHz}$ and a low cut-off frequency of $f_L = 10\,\text{kHz}$. The bifurcation diagram of the QW laser displays slow dynamic behaviour, yet it still shows several operational regions.

The simulated time traces of QD lasers shown in Fig. 3.57a and b, corresponding to the red vertical lines in Fig. 3.56a, replicate the 250 ns square waveform (feedback strength of $\eta = 1$) as well as the mixed waveform observed in Fig. 3.53, respectively. Switching focus to the QW lasers in Fig. 3.56b, it becomes evident that slow and fast oscillations emerge as the feedback strength increases. Subsequently, Fig. 3.58a

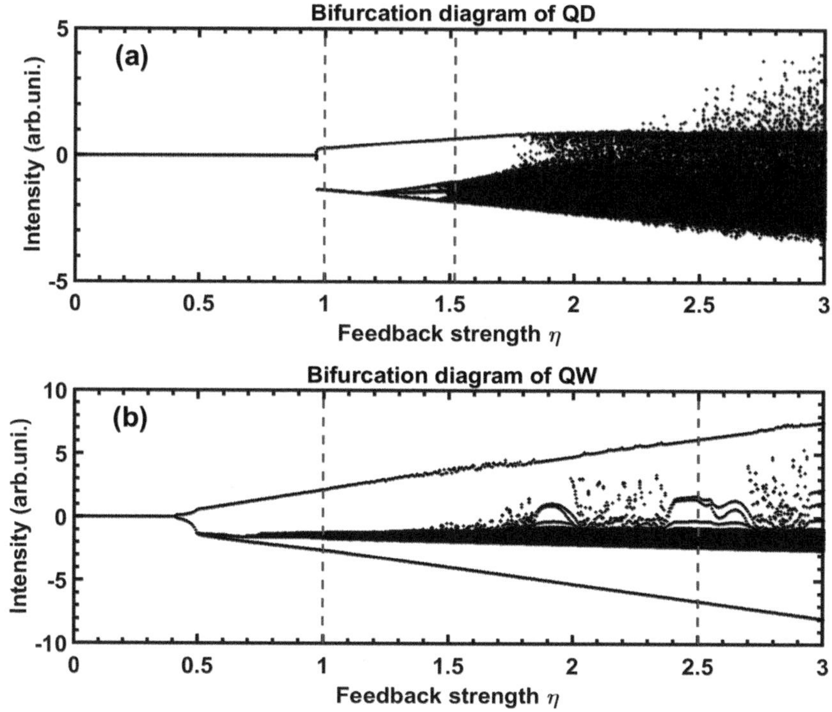

Fig. 3.56 Simulated bifurcation diagrams of **a** QD and **b** QW lasers

and b illustrate time traces of QW lasers at feedback strengths of $\eta = 1$ and $\eta = 2.5$, showing similar waveforms to the experimental data in Fig. 3.54.

To sum up, the findings presented in this section provide a compelling picture of the interplay between the nonlinearity transfer function of semiconductor lasers and the time delay within the framework of AC-coupled optoelectronic feedback. These results give rise to a rich spectrum of complex dynamical patterns. Especially, laser L-I characteristics play a pivotal role in shaping the nonlinear dynamics within the entire OEF loop. In general, QD lasers present a more complex L-I curve characterized by multiple segments and a higher susceptibility to saturation compared to QW lasers. This can be attributed not only to their higher gain compression coefficient ($\sim 10^{-16}$ cm^3), about 10 times that of QW lasers ($\sim 10^{-17}$ cm^3) [38, 103], but also to the potential for excited states emission.

Looking at the existing literature, however, it should be noted that there is just a few MHz high cut-off frequency with QW lasers [102]. In this study, when the QD laser was employed as the optical source, the OEF loop displayed oscillations at a cut-off frequency of 500 MHz. This frequency considerably exceeds what has been documented in previous research [74]. In addition, the absence of a modulation interface in the design of our QD laser could potentially constrain the upper band of

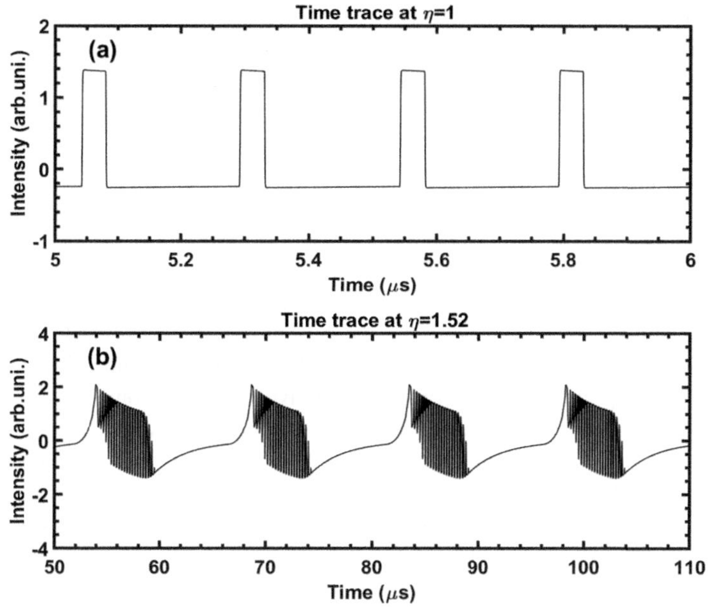

Fig. 3.57 Simulated time series of the QD lasers for a given feedback strength as indicated by the red arrows in Fig. 3.56a

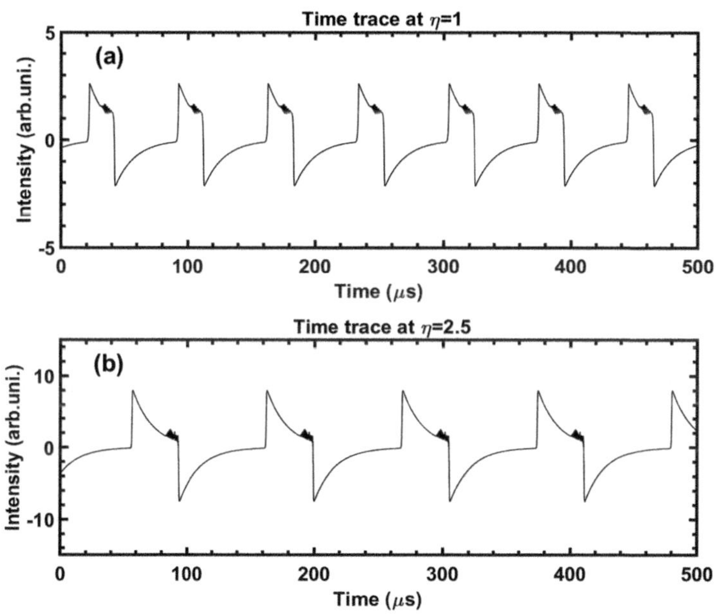

Fig. 3.58 Simulated time series of the QW lasers for a given feedback strength as indicated by the red arrows in Fig. 3.56b

the high cut-off frequency. Therefore, the progression towards future silicon-based on-chip OEF loops holds the promise of enabling the generation of nonlinear signals within the range of tens of GHz, which will decrease the time delay in the OEF loop.

Conclusion

QD lasers have positioned themselves as a promising choice for massive optical communication within data centres, capturing significant interest from major industry players. This is primarily due to the various advantages exhibited by QD lasers, including low relative intensity noise ($< 140\,$dB/Hz), narrow linewidth ($< 30\,$kHz), and a near-zero α_H-factor. Moreover, the QD active region demonstrates a substantial tolerance to threading dislocations when incorporated into the silicon substrate. In addition to this, QD lasers exhibit a robust immunity against unwanted optical feedback, which can be identified as an ideal feature for on-chip integration in silicon photonics. On the contrary, for those seeking nonlinear dynamics in optical computing structures or Ising machines, QD lasers prove to be eligible for generating complex waveforms when subject to the optoelectrical feedback loop.

In the case of mid-infrared ICLs, the results presented in this section indicate a likelihood of low relative intensity noise and narrow linewidth. Unlike QCLs, ICLs show a distinct relaxation oscillation, wide modulation bandwidth, and broadband chaos, owing to their class-B-like interband transition. These discoveries pave the way for secure and high-speed chaos transmission in free space.

References

1. Henry C (1982) Theory of the linewidth of semiconductor lasers. IEEE J Quantum Electron 18(2):259–264. https://doi.org/10.1109/JQE.1982.1071522
2. Petermann K (1991) Laser diode modulation and noise, vol 3. Springer Science & Business Media
3. Bachor H-A, Ralph TC (2004) A guide to experiments in quantum optics. Wiley
4. Yariv A (1991) Optical electronics. Saunders College Publishing
5. Coldren LA, Corzine SW, Mashanovitch ML (2012) Diode lasers and photonic integrated circuits. Wiley
6. Capua A, Rozenfeld L, Mikhelashvili V, Eisenstein G, Kuntz M, Laemmlin M, Bimberg D (2007) Direct correlation between a highly damped modulation response and ultra low relative intensity noise in an InAs/GaAs quantum dot laser. Optics Express 15(9):5388–5393
7. Lelarge F, Dagens B, Renaudier J, Brenot R, Accard A, Van Dijk F, Make D, Le Gouezigou O, Provost J-G, Poingt F et al (2007) Recent advances on InAs/InP quantum dash based semiconductor lasers and optical amplifiers operating at 1.55\μ m. IEEE J Sel Topics Quantum Electron 13(1):111–124
8. Grillot F, Duan J, Dong B, Huang H (2021) Uncovering recent progress in nanostructured light-emitters for information and communication technologies. Light: Sci & Appl 10(1):156. https://doi.org/10.1038/s41377-021-00598-3

9. Didier P, Spitz O, Cerutti L, Diaz-Thomas D, Baranov A, Carras M, Grillot F (2021) Relative intensity noise and intrinsic properties of RF mounted interband cascade laser. Appl Phys Lett 119(17):171107
10. Didier P, Knötig H, Spitz O, Cerutti L, Lardschneider A, Awwad E, Diaz-Thomas D, Baranov A, Weih R, Koeth J et al (2023) Interband cascade technology for energy-efficient mid-infrared free-space communication. Photonics Res 11(4):582–590. https://doi.org/10.1364/PRJ.478776
11. Yang C, Liang L, Qin L, Tang H, Lei Y, Jia P, Chen Y, Wang Y, Song Y, Qiu C et al (2023) Advances in silicon-based, integrated tunable semiconductor lasers. Nanophotonics 12(2):197–217
12. Schawlow AL, Townes CH (1958) Infrared and optical masers. Phys Rev 112(6):1940. https://doi.org/10.1103/PhysRev.112.1940
13. Lax M (1967) Classical noise. V. Noise in self-sustained oscillators. Phys Rev 160(2):290
14. Fleming MW, Mooradian A (1981) Fundamental line broadening of single-mode (GaAl) As diode lasers. Appl Phys Lett 38(7):511–513
15. Tran MA, Huang D, Bowers JE (2019) Tutorial on narrow linewidth tunable semiconductor lasers using Si/III-V heterogeneous integration. APL Photonics 4(11):111101
16. Jin W, Yang Q-F, Chang L, Shen B, Wang H, Leal MA, Wu L, Gao M, Feshali A, Paniccia M et al (2021) Hertz-linewidth semiconductor lasers using CMOS-ready ultra-high-Q microresonators. Nat Photon 15(5):346–353. https://www.nature.com/articles/s41566-021-00761-7
17. Septon T, Becker A, Gosh S, Shtendel G, Sichkovskyi V, Schnabel F, Sengül A, Bjelica M, Witzigmann B, Reithmaier JP et al (2019) Large linewidth reduction in semiconductor lasers based on atom-like gain material. Optica 6(8):1071–1077
18. Borri S, Siciliani de Cumis M, Viciani S, D'Amato F, De Natale P (2020) Unveiling quantum-limited operation of interband cascade lasers. APL Photonics 5(3):036101
19. Wang S, Lv Z-R, Yang Q-L, Wang S-L, Chai H-Y, Meng L, Lu D, Ji C, Yang X-G, Yang T (2023) High-power, narrow-linewidth, and low-noise quantum dot distributed feedback lasers. Laser & Photonics Rev 17(9):2200979
20. Osinski M, Buus J (1987) Linewidth broadening factor in semiconductor lasers-an overview. IEEE J Quantum Electron 23(1):9–29
21. Chow WW, Koch SW, Sargent MI (2012) Semiconductor-laser physics. Springer Science & Business Media
22. Chow WW, Jahnke F (2013) On the physics of semiconductor quantum dots for applications in lasers and quantum optics. Progr Quantum Electron 37(3):109–184. https://doi.org/10.1016/j.pquantelec.2013.04.001
23. Lingnau B (2015) Nonlinear and nonequilibrium dynamics of quantum-dot optoelectronic devices. Springer
24. Chow WW, Zhang Z, Norman JC, Liu S, Bowers JE (2020) On quantum-dot lasing at gain peak with linewidth enhancement factor $\alpha H= 0$. APL Photonics 5(2):026101
25. Provost J-G, Grillot F (2011) Measuring the chirp and the linewidth enhancement factor of optoelectronic devices with a Mach-Zehnder interferometer. IEEE Photonics J 3(3):476–488
26. Yariv A (1989) Quantum electronics. Wiley
27. Newkirk MA, Vahala KJ (1991) Amplitude-phase decorrelation: a method for reducing intensity noise in semiconductor lasers. IEEE J Quantum Electron 27(1):13–22
28. Karlsson A, Björk G (1991) Use of quantum-noise correlation for noise reduction in semiconductor lasers. Phys Rev A 44(11):7669
29. Villar AS (2008) The conversion of phase to amplitude fluctuations of a light beam by an optical cavity. Amer J Phys 76(10):922–929
30. Newkirk MA, Vahala KJ (1992) Large (14.5 dB) reduction of intensity noise from a semiconductor laser by amplitude-phase decorrelation. Appl Phys Lett 60(11):1289–1291
31. Vahala K, Yariv A (1984) Detuned loading in coupled cavity semiconductor lasers–Effect on quantum noise and dynamics. Appl Phys Lett 45(5):501–503

32. Chang T, Poizat J-P, Grangier P (1998) Intensity noise reduction using phase-amplitude coupling in a DFB diode laser. Optics Commun 148(1–3):180–186
33. Zavatta A, Marin F, Giacomelli G (2002) Quantum-state reconstruction of a squeezed laser field by self-homodyne tomography. Phys Rev A 66(4):043805
34. Richardson W, Yamamoto Y (1991) Quantum correlation between the junction-voltage fluctuation and the photon-number fluctuation in a semiconductor laser. Physical Rev Lett 66(15):1963
35. Berg TW, Bischoff S, Magnusdottir I, Mørk J (2001) Ultrafast gain recovery and modulation limitations in self-assembled quantum-dot devices. IEEE Photonics Technol Lett 13(6):541–543
36. Wang C, Osiński M, Even J, Grillot F (2014) Phase-amplitude coupling characteristics in directly modulated quantum dot lasers. Appl Phys Lett 105(22):221114
37. Miska P, Even J, Dehaese O, Marie X (2008) Carrier relaxation dynamics in InAs/ InP quantum dots. Appl Phys Lett 92(19)
38. Ludge K, Scholl E (2009) Quantum-dot lasers–desynchronized nonlinear dynamics of electrons and holes. IEEE J Quantum Electron 45(11):1396–1403. https://doi.org/10.1109/JQE.2009.2028159
39. Grillot F, Veselinov K, Gioannini M, Montrosset I, Even J, Piron R, Homeyer E, Loualiche S (2009) Spectral analysis of 1.55-μ m InAs-InP (113) B quantum-dot lasers based on a multipopulation rate equations model. IEEE J Quantum Electron 45(7):872–878
40. Gioannini M, Sevega A, Montrosset I (2006) Simulations of differential gain and linewidth enhancement factor of quantum dot semiconductor lasers. Opt Quantum Electron 38:381–394
41. Lingnau B, Lüdge K, Chow WW, Schöll E (2012) Failure of the α factor in describing dynamical instabilities and chaos in quantum-dot lasers. Phys Rev E 86(6):065201
42. Borri P, Langbein W, Schneider S, Woggon U, Sellin R, Ouyang D, Bimberg D (2002) Exciton relaxation and dephasing in quantum-dot amplifiers from room to cryogenic temperature. IEEE J Sel Topics Quantum Electron 8(5):984–991
43. Sugawara M, Mukai K, Nakata Y, Ishikawa H, Sakamoto A (2000) Effect of homogeneous broadening of optical gain on lasing spectra in self-assembled In x Ga 1–x A s/G a A s quantum dot lasers. Phys Rev B 61(11):7595
44. Duan J, Wang X-G, Zhou Y-G, Wang C, Grillot F (2018) Carrier-noise-enhanced relative intensity noise of quantum dot lasers. IEEE J Quantum Electron 54(6):1–7
45. Wang C, Zhuang J-P, Grillot F, Chan S-C (2016) Contribution of off-resonant states to the phase noise of quantum dot lasers. Optics Express 24(26):29872–29881
46. Melnik S, Huyet G, Uskov AV (2006) The linewidth enhancement factor α of quantum dot semiconductor lasers. Optics Express 14(7):2950–2955
47. Shang C, Feng K, Hughes ET, Clark A, Debnath M, Koscica R, Leake G, Herman J, Harame D, Ludewig P et al (2022) Electrically pumped quantum-dot lasers grown on 300 mm patterned Si photonic wafers. Light: Sci & Appl 11(1):299
48. Liang D, Bowers JE (2010) Recent progress in lasers on silicon. Nat Photon 4(8):511–517
49. Shang C, Hughes E, Wan Y, Dumont M, Koscica R, Selvidge J, Herrick R, Gossard AC, Mukherjee K, Bowers JE (2021) High-temperature reliable quantum-dot lasers on Si with misfit and threading dislocation filters. Optica 8(5):749–754
50. Shang C, Wan Y, Selvidge J, Hughes E, Herrick R, Mukherjee K, Duan J, Grillot F, Chow WW, Bowers JE (2021) Perspectives on advances in quantum dot lasers and integration with Si photonic integrated circuits. ACS Photonics 8(9):2555–2566
51. Chen S, Li W, Wu J, Jiang Q, Tang M, Shutts S, Elliott SN, Sobiesierski A, Seeds AJ, Ross I et al (2016) Electrically pumped continuous-wave III-V quantum dot lasers on silicon. Nat Photon 10(5):307–311
52. Gérard J, Cabrol O, Sermage B (1996) InAs quantum boxes: highly efficient radiative traps for light emitting devices on Si. Appl Phys Lett 68(22):3123–3125
53. Liu AY, Srinivasan S, Norman J, Gossard AC, Bowers JE (2015) Quantum dot lasers for silicon photonics. Photonics Res 3(5):B1–B9

54. Zhao S, Grillot F (2021) Effect of Shockley-Read-Hall recombination on the static and dynamical characteristics of epitaxial quantum-dot lasers on silicon. Phys Rev A 103(6):063521
55. Saldutti M, Tibaldi A, Cappelluti F, Gioannini M (2020) Impact of carrier transport on the performance of QD lasers on silicon: a drift-diffusion approach. Photonics Res 8(8):1388–1397
56. Duan J, Huang H, Jung D, Zhang Z, Norman J, Bowers J, Grillot F (2018) Semiconductor quantum dot lasers epitaxially grown on silicon with low linewidth enhancement factor. Appl Phys Lett 112(25):251111
57. Globisch B, Otto C, Schöll E, Lüdge K (2012) Influence of carrier lifetimes on the dynamical behaviour of quantum-dot lasers subject to optical feedback. Phys Rev E 86(4):046201
58. Ding S, Dong B, Huang H, Bowers JE, Grillot F (2022) Reflection sensitivity of InAs/GaAs epitaxial quantum dot lasers under direct modulation. Electron Lett 58(9):363–365
59. Yang RQ (1995) Infrared laser based on intersubband transitions in quantum wells. Superlattices Microstruct 17(1):77–83. https://doi.org/10.1006/spmi.1995.1017
60. Vurgaftman I, Bewley W, Canedy C, Kim C, Kim M, Merritt C, Abell J, Lindle J, Meyer J (2011) Rebalancing of internally generated carriers for mid-infrared interband cascade lasers with very low power consumption. Nat Commun 2(1):585. https://doi.org/10.1038/ncomms1595
61. Deng Y, Wang C (2020) Rate equation modeling of interband cascade lasers on modulation and noise dynamics. IEEE J Quantum Electron 56(2):1–9
62. Rana F, Ram RJ (2002) Current noise and photon noise in quantum cascade lasers. Phys Rev B 65(12):125313
63. Zhao S, Grillot F (2022) Modeling of amplitude squeezing in a pump-noise-suppressed interband cascade laser. IEEE Photonics J 14(3):1–8
64. Liu G, Chuang S-L (2002) Modeling of Sb-based type-II quantum cascade lasers. Phys Rev B 65(16):165220
65. Sciamanna M, Shore KA (2015) Physics and applications of laser diode chaos. Nat Photon 9(3):151–162
66. Soriano MC, García-Ojalvo J, Mirasso CR, Fischer I (2013) Complex photonics: dynamics and applications of delay-coupled semiconductors lasers. Rev Modern Phys 85(1):421
67. Uchida A (2012) Optical communication with chaotic lasers: applications of nonlinear dynamics and synchronization. Wiley
68. Lin F-Y, Liu J-M (2004) Chaotic lidar. IEEE J Sel Topics Quantum Electron 10(5):991–997
69. Kim K, Bittner S, Zeng Y, Guazzotti S, Hess O, Wang QJ, Cao H (2021) Massively parallel ultrafast random bit generation with a chip-scale laser. Science 371(6532):948–952
70. Cao H, Chriki R, Bittner S, Friesem AA, Davidson N (2019) Complex lasers with controllable coherence. Nat Rev Phys 1(2):156–168
71. Larger L, Baylón-Fuentes A, Martinenghi R, Udaltsov VS, Chembo YK, Jacquot M (2017) High-speed photonic reservoir computing using a time-delay-based architecture: Million words per second classification. Phys Rev X 7(1):011015
72. Böhm F, Verschaffelt G, Van der Sande G (2019) A poor man's coherent Ising machine based on opto-electronic feedback systems for solving optimization problems. Nat Commun 10(1):3538
73. Grillot F, Norman JC, Duan J, Zhang Z, Dong B, Huang H, Chow WW, Bowers JE (2020) Physics and applications of quantum dot lasers for silicon photonics. Nanophotonics 9(6):1271–1286
74. Chembo YK, Brunner D, Jacquot M, Larger L (2019) Optoelectronic oscillators with time-delayed feedback. Rev Modern Phys 91(3):035006
75. Lang R, Kobayashi K (1980) External optical feedback effects on semiconductor injection laser properties. IEEE J Quantum Electron 16(3):347–355
76. Schunk N, Petermann K (1988) Numerical analysis of the feedback regimes for a single-mode semiconductor laser with external feedback. IEEE J Quantum Electron 24(7):1242–1247
77. Mork J, Tromborg B, Mark J (1992) Chaos in semiconductor lasers with optical feedback: theory and experiment. IEEE J Quantum Electron 28(1):93–108

78. Fischer I, Van Tartwijk G, Levine A, Elsässer W, Göbel E, Lenstra D (1996) Fast pulsing and chaotic itinerancy with a drift in the coherence collapse of semiconductor lasers. Phys Rev Lett 76(2):220
79. Heil T, Fischer I, Elsäßer W, Gavrielides A (2001) Dynamics of semiconductor lasers subject to delayed optical feedback: the short cavity regime. Phys Rev Lett 87(24):243901
80. Otto C, Lüdge K, Schöll E (2010) Modeling quantum dot lasers with optical feedback: sensitivity of bifurcation scenarios. Physica Status Solidi (b) 247(4):829–845
81. Lin L-C, Chen C-Y, Huang H, Arsenijević D, Bimberg D, Grillot F, Lin F-Y (2018) Comparison of optical feedback dynamics of InAs/GaAs quantum-dot lasers emitting solely on ground or excited states. Optics Lett 43(2):210–213
82. Dong B, Chen J-D, Lin F-Y, Norman JC, Bowers JE, Grillot F (2021) Dynamic and nonlinear properties of epitaxial quantum-dot lasers on silicon operating under long-and short-cavity feedback conditions for photonic integrated circuits. Phys Rev A 103(3):033509
83. Tang S, Liu J (2001) Chaotic pulsing and quasi-periodic route to chaos in a semiconductor laser with delayed opto-electronic feedback. IEEE J Quantum Electron 37(3):329–336
84. Kouomou YC, Colet P, Larger L, Gastaud N (2005) Chaotic breathers in delayed electro-optical systems. Phys Rev Lett 95(20):203903
85. Islam MS, Kovalev A, Viktorov E, Citrin D, Locquet A (2021) Optical square-wave generation in a semiconductor laser with optoelectronic feedback. Optics Lett 46(24):6031–6034
86. Ikeda K (1979) Multiple-valued stationary state and its instability of the transmitted light by a ring cavity system. Optics Commun 30(2):257–261
87. Zhou Z, Ou X, Fang Y, Alkhazraji E, Xu R, Wan Y, Bowers JE (2023) Prospects and applications of on-chip lasers. Elight 3(1):1–25. https://doi.org/10.1186/s43593-022-00027-x
88. Stojanović V, Ram RJ, Popović M, Lin S, Moazeni S, Wade M, Sun C, Alloatti L, Atabaki A, Pavanello F et al (2018) Monolithic silicon-photonic platforms in state-of-the-art CMOS SOI processes. Optics Express 26(10):13106–13121. https://doi.org/10.1364/OE.26.013106
89. IEEE Standard for Information Technology– Local and metropolitan area networks– Part 3: CSMA/CD access method and physical layer specifications amendment: media access control parameters, Physical Layers, and Management Parameters for Subscriber Access Networks. IEEE Std 802.3ah-2004 (2004), pp 1–640
90. White AD, Ahn GH, Gasse KV, Yang KY, Chang L, Bowers JE, Vučković J (2023) Integrated passive nonlinear optical isolators. Nat Photon 17(2):143–149
91. Yu M, Cheng R, Reimer C, He L, Luke K, Puma E, Shao L, Shams-Ansari A, Ren X, Grant HR et al (2023) Integrated electro-optic isolator on thin-film lithium niobate. Nat Photon 1–6
92. Bi L, Hu J, Jiang P, Kim DH, Dionne GF, Kimerling LC, Ross C (2011) On-chip optical isolation in monolithically integrated non-reciprocal optical resonators. Nat Photon 5(12):758–762
93. Duan J, Huang H, Dong B, Jung D, Norman JC, Bowers JE, Grillot F (2019) 1.3-μm reflection insensitive InAs/GaAs quantum dot lasers directly grown on silicon. IEEE Photonics Technol Lett 31(5):345–348
94. Hohl A, Gavrielides A (1999) Bifurcation cascade in a semiconductor laser subject to optical feedback. Phys Rev Lett 82(6):1148
95. Lenstra D, Verbeek B, Den Boef A (1985) Coherence collapse in single-mode semiconductor lasers due to optical feedback. IEEE J Quantum Electron 21(6):674–679
96. Spitz O, Herdt A, Wu J, Maisons G, Carras M, Wong C-W, Elsäßer W, Grillot F (2021) Private communication with quantum cascade laser photonic chaos. Nat Commun 12(1):3327
97. Jumpertz L, Schires K, Carras M, Sciamanna M, Grillot F (2016) Chaotic light at mid-infrared wavelength. Light: Sci & Appl 5(6):e16088–e16088
98. Deng Y, Fan Z-F, Zhao B-B, Wang X-G, Zhao S, Wu J, Grillot F, Wang C (2022) Mid-infrared hyperchaos of interband cascade lasers. Light: Sci & Appl 11(1):7
99. Grillot F, Thedrez B, Gauthier-Lafaye O, Martineau M, Voiriot V, Lafragette J, Gentner J, Silvestre L (2003) Coherence-collapse threshold of 1.3-μm semiconductor DFB lasers. IEEE Photonics Technol Lett 15(1):9–11

100. Sukow DW, Gardner JR, Gauthier DJ (1997) Statistics of power-dropout events in semiconductor lasers with time-delayed optical feedback. Phys Rev A 56(5):R3370
101. Han H, Cheng X, Jia Z, Shore KA (2021) Nonlinear dynamics of interband cascade laser subjected to optical feedback. Photonics 8(9):366. MDPI
102. Chengui GRG, Jacques K, Woafo P, Chembo YK (2020) Nonlinear dynamics in an optoelectronic feedback delay oscillator with piecewise linear transfer functions from the laser diode and photodiode. Phys Rev E 102(4):042217
103. Nielsen TR, Gartner P, Jahnke F (2004) Many-body theory of carrier capture and relaxation in semiconductor quantum-dot lasers. Phys Rev B 69(23):235314. https://doi.org/10.1103/PhysRevB.69.235314

Chapter 4
Quantum Fluctuations and Amplitude Squeezing in Quantum-Confined Lasers

This chapter is devoted to the investigation of quantum fluctuations and amplitude squeezing in quantum-confined lasers. In semiconductor lasers, the presence of quantum fluctuations has been a longstanding concern. The presence of intrinsic constraints on electromagnetic field measurements can be related to the principles of quantum mechanics. The emergence of quantum technologies has rekindled interest in generating quantum states of light using semiconductor lasers. A significant factor that distinguishes semiconductor lasers from other laser types is their pumping mechanism. Specifically, semiconductor lasers are pumped by injection current delivered through an electrical circuit. Optical pumping, which relies on photon absorption, follows a Poisson point process and is constrained by shot noise limit. However, electrical pumping doesn't strictly conform to a Poisson point process due to Coulomb interactions [1]. Interestingly, in the case of semiconductor lasers, the inherent noise stemming from pumping can be effectively decreased below the shot-noise limit. This reduction in pump noise leads to the phenomenon of amplitude squeezing in the output of a semiconductor laser.

The chapter commences by introducing the foundational notion of quantum fluctuations and quantum states of light. The subsequent exploration delves into the theoretical analysis of the potential generation of squeezed light in interband cascade lasers and quantum dot lasers. To validate the theoretical framework, a well-established experiment is conducted, showcasing the manifestation of broadband amplitude squeezing in quantum dot lasers.

4.1 Introduction to the Quantum Fluctuations

This century has witnessed a transformative change in the concept of light. This revolution traces its origins back to the early days of the century, ignited by the groundbreaking work of Max Planck, and subsequently advanced through the pioneering contributions of Albert Einstein, Arthur Compton, Paul Dirac, Niels Bohr,

S. Zhao, *Noise, Dynamics and Squeezed Light in Quantum Dot and Interband Cascade Lasers*, Springer Theses, https://doi.org/10.1007/978-3-031-66818-0_4

and other luminaries. While classical optics has proven adept at explaining numerous aspects of light, its coverage is not exhaustive. Many captivating effects arise from the inherent quantum nature of light, forming the cornerstone of quantum optics. Indeed, the marvel of quantum optics lies in the coexistence of seemingly contradictory perspectives: the localized, particle-like emission and detection, alongside the delocalized, wave-like propagation, often referred to as wave-particle duality.

Every optical field found in nature has certain fluctuations associated with it. Fluctuations assume a distinctive and illuminating role in the context of particle-wave duality. It was in 1909 that Albert Einstein first recognized the disparity in electromagnetic fluctuations based on whether energy is carried by waves or particles [2]. Within the domain of classical waves, the amplitude of energy fluctuations exhibits a linear relationship with the mean energy, while in the case of classical particles, this amplitude scales proportionally to the square root of the mean energy. Given that a photon defies categorization as either a classical wave or particle, it possesses both linear and square-root contributions within its behaviour. Typically, at optical or higher frequencies, the dominant contribution is the square root (particle) aspect, while at radio or lower frequencies, the linear (wave) aspect becomes prominent. In his article, the energy fluctuation ε of blackbody radiation can be described within a frequency range spanning from ν to $\nu + d\nu$ inside a volume V and for a spectral density $\rho(n, T)$ (energy per unit volume and unit frequency interval) [3]:

$$\left\langle \varepsilon^2(\nu, T) \right\rangle = \left(h\nu\rho + \frac{c^3}{8\pi\nu^2}\rho^2 \right) V d\nu \tag{4.1}$$

Einstein noted that this relationship, deduced from Planck's law, stands in contradiction to the classical wave theory of light. In fact, the classical theory, which results in the Rayleigh-Jeans law for spectral density, would exclusively generate the second term on the right-hand side of (4.1), which can be labelled as the "wave term". The first term, on the other hand, represents the characteristic of pointlike quanta moving independently, each with energy $h\nu$. This term, often referred to as "shot noise", is the sole contribution to the variance of a flux of independent particles with random arrival times (for instance, raindrops falling on a roof: in this context, ε^2 signifies the variance in the count of raindrops hitting the roof during a fixed time interval). Therefore, (4.1) effectively displays both the particle and wave aspects of radiation.

For a long period of time, only thermal sources were accessible, and Planck's distribution held as the foundational framework for statistically describing light. This paradigm underwent a transformation with the invention of the first laser in 1960 by Theodore Maiman [4], which brought about a second revolution in the concept of light. Lasers operate far from thermal equilibrium, resulting in the emission of radiation characterized by a statistical distribution that deviates markedly from Planck's formula. Well above the threshold, the statistics of the emitted light approaches that of an ensemble of independent particles "photon", so that only the shot-noise term is present in (4.1), corresponding to a Poissonian distribution for the number of photons. The statistical difference between laser and thermal light was experimentally demonstrated in 1965 [5].

The advent of the laser light source has motivated a series of subsequent theoretical efforts, with the goal of constructing a general framework to describe diverse states of radiation. Traditionally, optical coherence theory has centred on the statistical representation of fluctuations. The concept of optical coherence has commonly been associated with interference, likely because interference is the simplest phenomenon that reveals the correlation between light beams. Nevertheless, as modern light detectors and ultra-fast electronic circuitry became available, investigations into various types of correlations in optical fields gained traction. Hanbury Brown and Twiss [6] demonstrated that correlations could extend to quantities reliant on field variables in a quadratic manner. This observation prompted the study of higher-order correlation effects in optical fields. Around 1963, Roy Glauber and George Sudarshan [7–9] were instrumental in introducing phase-space distributions. Their work established a formal correspondence between quantum-mechanical and classical statistical descriptions of optical coherence functions, closely associated with the measurements made using photoelectric detectors. This correspondence essentially asserts that "any statistical state of the quantum mechanical system may be described by a classical probability distribution over a complex plane, provided all field operators are written in the normal-ordered form". These developments embody the two fundamental pillars of quantum fluctuations: the intrinsic statistical nature of quantum mechanics and the statistical aspect arising from our ability to define each of the infinite degrees of freedom within the light field. Thus, the very notion of "quantum fluctuations" implies a unification between statistical principles and quantum mechanics.

4.1.1 Coherent State of Light

A robust connection between classical variables and quantum operators is established through the process of canonical quantization, initially pioneered by Paul Dirac. If a classical system is described by canonically conjugate variables, the corresponding quantum mechanical description can be derived by introducing operators and substituting classical Poisson brackets with commutation relation. In this context, the classical depiction of light introduced in Sect. 3.1 can be transformed into a quantum mechanical representation. As per (3.2), the classical electric field of light $\mathbf{E}(t)$ is expressed in terms of the electric field operator $\hat{\mathbf{E}}(t)$ associated with an electromagnetic field mode with angular frequency ω and at time t:

$$\hat{\mathbf{E}}(t) = \frac{E_0}{2}\left[\hat{\alpha}e^{-i\omega t} + \hat{\alpha}^{\dagger}e^{i\omega t}\right] \tag{4.2}$$

where once more, $E_0 = \sqrt{\frac{\hbar\omega}{\varepsilon_0 V}}$ represents the one-photon field within volume V. $\hat{\alpha}$ and $\hat{\alpha}^{\dagger}$ are the photon annihilation and creation operators, satisfying $\left[\hat{\alpha}, \hat{\alpha}^{\dagger}\right] = 1$. The coherent states $|\alpha\rangle$ correspond to the quantized electromagnetic field states that closely resemble the behaviour of a classical harmonic oscillator, and these states are eigenstates of the annihilation operator $\hat{\alpha}$ [8]:

$$\hat{\alpha} |\alpha\rangle = \alpha |\alpha\rangle$$

The parameter $\alpha = |\alpha| \exp(i\phi)$ represents a complex number depicting the phasor diagram of the electromagnetic field. The electric field of (4.2) can be reexpressed with Euler's formula ($e^{jx} = \cos(x) + j \sin(x)$):

$$\hat{\mathbf{E}}(t) = E_0(\hat{X}_1 \cos \omega t + \hat{X}_2 \sin \omega t) \tag{4.3}$$

where $\hat{X}_1 = \frac{1}{2}(\hat{\alpha} + \hat{\alpha}^\dagger)$ and $\hat{X}_2 = \frac{1}{2i}(\hat{\alpha} - \hat{\alpha}^\dagger)$ represent the two quadrature operators, corresponding to the real and imaginary parts of the complex amplitude α.

$$\langle\alpha| \hat{X}_1 |\alpha\rangle = \langle\alpha| \frac{1}{2}(\hat{\alpha} + \hat{\alpha}^\dagger) |\alpha\rangle = \frac{1}{2}(\alpha + \alpha^*) = \Re\mathrm{e}(\alpha)$$
$$\langle\alpha| \hat{X}_2 |\alpha\rangle = \langle\alpha| \frac{1}{2i}(\hat{\alpha} - \hat{\alpha}^\dagger) |\alpha\rangle = \frac{1}{2i}(\alpha - \alpha^*) = \Im\mathrm{m}(\alpha)$$

The commutation relation between these two quadrature operators can be thus calculated as:

$$\left[\hat{X}_1, \hat{X}_2\right] = \frac{1}{4i}\left[\hat{\alpha} + \hat{\alpha}^\dagger, \hat{\alpha} - \hat{\alpha}^\dagger\right] = \frac{-1}{4i}\left(\left[\hat{\alpha}^\dagger, \hat{\alpha}^\dagger\right] - \left[\hat{\alpha}^\dagger, \hat{\alpha}\right] + \left[\hat{\alpha}, \hat{\alpha}^\dagger\right] - \left[\hat{\alpha}, \hat{\alpha}\right]\right) = \frac{i}{2}$$

The commutation relation between the two quadrature operators bears a strong resemblance to the commutation relation of the position and momentum in classical mechanics. Therefore, it's advantageous to treat $\hat{X}_1$ and $\hat{X}_2$ as their counterparts in the "*de facto* standard" of quantum optics, characterized by the Hamiltonian $\hat{H} = \hbar\omega(\hat{\alpha}\hat{\alpha}^\dagger + 1/2)$. Heisenberg's uncertainty principle postulates that the simultaneous measurement of two canonically conjugate variables (i.e., position and momentum), cannot be accomplished with unrestricted precision. According to the variances $V(\hat{X}_1) = \mathbb{E}\left[(X_1 - \left\langle\hat{X}_1\right\rangle)^2\right]$ and $V(\hat{X}_2) = \mathbb{E}\left[(X_2 - \left\langle\hat{X}_2\right\rangle)^2\right]$ [10, 11]:

$$V(\hat{X}_1) = \langle\alpha| \hat{X}_1^2 |\alpha\rangle - \langle\alpha| \hat{X}_1 |\alpha\rangle^2 = \Re\mathrm{e}(\alpha)^2 + \frac{1}{4} - \Re\mathrm{e}(\alpha)^2 = \frac{1}{4}$$
$$V(\hat{X}_2) = \langle\alpha| \hat{X}_2^2 |\alpha\rangle - \langle\alpha| \hat{X}_2 |\alpha\rangle^2 = \Im\mathrm{m}(\alpha)^2 + \frac{1}{4} - \Im\mathrm{m}(\alpha)^2 = \frac{1}{4}$$

Hence, the uncertainty product in the measurements of the two quadrature operators $\hat{X}_1$ and $\hat{X}_2$ follows the Schwartz inequality:

$$V(\hat{X}_1)V(\hat{X}_2) \geq \frac{1}{4}\left|\left\langle\left[\hat{X}_1, \hat{X}_2\right]\right\rangle\right|^2 = \frac{1}{16} \tag{4.4}$$

Defining the standard deviations $\Delta X_1 = \sqrt{V(\hat{X}_1)}$ and $\Delta X_2 = \sqrt{V(\hat{X}_2)}$, then (4.4) can be rewritten by

$$\Delta X_1 \Delta X_2 \geq \frac{1}{2} \left| \left\langle \left[\hat{X}_1, \hat{X}_2 \right] \right\rangle \right| = \frac{1}{4} \tag{4.5}$$

In a coherent state, the mean values correspond to those of the real and imaginary components of the complex amplitude α. Furthermore, the standard deviations in (4.5) are the same regardless of the complex amplitude, and both become equal to 1/2. Consequently, a coherent state in phase space is described in Fig. 4.1a. A coherent state $|\alpha\rangle$ can be described by a superposition of the photon-number (Fock) states $|n\rangle$ (eigenstates of the photon number operator $\hat{n} = \hat{\alpha}^\dagger \hat{\alpha}$):

$$|\alpha\rangle = e^{-|\alpha|^2/2} \sum_{n=0}^{\infty} \frac{\alpha^n}{\sqrt{n!}} |n\rangle$$

corresponding to a Poisson distribution of the photon number n:

$$p(n) = |\langle n|\alpha\rangle|^2 = e^{-|\alpha|^2} \frac{|\alpha|^{2n}}{n!} = e^{-\langle n\rangle} \frac{\langle n\rangle^n}{n!}$$

with mean value and variance such that $\left\langle \Delta n^2 \right\rangle = \langle n\rangle = |\alpha|^2$. Let's think about the time evolution of a coherent state with a Hamiltonian $\hat{H} = \hbar\omega(\hat{\alpha}\hat{\alpha}^\dagger + 1/2)$ and the

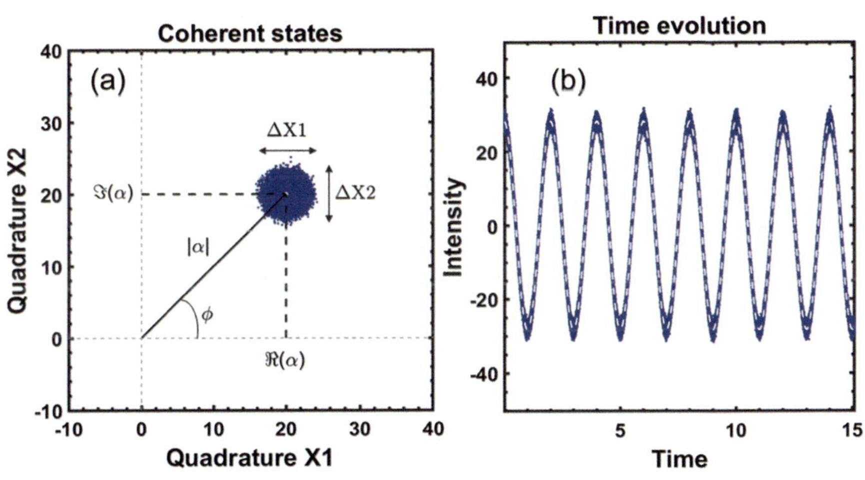

Fig. 4.1 A coherent state **a** in the phase diagram and **b** in the time evolution. The white dashed line is the corresponding sine function without any fluctuations

following Schrödinger equation with a time-invariant Hamiltonian with the initial state as $|\alpha\rangle$ and the final state as $|\alpha(t)\rangle$ [11]:

$$|\alpha(t)\rangle = e^{-j\frac{\hat{H}}{\hbar}t}\,|\alpha\rangle = e^{-j\frac{\omega}{2}t}\,\left|\alpha e^{-j\omega t}\right\rangle$$

Since the phase factor for the overall state has no physical meaning, the initial coherent state with a complex amplitude of α transforms into one with a complex amplitude of $\alpha e^{-j\omega t}$ after time t. This implies that a coherent state retains its coherence and only undergoes a change in phase. Therefore, the time evolution of a coherent state can be represented as the time dependence of either the real or imaginary part of the optical field, as illustrated in Fig. 4.1b, which is just a simple wave! In summary, the coherent states exhibit properties similar to those of classical coherent light and represent the closest quantum mechanical approximation to it. Much like laser light can be approximated as idealized classical coherent light under specific conditions, in situations demanding a quantum description, laser light can often be effectively approximated using coherent states [8].

As discussed in Sect. 3.1, phasor diagram is widely used in quantum optics due to their utility in representing any state of light. A phasor diagram plots $\hat{X}_1$ versus $\hat{X}_2$, offering a concise visualization of quantum states in Fig. 4.2. In this representation, each coherent state $|\alpha\rangle$ corresponds to an area centred around the point defined by the mean values for that particular state. The size of this area is determined by the variances ΔX_1^2 and ΔX_2^2, often depicted as a circle with a diameter equivalent to the variance. The area of the circle conveys the extent of the uncertainty distribution. This qualitative phasor diagram provides several key insights:

- The uncertainty area maintains symmetry. When measuring fluctuations in a coherent state, the outcome remains consistent, regardless of the chosen projection angle.
- The uncertainty area's size remains constant and is unaffected by the state's intensity $|\alpha|^2$.

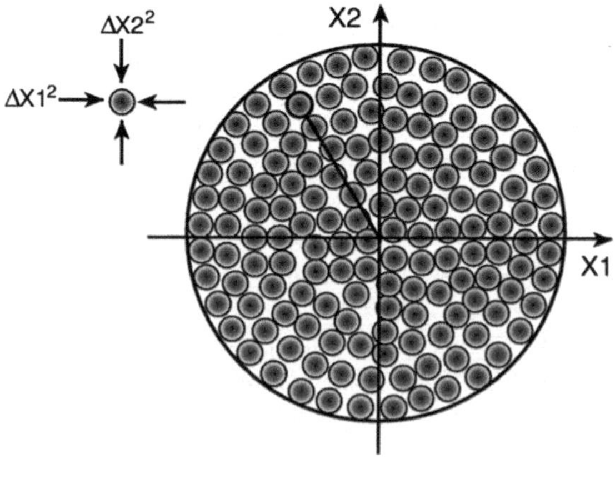

Fig. 4.2 Uncertainty areas of the coherent state can span the entire parameter space like tiles. (*Source* On p. 70 in Ref. [12])

- Two states are distinguishable from each other only if their average positions in the $\hat{X}_1$, $\hat{X}_2$ plane are spaced by more than ΔX.

The last insight is so important that it provides a practical guideline for determining when two states can be distinguished through measurement. The uncertainty area sets a minimum requirement for a change in amplitude or phase that is needed before two states can be differentiated. While in classical systems, such an uncertainty region might be attributed to noise arising from technical limitations, in the quantum context, the uncertainty area arises from fundamental quantum principles rather than technical imperfections. As a result, the uncertainty area embodies the influence of quantum fluctuations on any form of light measurement, whether it pertains to amplitude, phase, or any other property. Ultimately, experimental measurements cannot surpass the precision imposed by quantum fluctuations. Another convenient approach to representing the state of the field is through quasiprobability distributions. These distributions enable the expectation values of quantum operators to be expressed as integrals of corresponding classical quantities, with the integration weighted by the distribution function associated with the state. Additional details can be found in Appendix C.

4.1.2 Squeezed States of Light

Apart from coherent states, there exist other types of light states as well. In certain applications, such as communication [13] or precision measurements [14], it's often necessary to consider only one of the quadratures of the field. This approach offers a way to work around the quantum limitation presented by the Heisenberg uncertainty principle. By reducing the fluctuations in one of the quadratures below that of a coherent state, it becomes possible to achieve enhanced precision in that particular quadrature. However, this comes at the cost of increased noise in the other quadrature, ensuring that the Heisenberg uncertainty principle is upheld and not violated. Such states are called squeezed states of light [15, 16]. Here is a brief overview of the mathematical characteristics of different types of squeezed states.

A coherent state $|\alpha\rangle$ can be created by applying the displacement operator $D(\alpha)$ to the vacuum state $|0\rangle$ [8, 10]:

$$|\alpha\rangle = D(\alpha)\,|0\rangle$$

where

$$D(\alpha) = e^{-\frac{1}{2}|\alpha|^2} e^{\alpha\hat{a}^\dagger} e^{-\alpha^*\hat{a}}$$

A general quadrature-squeezed state $|\alpha, \zeta\rangle$ may be generated by first acting with the squeeze operator $S(\zeta)$ on the vacuum state $|0\rangle$ followed by the displacement operator $D(\alpha)$ [16]:

$$|\alpha, \zeta\rangle = D(\alpha)S(\zeta)\,|0\rangle$$

where the squeeze operator $S(\zeta)$

$$S(\zeta) = e^{(\frac{1}{2}\zeta^*\hat{a}^2 - \frac{1}{2}\zeta(\hat{a}^\dagger)^2)}, \quad \zeta = re^{i\theta}$$

The function of the squeeze operator can be broken down into a sequence of steps: first, rotating the angle by $\theta/2$ counterclockwise, then implementing the squeeze operation with the parameter r, and eventually rotating back to the original configuration [11]. By defining a rotation angle of $\theta/2$, satisfying $\langle Y_1 + iY_2\rangle = \langle X_1 + iX_2\rangle\, e^{-i\theta/2}$, as depicted in Fig. 4.3, the expectation values and variances for a general quadrature-squeezed state $|\alpha, \zeta\rangle$ are provided by [10]:

$$\Delta Y_1 = \frac{1}{4}e^{-2r}, \quad \Delta Y_2 = \frac{1}{4}e^{2r}$$
$$\langle\hat{n}\rangle = |\alpha|^2 + \sinh^2 r$$
$$\langle\hat{n}^2\rangle = (|\alpha|^2 + \sinh^2 r)^2 + 2\sinh^2 r\cosh^2 r + |\alpha\cosh r - \alpha e^{i\theta}\sinh r|^2$$
$$V(\hat{n}) = \langle\hat{n}^2\rangle - \langle\hat{n}\rangle^2 = |\alpha|^2\left[e^{-2r}\cos^2(\phi - \frac{\theta}{2}) + e^{2r}\sin^2(\phi - \frac{\theta}{2})\right] + 2\sinh^2 r\cosh^2 r$$

For the vacuum squeezed state ($\alpha = 0$), it consequently exhibits a super-Poissonian distribution:

$$V(\hat{n}) = 2\sinh^2 r\cosh^2 r \geqslant \langle\hat{n}\rangle = \sinh^2 r$$

For the quadrature-squeezed state with the coherent amplitude greatly exceeding the squeezing ($\alpha \neq 0$ and $|\alpha|^2 \gg e^{2r}$) when the rotation angle $\theta = 2\phi$

$$V(\hat{n}) = |\alpha|^2 e^{-2r} + 2\sinh^2 r\cosh^2 r$$

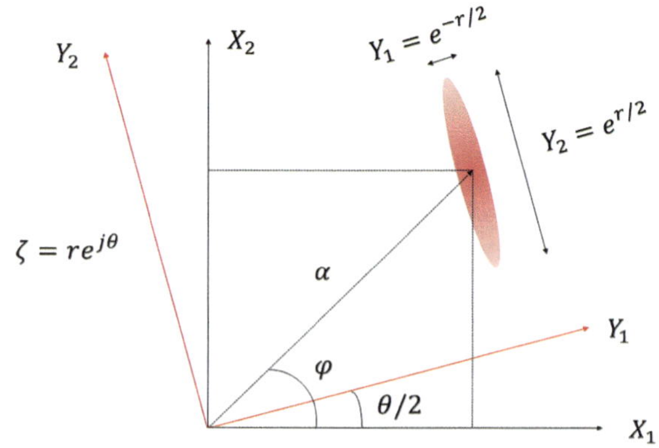

Fig. 4.3 The phase diagram of a general quadrature-squeezed state $|\alpha, re^{i\theta}\rangle$ ($r > 0$) in the Cartesian coordinate system

It has a sub-Poissonian distribution. On the contrary, when the rotation angle $\theta = 2\phi + \pi$,

$$V(\hat{n}) = |\alpha|^2 e^{2r} + 2\sinh^2 r \cosh^2 r$$

It has a super-Poissonian distribution. More specific cases are obtained when the rotation angle θ equals 0 or π in the Cartesian coordinate system, as depicted in Fig. 4.4. Squeezing is present in the $\hat{X}_1$ quadrature when $\theta = 0$, and squeezing is evident in the $\hat{X}_2$ quadrature when $\theta = \pi$.

A quadrature-squeezed state is created through a phase-sensitive amplification process occurring in many nonlinear optical interactions. Among the successful methods for producing such states, the employment of a four-wave mixer and a degenerate parametric amplifier has emerged as the two most effective methods [17, 18]. This preference arises because, while a coherent state is generated via linear terms involving $\hat{a}$ and $\hat{a}^\dagger$ in the exponent, a squeezed state necessitates the inclusion of quadratic terms. The prototype governing these interactions is characterized by the Hamiltonian [16]:

$$\mathcal{H} = \hbar\left[\chi^{(n)*}(\varepsilon)\hat{a}^2 + \chi^{(n)}(\varepsilon)(\hat{a}^\dagger)^2\right]$$

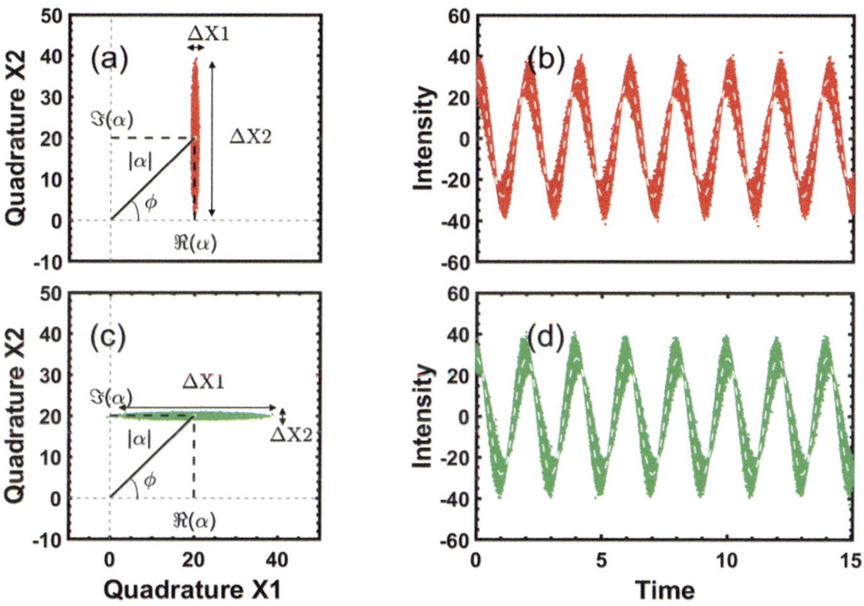

Fig. 4.4 The phase diagram and the time evolution of specific quadrature-squeezed states, **a**–**b** in the $\hat{X}_1$ quadrature, and **c**–**d** in the $\hat{X}_2$ quadrature. The white dashed line is the corresponding sine function without any fluctuations

where $\chi^{(n)}$ is the nonlinear susceptibility of the optical medium and ε is the amplitude of the pump field which has been treated classically.

$$\chi^{(2)}(\varepsilon) = \varepsilon\chi^{(2)} \quad \text{(degenerate parametric amplifier)}$$
$$\chi^{(3)}(\varepsilon) = \varepsilon^2\chi^{(3)} \quad \text{(four wave mixing)}$$

A related yet not entirely equivalent class of quantum states is referred to as photon-number squeezed states or amplitude-squeezed states with the number-phase minimum uncertainty condition ($\langle\Delta\hat{n}^2\rangle\langle\Delta\hat{\phi}^2\rangle \approx \frac{1}{4}$). The precise definition of these states necessitates the introduction of an operator $\hat{\phi}$ corresponding to the phase of an electromagnetic field, which has been a longstanding challenge in physics [19]. Squeezing in photon number more readily occurs when the photon number fluctuations become smaller than those for a coherent state. These states are also classified as "nonclassical," akin to quadrature-squeezed states, but the error contour assumes a "crescent" shape in the polar coordinate system, as shown in Fig. 4.5. Specifically, concerning the photon-number variance, the relationship is as follows [3]:

$$V(\hat{n}) = \langle\Delta\hat{n}^2\rangle = \langle\hat{n}\rangle + \left\langle(\hat{a}^\dagger\hat{a}^\dagger\hat{a}\hat{a} - \langle\hat{a}^\dagger\hat{a}\rangle^2)\right\rangle$$

or, using the Glauber-Sudarshan P function (see Appendix C)

$$\langle\Delta\hat{n}^2\rangle = \langle\hat{n}\rangle + \int \mathrm{d}^2\alpha\, P(\alpha,\alpha^*)(|\alpha|^2 - \langle|\alpha|^2\rangle)^2$$

As previously discussed, $\langle\Delta\hat{n}^2\rangle < \langle\hat{n}\rangle$ only if the probability distribution $P(\alpha,\alpha^*)$ is not positive definite. The integral associated with the amplitude-squeezed state takes

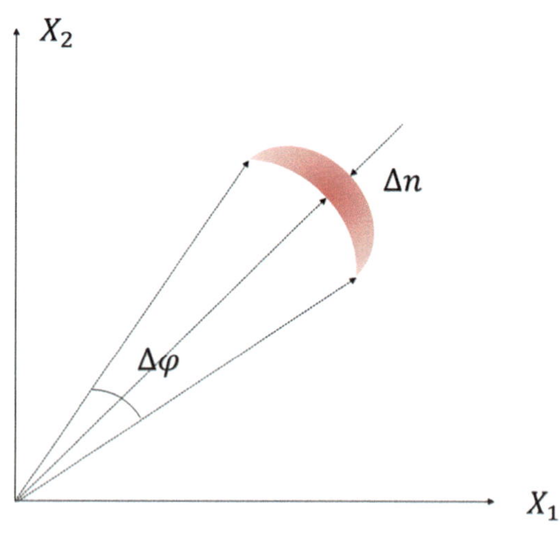

Fig. 4.5 The phase diagram of a "crescent" amplitude-squeezed state in the polar coordinate system

on a negative value, thereby indicating the quantum nature inherent to the state. An amplitude-squeezed state can be also generated by a third-order nonlinear medium [20] or by semiconductor lasers [21].

4.1.3 Measurement of Quantum States of Light

The straightforward method of detecting quantum fluctuations involves directly detecting the light with a single photodetector and then analyzing the resulting photocurrent using an electronic spectrum analyzer or oscilloscope, as depicted in Fig. 4.6. Through direct detection, the acquired electrical noise power should primarily arise from the quantum fluctuations of the light in an ideal scenario with a perfect laser mode and detection system. Nevertheless, some technical challenges must be circumvented [12]. This measurement is not particularly sensitive to squeezing but is rather sensitive to effects like sub- or super-Poisson statistics, which can also manifest in unsqueezed fields. Additionally, direct detection cannot differentiate between the quadrature of the optical field.

Since the detection of squeezed states requires a phase-sensitive scheme that measures the variance of a field quadrature component, Yuan and Shapiro [22] proposed that homodyne detection measures a single quadrature of the field, whereas heterodyne detection entails measuring two quadratures of the field. Both of these schemes involve the interference of the light in a squeezed state with a coherent field (often referred to as a *local oscillator*, LO), thereby adding a large coherent component to it. The current discussion is focused exclusively on the approach of balanced homodyne detection, which is outlined in Fig. 4.7.

In this method, the term "homodyne" indicates that both the unknown beam α_{in} and the LO beam α_{lo} share the same frequency and originate from the same laser source. The process begins by dividing the incoming light into two paths after the 90/10 beamsplitter M_1. One of these paths contains the unknown beam to be detected, while the other serves as the reference beam or LO beam, which can provide a phase reference to distinguish between the quadratures. The phase shifter in the LO path allows for adjustment of the phase difference ϕ_{lo} between α_{in} and α_{lo}. Subsequently, these two paths are combined using an exact 50/50 beamsplitter M_4, maintaining equal optical power. This balanced configuration is crucial. The resulting beams are detected by identical photodetectors D_1 and D_2. The photocurrents generated by

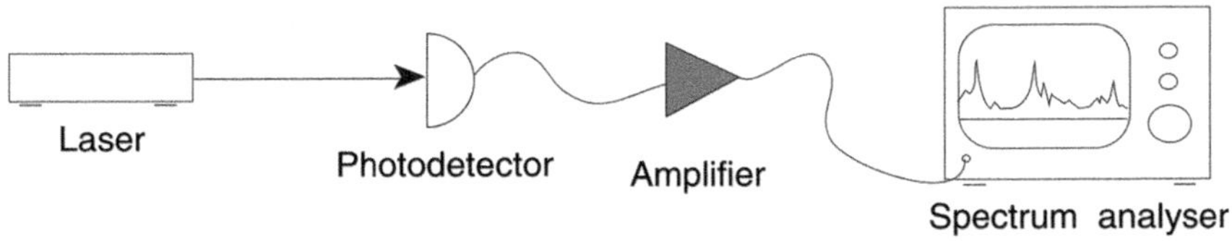

Fig. 4.6 Schematic diagram of direct detection. (*Source* On p. 200 in Ref. [12])

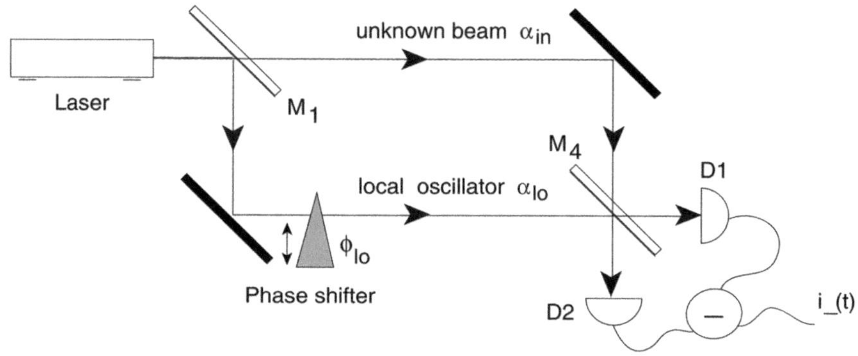

Fig. 4.7 Schematic diagram of balanced homodyne detection. (*Source* On p. 206 in Ref. [12])

these detectors can be either added or subtracted using a differential amplifier. Finally, an electronic spectrum analyzer is employed to create and record two noise spectra: one for the sum of the photocurrents and another for their difference. This technique effectively distinguishes between quantum noise and classical noise. Classical noise generates correlated photocurrents, allowing them to be subtracted to eliminate each other. When added, the complete noise power is restored. In contrast, for quantum noise, the photocurrents are uncorrelated, leading to both the sum and difference channels adding in quadrature. Therefore, the subtracted spectrum shows quantum noise only. A more formal derivation of these results is provided in Appendix D. It's important to note that homodyne detection differs significantly from direct detection in two key aspects: (a) Intensity fluctuations in homodyne detection directly evaluate the fluctuations in a specific quadrature of the input field. (b) The signal and its variance in homodyne detection are influenced by the phase angle of the LO, which is an external parameter.

Another essential detection scheme is a measurement of intensity correlation, known as the Hanbury-Brown and Twiss experiment [6]. In this experiment, the incoming light is divided into two beams using a 50–50% beam splitter. These split beams are then captured by two (usually single-photon) detectors, as depicted in Fig. 4.8a. The joint photon-counting rate at these two detectors is proportional to the second-order coherence function $g^{(2)}(\tau)$ [23]. The normalized second-order correlation function for photons generated at times $t = 0$ and $t = \tau$:

$$g^{(2)}(\tau) = \frac{\langle \hat{a}^\dagger(t)\hat{a}^\dagger(t+\tau)\hat{a}(t+\tau)\hat{a}(t)\rangle}{\langle \hat{a}^\dagger(t)\hat{a}(t)\rangle\langle \hat{a}^\dagger(t+\tau)\hat{a}(t+\tau)\rangle}; \quad g^{(2)}(0) = \frac{\langle (\hat{a}^\dagger)^2\hat{a}^2\rangle}{\langle \hat{a}^\dagger\hat{a}\rangle^2} = \frac{\langle \hat{n}(\hat{n}-1)\rangle}{\langle \hat{n}\rangle^2} \tag{4.6}$$

The normalized second-order coherence function $g^{(2)}(\tau)$ as represented by (4.6), provides insights into the joint probability of detecting one photon followed by another within the time delay τ at a fixed position. This metric helps determine

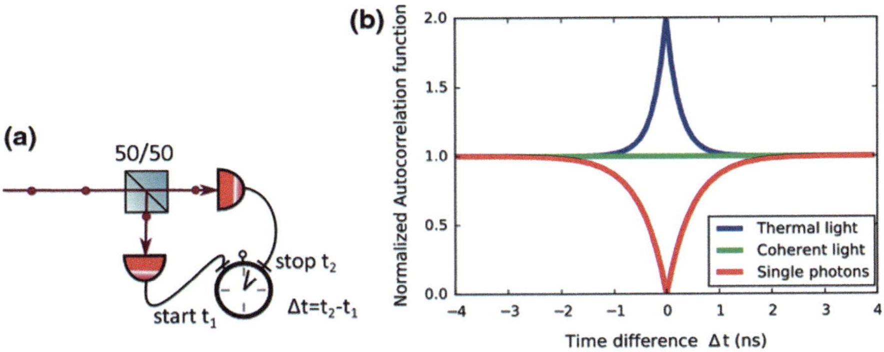

Fig. 4.8 **a** Schematic diagram of the Hanbury Brown-Twiss experiment for measuring the normalized second-order coherence function $g^{(2)}(\tau)$ with $\Delta t = t_2 - t_1 = \tau$. **b** Schematic of the $g^{(2)}(0)$ function for the three possible light statistics. The value around zero delay carries crucial information (*Source* in the thesis of Markus Kantner)

whether photons within the beam tend to cluster together or remain separate. When $g^{(2)}(\tau) = 1$, it means that the probability of joint detection is equivalent to the probability of independent detection. This situation is expected to occur as $\tau \to \infty$ since the "memory" of the first detected photon fades after a sufficiently long time. If $g^{(2)}(\tau) < g^{(2)}(0)$, it suggests that the probability of detecting the second photon decreases with the time delay, indicating a phenomenon known as "*bunching*" of photons. On the contrary, if $g^{(2)}(\tau) > g^{(2)}(0)$, it means that the probability of detecting the second photon increases with the time delay, illustrating the phenomenon of "*antibunching*" of photons.

Considering that the photodetection process involves only normal-ordered correlation functions, the normalized second-order coherence function $g^{(2)}(0)$ can be derived from the Glauber-Sudarshan P function [24]:

$$g^{(2)}(0) = 1 + \frac{\int P(\alpha, \alpha^*) \left[|\alpha|^2 - \langle \hat{a}^\dagger \hat{a} \rangle\right]^2 \mathrm{d}^2\alpha}{\langle \hat{a}^\dagger \hat{a} \rangle^2} \tag{4.7}$$

For classical states where the Glauber-Sudarshan P function is nonnegative ($P(\alpha, \alpha^*) \geq 0$), thus $g^{(2)}(0) \geq 1$. But for a nonclassical field state, it is possible to have $g^{(2)}(0) < 1$. For example, the Glauber-Sudarshan P function $P(\alpha, \alpha^*)$ of a single-mode thermal field is described by Bose-Einstein statistics [24]:

$$P(\alpha, \alpha^*)_{\text{Thermal}} = \frac{1}{\pi \langle \hat{n} \rangle} e^{-|\alpha|^2/\langle \hat{n} \rangle}$$

Then

$$g^{(2)}(0)_{\text{Thermal}} = \frac{\int P(\alpha, \alpha^*)|\alpha|^4 d^2\alpha}{\left[\int P(\alpha, \alpha^*)|\alpha|^2 d^2\alpha\right]^2} = 2$$

For the coherent state $|\beta\rangle$ [24]

$$P(\alpha, \alpha^*)_{\text{Coherent}} = \delta^{(2)}(\alpha - \beta)$$

Then

$$g^{(2)}(0)_{\text{Coherent}} = 1$$

The coherent state exhibits coherence in all orders of correlation functions ($g^{(n)}(0) = 1$) so that photon detection events are entirely uncorrelated. Hence, the detection of a photon at time $\tau = 0$ does not provide any information about the timing of subsequent photon detections. Additionally, for a single-mode photon number state $|\hat{n}\rangle$ [24],

$$g^{(2)}(0)_{\text{Fock}} = 1 - \frac{1}{n} < 1$$

All these three states are shown in Fig. 4.8b. The relation between antibunching and sub-Poissonian statistics is a subtle one. In fact, by utilizing (4.7), one can establish the following connection between $g^{(2)}(0)$ and the variance of photon numbers for a single-mode field:

$$g^{(2)}(0) = 1 + \frac{\langle \Delta\hat{n}^2\rangle - \langle \hat{n}\rangle}{\langle \hat{n}\rangle^2} \tag{4.8}$$

Therefore, in this case, sub-Poissonian statistics imply $g^{(2)}(0) < 1$. On the other hand, one cannot say in general that antibunching leads to sub-Poissonian statistics, since $g^{(2)}(\tau) > g^{(2)}(0)$ does not necessarily require $g^{(2)}(0) < 1$. Using (4.8), a more straightforward approach to gauge the photon statistics of any state is by computing the *Mandel Q-parameter* [25]:

$$Q = \frac{\langle \Delta\hat{n}^2\rangle - \langle \hat{n}\rangle}{\langle \hat{n}\rangle} = \langle \hat{n}\rangle \left[g^{(2)}(0) - 1\right] \tag{4.9}$$

For a state with Q in the range of $-1 \leq Q < 0$, the photon statistics are sub-Poissonian. Conversely, if $Q > 0$, the statistics are super-Poissonian. Notably, a coherent state is characterized by $Q = 0$.

Additionally, there exists a technique known as quantum tomography or quantum state tomography through the Wigner function, which involves the reconstruction of a quantum state through measurements on a collection of identical quantum states. Further information can be found in the pioneering article by Smithey et al. [26].

4.1.4 Quantum Noise from Electronics

In the realm of electronics, another facet of quantum noise has equally undergone development since electrons share the particle-wave duality with photons. Exploring the quantum noise of electronics leads to a historical review. In 1918, Walter H. Schottky observed that even within ideal vacuum tubes, where all extraneous noise sources were eliminated, two types of intrinsic noise persisted, which he termed the *Wärmeeffekt* and the *Schroteffekt* [27]. The former is known as Johnson-Nyquist noise or "thermal noise" [28]. Thermal noise arises from the thermal agitation or Brownian motion of electrons within an electrical conductor. As temperature increases, thermal agitation grows, causing a corresponding increase in thermal noise power. Moreover, thermal noise power is independent of applied voltage. The latter noise, often referred to as "shot noise," emerges due to the discrete nature of electrical charges. Shot noise power is proportional to the average current, rather than the square of the current as one would anticipate from Ohm's law. This deviation occurs because, in the case of vacuum tubes, electrons are emitted by the cathode independently and reach the anode at random intervals, following a Poisson process. These intrinsic noises cannot be completely removed, even in perfectly designed and operated semiconductor systems [29].

A conductor possessing finite resistance R and operating at temperature T serves as the simplest system that demonstrates both types of intrinsic noise. Shifting the focus to thermal noise, Nyquist's analysis of thermal noise [28] followed shortly after Johnson's initial findings [27]. Nyquist's work revealed that the power spectral density of open-circuit voltage noise $S_\nu(\omega)$ remains unaffected by the conductor's material composition and the measurement frequency ω. Instead, it is only determined by the temperature and resistance:

$$S_\nu(\omega) = 4k_B T R \tag{4.10}$$

where k_B is the Boltzmann constant. Back in 1928, Nyquist [28] even considered the scenario at high frequencies, where $\hbar\omega \gg k_B T$, since it was known that in the case of the black body radiation, the effect of quantization was to yield the Planck spectrum, which is equivalent to

$$S_\nu(\omega) = \frac{4\hbar\omega R}{e^{\hbar\omega/k_B T} - 1} \tag{4.11}$$

This function remains flat when $\hbar\omega \ll k_B T$, but as $\hbar\omega > k_B T$, it rapidly diminishes towards zero. As a result, Nyquist concluded that the physically unrealistic phenomenon of a flat noise spectrum (white noise) would be eliminated through the integration of quantum mechanics. A more rigorous quantum treatment emerged in 1951, courtesy of Callen and Welton [30]. They provided a line of reasoning that suggested the generalized Nyquist noise should have a fully quantum mechanical expression when accounted for the zero-point fluctuation contribution to (4.11):

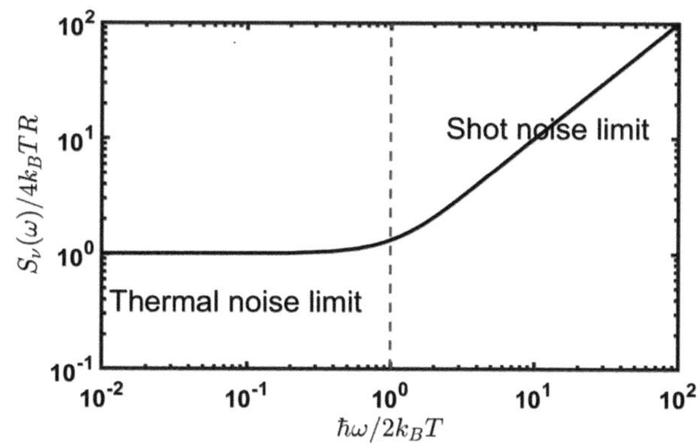

Fig. 4.9 The generalized open-circuit voltage noise spectrum including the zero-point fluctuation according to the (4.12) [31]

$$S_v(\omega) = 2\hbar\omega R \coth\left(\frac{\hbar\omega}{2k_BT}\right) = 4R\left[\frac{1}{2}\hbar\omega + \frac{\hbar\omega}{e^{\hbar\omega/k_BT} - 1}\right] \tag{4.12}$$

This result confirms Nyquist's insight that the thermal spectrum would be cut off when $\hbar\omega > k_BT$. However, the entire spectrum doesn't actually get cut off; rather, it exhibits a linear rise with increasing ω. This behaviour arises from the first term in (4.12), which originates from the zero-point fluctuations inherent in the microscopic structure of the resistive element, modelled by harmonic oscillators. Meanwhile, the second term in (4.12) corresponds to Planck's blackbody radiation spectrum. As depicted in Fig. 4.9, thermal noise ($4k_BTR$) prevails as a dominant noise source at high temperatures and/or low frequencies ($k_BT \gg \hbar\omega$), whereas shot noise ($2\hbar\omega R$) dominates at low temperatures and/or high frequencies ($k_BT \ll \hbar\omega$).

The generalized Nyquist formula holds some noteworthy implications. Thermal noise is characterized by actual fluctuation energy that can be directly measured. It serves as an intrinsic noise source, inevitably present when a system operates at a finite temperature T. In contrast, shot noise is a virtual fluctuation energy that cannot be measured directly. It manifests itself only indirectly in the form of photon shot noise in a laser or mechanical (Casimir) force in metallic parallel plates [1]. Another implication is that in certain low-frequency regions, the thermal noise effect can exhibit fewer fluctuations than the shot noise effect [32].

4.1.5 Generation of Quantum States of Light

Quantum states of light have a purpose that extends beyond being solely a mathematical foundation for quantum optics. They can indeed be realized through practical experiments. The first experimental observation of squeezing, conducted by Slusher et al. [17], harnessed the four-wave mixing in sodium vapour. The measured squeezing level was 0.3 dB below the shot noise limit. A nondegenerate four-wave-mixing experiment was performed by Shelby et al. [33] in a 114 m single-mode silica fibre, cooled to 4.2 K. They observed a squeezing level of 0.6 dB below shot noise. A subsequent experiment by Wu et al. [18] used the parametric down-conversion in a magnesium-doped lithium niobate crystal embedded in a standing-wave cavity. The squeezing level was about 3.5 dB below the shot noise limit. Figure 4.10 summarizes the fundamental principles of those nonlinear interactions. In the context of parametric down-conversion, which relies on $\chi^{(2)}$ nonlinear processes, a single pump photon at frequency ω_p is destroyed, resulting in the production of a photon pair. This pair consists of a signal photon with frequency ω_s and an idler photon with frequency ω_i with the relationship $\omega_s + \omega_i = \omega_p$. An analogous process in four-wave mixing operates through $\chi^{(3)}$ nonlinear interactions. In this case, photon pairs emerge when two pump photons are simultaneously annihilated, conserving energy and momentum. This is governed by the relationship $\omega_s + \omega_i = 2\omega_p$, providing enhanced flexibility in adjusting the signal and idler photon energies.

Since the landmark experiments on squeezed light generation in the 1980s, various research groups worldwide have pursued the goal of achieving stronger squeezing using diverse technologies and a comprehensive review is available in Ref. [34]. For the past few decades, the workhorse for generating squeezed light has been an optical parametric oscillator (OPO) cavity. This configuration involves placing a $\chi^{(2)}$ or $\chi^{(3)}$ nonlinear crystal inside the cavity to enhance the nonlinear coupling, as their intrinsic values are relatively small. A number of nonlinear materials have been applied, including $LiNbO_3$ (PPLN) [35], $KTiOPO_4$ (PPKTP) [36], silicon [37], and Si_3N_4 [38]. Recent advancements as highlighted in Ref. [39] also have enabled the transition from traditional benchtop instruments to a compact single-chip design, maintaining efficiency and preserving key performance metrics. Overall, the domain

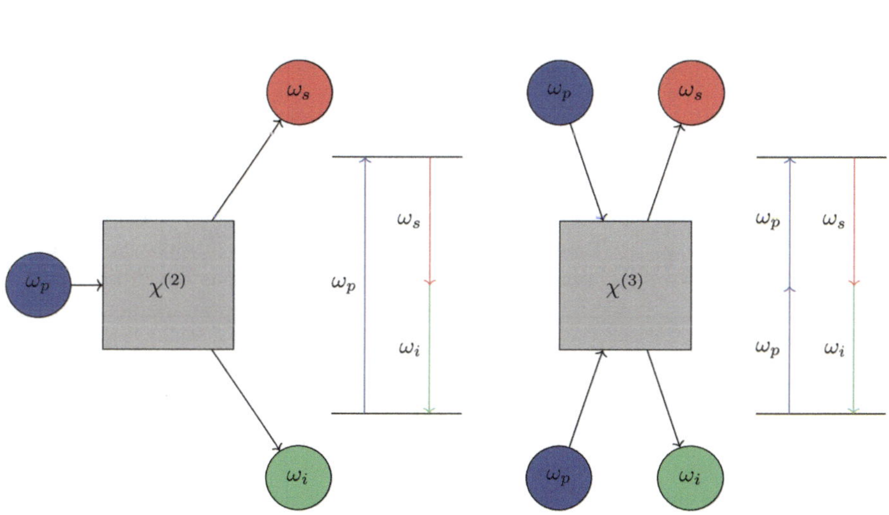

Fig. 4.10 Quantum states of light generation with nonlinear materials relies on $\chi^{(2)}$ (parametric down-conversion) or $\chi^{(3)}$ (four-wave mixing) processes in which one or two pump photons (ω_p) are converted into correlated signal (ω_s) and idler (ω_i) photons, respectively. In both cases, the systems are operated in the degenerate regime ($\omega_s \approx \omega_i$)

of squeezing technology has witnessed ongoing advancements driven by the incorporation of low-loss optical components, high-efficiency detectors, and low-noise electronics [40].

The preceding discussion is concerned with quadrature-squeezed states and now transitions to the topic of generating photon-number squeezed states or amplitude-squeezed states. An amplitude-squeezed state shows reduced photon-number fluctuations (sub-Poisson statistics instead of Poisson statistics) while phase fluctuations are enhanced, as illustrated in Fig. 4.5. In essence, the statistical characteristics of a light beam are reflected in the stochastic process governing the arrival of photons at a detector. Light with a regular photon arrival pattern displays lower photon-number fluctuations than light with statistically independent photon arrivals. Consequently, the production of amplitude-squeezed states involves inducing anticorrelations among closely spaced photon occurrences. This concept can be visualized using the analogy of a sequence of photons being "shot" from a hypothetical gun. Initially, these photons follow a Poisson distribution in terms of their numbers. However, by selectively removing photons that appear within a certain specified time interval of each other, the sequence becomes more regular with sub-Poisson behaviour. To induce such anticorrelations, different mechanisms can be employed as shown in Fig. 4.11. These include triggering photon emissions at regular time intervals (excitation control), incorporating constraints related to the emission's reset time (emission control), or using information about photon occurrence times to influence future emissions or excitations (feedback control).

Several excitation control methods have used the inherent sub-Poisson nature of an electron current. The genesis of amplitude-squeezed states found its roots in a Franck-Hertz-like experiment [41]. In this context, Coulomb repulsion plays a crucial role in potentially constraining the electron stream due to inelastic collisions. As

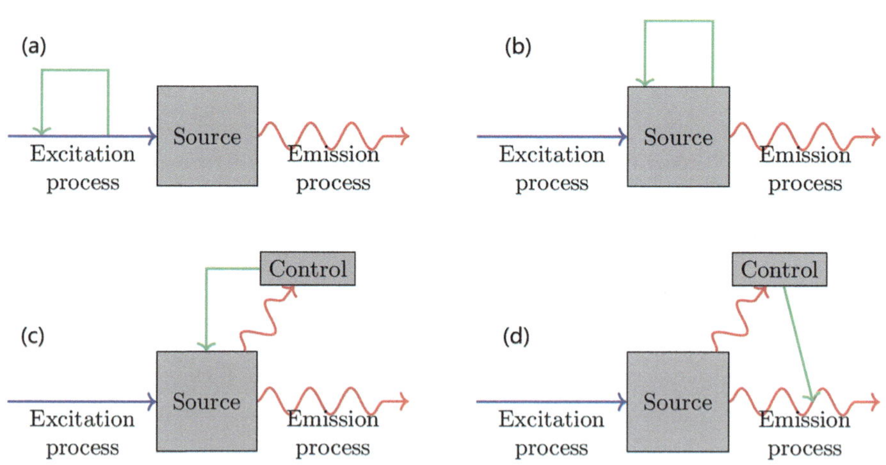

Fig. 4.11 Schematic diagram illustrating the generation of amplitude-squeezed light by **a** excitation control, **b** emission control, **c** and **d** feedback control

a result, an element of regularization is introduced into the flow of electrons, which then translates to regularization in the number of atoms they excite and, by extension, in the number of photons spontaneously emitted by these atoms. The outcome of such an excitation control method is the emission of spontaneously fluorescing photons displaying amplitude squeezing. However, it is essential to acknowledge that the sub-Poisson behaviour can be randomized due to photon losses stemming from imperfect photon generation, collection, and detection. In cases where these losses are substantial, random (Poisson distribution) photons could be regenerated. Effects such as attenuation, scattering, and the presence of background photons can significantly diminish the degree of amplitude squeezing, necessitating diligent efforts to mitigate these influences. To overcome these challenges, several high-efficiency configurations have been developed. For instance, current supplied from a DC source like a battery, inherently exhibits sub-Poisson characteristics owing to the intrinsic Coulomb repulsion of electrons, with Johnson noise being the principal noise source. In such cases, driving a light emitter operating through photon emission with sub-Poisson current is sufficient. Thus, a simple light-emitting diode (LED) powered by a constant current source can emit amplitude-squeezed light [42]. This phenomenon has been demonstrated by Tapster et al. who carried out an experiment similar to a solid-state version of the space-charge-limited Franck-Hertz experiment. Similarly, a semiconductor injection laser driven by a constant current [21], analogous to a stimulated-emission version of the space-charge-limited Franck-Hertz experiment, also behaves in this manner.

Emission control methods can also yield amplitude squeezing. For example, the introduction of a "dead time" can prevent a second event from occurring within a fixed time interval after a previous event. This constraint effectively prevents the events from being arbitrarily close to each other and regularizes them. This technique has found particular application in resonance fluorescence emissions [43]. In these experiments, isolated atoms subject to Poisson-like excitations are unable to emit photons while they are undergoing re-excitation.

Controlling excitations or emissions by feedback offers an alternative approach. By monitoring arriving photons without destroying them, their arrival times can be utilized to modify subsequent excitations or emissions. This form of feedback control method becomes practicable when photons are observed using quantum-nondemolition measurements, allowing for observations without introducing perturbations. Yamamoto and his colleagues have proposed strategies for implementing quantum-nondemolition measurements based on self-phase modulation in a Kerr medium [20]. It is important to note that if conventional destructive photon detectors are used instead of the quantum-nondemolition measurement of photon number, the extraction of amplitude-squeezed states from the system is not feasible [44]. Furthermore, an additional avenue for achieving feedback control is to use twin photon streams. In this scenario, one of the streams can be annihilated to generate a control signal, while the twin stream persists unaffected. In conclusion, a compilation of early experiments in which squeezed light was generated is presented in Table 4.1.

Table 4.1 Early squeezed light experiments

Year	Effect	Experiment	Reference
1983	Amplitude-squeezed	Single-atom resonance fluorescence	[43]
1985	Amplitude-squeezed	Space-charge-limited Franck-Hertz effect	[41]
1985	Quadrature squeezing	Nondegenerate four-wave mixing in Na atoms	[17]
1986	Quadrature squeezing	Nondegenerate four-wave mixing in optical fibre	[33]
1986	Quadrature squeezing	Parametric downconversion in $MgO:LiNbO_3$	[18]
1986	Amplitude-squeezed	Quantum nondemolition measurement	[20, 44]
1987	Amplitude-squeezed	Constant-current-driven LED	[42]
1987	Amplitude-squeezed	Constant-current-driven semiconductor laser	[21]

4.2 Introduction to Amplitude Squeezing in Semiconductor Lasers

The study of quantum fluctuations properties in lasers gained significant attention shortly after their invention in the early 1960s. The resulting consensus from these investigations was that, when a laser is operated well above its threshold and only short time intervals (high frequencies) are taken into account, its output closely resembles the noise characteristics of a coherent state [8]. This implies that, for sufficiently high Fourier frequencies, the spectral variance of photon number and phase fluctuations will approach the Poissonian limit or standard quantum limit, also known as the shot noise limit (SNL). However, at lower frequencies, the presence of excess intensity noise, encompassing both intrinsic and technical factors, is anticipated. In most types of lasers pumped well above the threshold, the quantum noise is often overwhelmed by extrinsic noises, making its measurement difficult in practical experiments.

Semiconductor lasers stand out as an exceptional case, with their intensity noise converging towards the SNL for all frequencies under optimal conditions. Their compact size and stable operation make quantum noise more pronounced compared to other extrinsic noises, allowing it to remain observable across varying pump rates. In contrast, the laser's phase is not constrained and undergoes temporal drift, known as laser phase diffusion. As a consequence, the semiconductor laser is a unique device that necessitates the application of quantum mechanics for a comprehensive understanding of its performance. Therefore, it provides a practical experimental platform to test the principles of the quantum theory of light. This conclusion was reached based on the consideration of three fundamental quantum noise sources [45]:

1. The quantum statistical nature of the fundamental interaction between light and atoms.
2. The stochastic nature of processes that lead to photon losses.
3. The statistical characteristics of the atomic excitation processes within the active medium.

Of particular relevance is the observation that the excitation process can be effectively approximated by a Poissonian distribution for the number of atoms excited during a given time interval. This distribution assumption underlies much of the theoretical work on lasers, either implicitly or explicitly. In the early 1980s, Golubev and Sokolov [45] made pioneering contributions to the study of lasers pumped in a regular manner (thus eliminating the third noise source), where equal numbers of atoms are pumped within equal time intervals.

The experimental validation of the reduction in intensity fluctuations through the regularization of a light source's pumping was initially demonstrated by Teich and Saleh in their work [41]. They observed the emission of sub-Poissonian light from a mercury-vapor lamp that was being pumped by a space-charge-limited electron beam. This phenomenon was termed the "inverse" of Einstein's photoelectric effect, as the average photon flux exhibited proportionality to the average electron current. Subsequent experiments by Tapster et al. in 1987 [42] revealed sub-Poissonian light emission from a light-emitting diode (LED) driven by a constant-current source. In a similar vein, S. Machida et al. in 1987 [21] demonstrated sub-Poissonian light emission from semiconductor lasers, including both 852 nm GaAs and 1550 nm InGaAsP lasers. These results provided strong evidence for the phenomenon of sub-Poissonian light emission due to the regularization of the pumping process. The work of Machida and Yamamoto in 1988 and 1989 further expanded on these findings. In these experiments, a reduction in noise from 9% to 19% was achieved over a frequency range spanning from near DC to 1.1 GHz [46]. The primary limitation stemmed from the measurement system, attributed to optical feedback and limited light-collection efficiency. These challenges were successfully addressed by implementing a direct "face-to-face" coupling between the laser and the detector, both situated within a cryostat at liquid-helium temperature. This innovation resulted in an impressive 85% reduction in noise [47]. To attain sub-Poissonian behaviour from commercial semiconductor lasers operating at room temperature, the suppression of spurious modes is essential to operate them in a monomode regime [48]. This goal has been achieved through diverse methods such as injection locking [49, 50], optical feedback from a grating [51–54], or active stabilization of the laser temperature [55]. The field has witnessed a wealth of experimental studies, as summarized in Table 4.2, particularly in the domain of semiconductor lasers.

Table 4.2 The well-known experimental observation of amplitude-squeezed states in semiconductor lasers. RT means room temperature, TJS means "transverse junction stripe", and VCSEL means "vertical-cavity surface-emitting laser". × means no extra technique

Number	Year	Laser type	Temperature	+Technique	Reference
1	1987	1560 nm DFB QW	RT	×	[21]
2	1988	810 nm single-mode TJS QW	77 K	×	[46]
3	1989	TJS QW	77 K	×	[56]
4	1990	780 nm TJS QW	4 K	×	[55]
5	1991	786 nm TJS QW	66 K	×	[57]
6	1993	788 nm TJS QW	77 K	Optical injection	[49]
7	1993	810 nm QW	10 K	×	[58]
8	1993	815 nm QW	10 K, 78K	Optical injection	[50]
9	1993	830–840 nm QW	RT	Optical feedback	[51]
10	1995	852 nm QW	RT	Optical feedback	[52]
11	1995	852 nm QW	RT	×	[48]
12	1997	960 nm VCSEL	RT	×	[59]
13	1998	852 nm QW	80 K	Optical feedback	[53]
14	1999	1550 nm DFB QW	RT	Optical feedback	[54]
15	2001	840 nm single-mode VCSEL	RT	×	[60]

4.2.1 Suppressed-Pump-Noise Mechanism

The preparation of a sub-Poissonian light or amplitude-squeezed state is not an easy job. Nevertheless, generating a sub-Poissonian electron beam or achieving an electron-number state becomes achievable through mutual electron interactions within a space-charge-limited beam or a simple resistor. In fact, a noteworthy characteristic setting semiconductor lasers apart from other lasers is their pumping mechanism. Specifically, semiconductor lasers can be pumped by injection current through an electrical circuit. The implementation of such an electron excitation scheme is very promising to significantly reduce pump noise, along with the facile generation of amplitude-squeezed states in semiconductor lasers.

The left panel of Fig. 4.12 illustrates the configuration for generating a sub-Poissonian electron beam within a space-charge-limited vacuum tube. As the electron emission rate surpasses its average, a rise in the count of space charges near the cathode ensues. This, in turn, induces a more negative potential minimum (U2) between the cathode and anode due to the Coulomb repulsion among the "crowded" space electrons. Consequently, an increased number of electrons experience reflection back towards the cathode due to their insufficient initial velocity, $v \leqslant (2q|\mathrm{U}|/m)^{\frac{1}{2}}$ [31]. On the contrary, when the electron emission rate falls below its average, the potential minimum becomes less negative (U1), resulting in fewer reflected electrons. This modulation of the potential minimum effectively governs the flow of electrons traversing through this region and naturally establishes a feedback stabilization mechanism. The electron arrival process at the anode exhibits a strong departure from a Poissonian distribution, leading to the anode current fluctuation at a sub-shot-noise level. Utilizing this principle, Teich and Saleh achieved the generation of sub-Poissonian light from a space-charge-limited vacuum tube containing mercury atoms [41], despite the light not being a product of laser emission but rather arising from weak spontaneous emission.

Applying the same principle, the previous regularization concept can be extended to a laser oscillator [32]. The right panel of Fig. 4.12 shows an injection-current-pumped semiconductor laser, wherein the carrier-injection process is controlled by a similar mechanism. The carrier injection rate, namely, the diode junction current, is determined by the forward-biased voltage across a p-n junction. This voltage opposes the inherent potential within a depletion layer, enabling the predominance of the carrier-diffusion process over the reverse-oriented drift process. When the junction current surpasses its average level, the junction voltage diminishes due to the increased voltage across the series resistance R_s. As a consequence, a greater number of minority carriers (in this context, electrons as depicted in Fig. 4.12) return to the n-type bulk layer due to the dominating built-in field within the depletion region. Oppositely, as the injection current decreases below its average value, the junction voltage increases, facilitating the diffusion of more electrons into an active

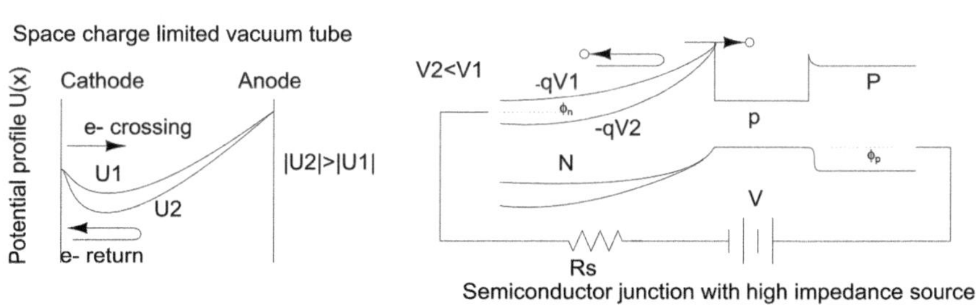

Fig. 4.12 Left panel: Regulated electron emission process in a space-charge-limited vacuum tube obtained by self-modulation of the potential field. Right panel: Semiconductor junction diode driven by a high impedance current source which also shows the regulated electron emissions through junction voltage modulation (*Source* In Ref. [31])

layer. This modulation of the junction voltage, induced by the presence of the series resistance R_s, plays a role in regulating the injection of electrons into the active layer. Hence, it becomes evident that the current noise produced within the series resistance R_s is not dominated by shot noise $(2qI)$ but instead, mostly constitutes thermal noise $(4k_BT/R)$ when $\hbar\omega < k_BT$ in (4.12), a factor unaffected by the current I due to such a modulation effect. If the applied voltage $V = IR_s$ is larger than the equivalent thermal voltage $V_T = 2k_BT/e$=42 mV at room temperature, as commonly encountered, the resulting current noise becomes lower than the shot noise level.

4.2.2 A Noise Equivalent Circuit of the Semiconductor Laser

Generally speaking, the origin of current noise in laser diodes can be ascribed to two fundamental processes: thermal diffusion and recombination of minority carriers [31]. Therefore, the relaxation current is directed into the diode's junction from the external circuit in response to these internal events, aiming to restore the equilibrium distribution of minority carriers. The equivalent noise circuit for laser diodes needs to incorporate this external circuit relaxation current in parallel with the diode's differential resistance (R_d). In fact, this noise-equivalent circuit method provides an ideal framework for elucidating the regulation mechanisms, as previously discussed. As is well-known, a laser diode can be simplified to a p-n junction structure. Consequently, a noise-equivalent circuit for this p-n junction is illustrated in Fig. 4.13 [61, 62]. vth is the thermal noise voltage associated with R_s and v_{sn} is the shot noise voltage associated with R_d. Alternatively, in their Norton equivalents, $i_{sn} = v_{sn}/R_d$ is the shot noise current and ith $= v$th$/R_s$ is the thermal noise current. The Kirchhoff circuit equation for this noise-equivalent circuit is as follows:

$$\frac{\mathrm{d}}{\mathrm{d}t}v_{jn} = -\frac{v_{jn}}{CR_s} + i_{th} - \frac{v_{jn}}{CR_d} + i_{sn} \tag{4.13}$$

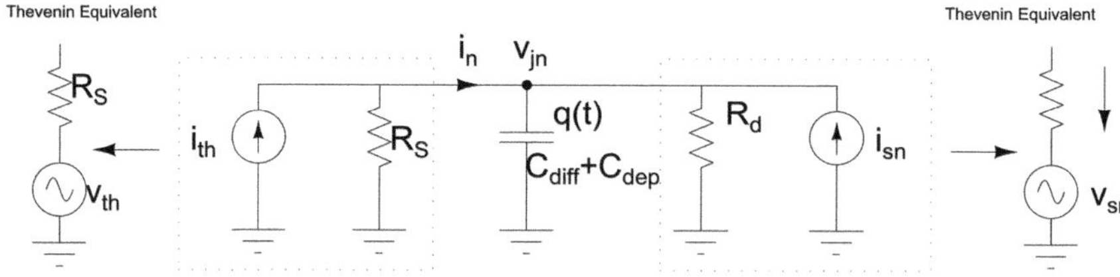

Fig. 4.13 Noise-equivalent circuit of a p-n junction. The circuit shows the resistance R_s, diode resistance R_d, total capacitance $C = C_{\text{dep}} + C_{\text{diff}}$ with C_{dep} the depletion layer capacitance and C_{diff} the diffusion capacitance, stored charge fluctuation $q(t)$, junction voltage fluctuation v_{jn}, recombination current i_{jn}, junction current i_n, and the noise generators v_{sn} and vth

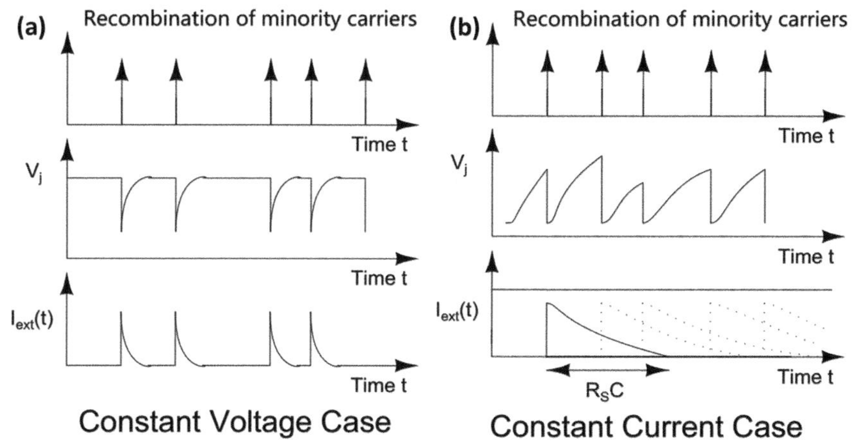

Fig. 4.14 The **a** Constant voltage operation and **b** Constant current operation on a p-n junction diode

The first term on the right-hand side of Eq. (4.13) represents the relaxation rate of v_{jn}, which is associated with the relaxation current in the external circuit. The second term corresponds to Johnson-Nyquist thermal noise resulting from R_s, with a spectral density denoted as $S_{i\text{th}}(\omega) = 4k_BT/R_s$. The third term on the right-hand side of Eq. (4.13) signifies the relaxation rate of v_{jn} due to the minority carriers across the depletion layer or minority carriers' recombination. Finally, the fourth term relates to the shot noise current associated with R_d, with a spectral density denoted as $S_{i_{sn}}(\omega) = 2qI$.

There are two operational modes of p-n junction diodes, as illustrated in Fig. 4.14.

- Constant voltage operation ($R_s \ll R_d$): Even though the junction voltage changes from the steady state value due to the internal events, it is immediately relaxed by the external circuit within the CR_s time constant which is much smaller than the internal time constant CR_d. As a result, the junction voltage is almost constant and the whole system lacks a memory effect for events related to thermal diffusion and the recombination of minority carriers. Then, the noise current associated with these occurrences exhibits full-shot noise behaviour, represented as $S_i(\omega) = 2qI$.
- Constant current operation ($R_s \gg R_d$): Every recombination of minority carriers triggers an external relaxation pulse. However, since the time constant for the external circuit CR_s is much larger than the internal time constant CR_d, the junction voltage is now allowed to fluctuate and the whole system can not be back to equilibrium distribution in time, which decreases the chances of the next minority carriers' recombination (photon emission). Also, the self-modulation mechanism previously discussed works in this situation. As a result, it is anticipated that the junction current fluctuation will exhibit a sub-shot-noise nature. However, it's worth noting that the junction voltage fluctuation v_{jn} is not suppressed by this process.

When a p-n junction diode is subjected to a strong forward bias, its differential resistance (R_d) often becomes negligible when compared to the source resistance (R_s). This situation is particularly evident in semiconductor light-emitting diodes and lasers [1]. Within such devices, the noise stemming from junction current is less than the standard full shot noise, and the emitted photon flux exhibits a sub-shot noise characteristic.

Taking the Fourier transform of (4.13), one can obtain

$$(i\omega C + \frac{1}{R_s} + \frac{1}{R_d})V_{jn}(\omega) = I_{th}(\omega) + I_{sn}(\omega) \tag{4.14}$$

where $V_{jn}(\omega)$, $I_{th}(\omega)$ and $I_{sn}(\omega)$ are now in their Fourier transforms. The spectral density of the junction voltage fluctuations is obtained by calculating $\left\langle V^*_{jn}(\omega)V_{jn}(\omega)\right\rangle$ and is

$$S_{V_{jn}} = \frac{2qIR_d^2 + 4k_BTR_d^2/R_s}{(1+\frac{R_d}{R_s})^2 + (\omega CR_d)^2} \tag{4.15}$$

The junction current fluctuations are obtained from Kirchhoff's law

$$i_n = i_{th} - \frac{v_{jn}}{R_s}$$

Fourier transforming the above equations and substituting for $V_{jn}(\omega)$, one gets

$$I_n(\omega) = I_{th}(\omega) - \frac{V_{jn}(\omega)}{R_s} = \frac{-I_{sn}(\omega) + I_{th}(\omega)(i\omega CR_s + \frac{R_s}{R_d})}{1 + \frac{R_s}{R_d} + i\omega CR_s} \tag{4.16}$$

The spectral density of the junction current fluctuations is expressed as

$$S_{I_n}(\omega) = \left\langle I_n^*(\omega)I_n(\omega)\right\rangle = \frac{2qI + \left[(\frac{R_s}{R_d})^2 + (\omega CR_s)^2\right]\frac{4k_BT}{R_s}}{(1+\frac{R_s}{R_d})^2 + (\omega CR_s)^2} \tag{4.17}$$

When discussing the noise spectral density, it typically refers to the spectral density of fluctuations in the junction current. Let's consider two operational modes:

- **Constant voltage operation** ($R_s \ll R_d$): Under this condition, the noise spectral density is reduced to

$$S_{I_n}(\omega) = \left\langle I_n^*(\omega)I_n(\omega)\right\rangle = \frac{2qI + (\omega CR_s)^2\frac{4k_BT}{R_s}}{1 + (\omega CR_s)^2}$$

  In the low-frequency limit ($\omega CR_s \ll 1$), the complete shot noise is recovered as $S_{I_n}(\omega) = 2qI$. However, in the opposite limit ($\omega CR_s \gg 1$), the thermal noise is recovered as $S_{I_n}(\omega) = \frac{4k_BT}{R_s}$. This divergence is because, at high frequencies, the p-n junction is effectively short-circuited by the junction capacitance C, making it impossible to extract the internal current noise I_n to the external circuit.

- **Constant current operation** ($R_s \gg R_d$): Assuming that $2qI = \frac{2k_BT}{R_d}$, in this scenario, the noise spectral density always corresponds to the thermal noise limit, given by $S_{I_n}(\omega) = \frac{4k_BT}{R_s} < 2qI$.

4.2.3 Limitation of Squeezing Level

The squeezing level $\mathcal{R}$, or the reduction in intensity noise, is solely determined by the laser's internal quantum efficiency η_{in}, which characterizes the conversion from injected electrons to emitted photons.

$$\mathcal{R} = 1 - \eta_{in}$$

Achieving photon number squeezing of more than 10 dB is feasible, especially considering that an internal quantum efficiency exceeding 90% is readily attainable in modern semiconductor laser technology. However, as mentioned in Sect. 4.1.5, it is crucial to acknowledge that sub-Poisson behaviour can be disrupted by photon losses arising from different processes, such as imperfect photon generation, collection, and detection. When these losses are large, the Poisson distribution may be recovered.

For instance, in Ref. [46], the photocurrent noise spectral density versus external quantum efficiency η is shown in Fig. 4.15 and a small amount of optical loss quickly destroys the squeezing. The ordinate is normalized relative to the corresponding shot-noise level. As several sources of optical loss appear, the noise level gradually

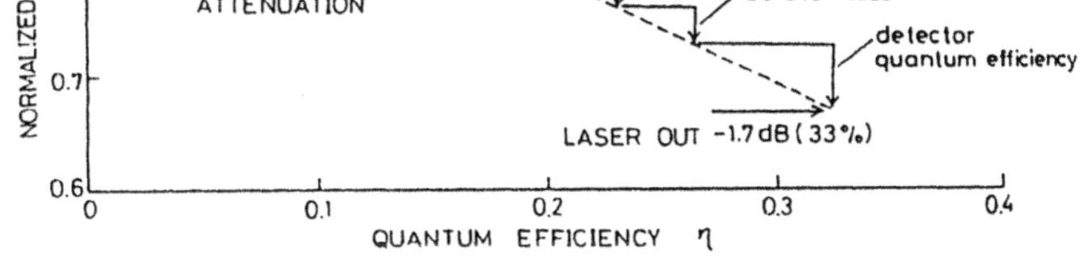

Fig. 4.15 The squeezing level versus external quantum efficiency (*Source* On p. 533 in Ref. [1])

rises towards the shot-noise level. In the limit of infinite loss, amplitude squeezing is entirely lost. With such substantial optical loss, the original quantum noise of the laser is completely replaced by the vacuum field fluctuation, resulting in the emergence of shot noise. This apparent increase in normalized photocurrent noise is another distinct signature of amplitude squeezing. The external quantum efficiency, from the increment in laser injection current to the increment in photodetector current, is 22% in this experiment when the optical loss is eliminated [46]. The overall detection quantum efficiency of 55% is composed of several factors: a photodetector quantum efficiency of 0.93, focusing-lens loss of 0.90, isolator insertion loss of 0.81, mirror reflection loss of 0.95, cryostat window loss of 0.93, and laser collimating-lens loss of 0.93. After correcting for these factors, the observed noise level indicates amplitude squeezing of −1.7 dB (33%) below the shot noise limit at the output of the laser facet, as shown in Fig. 4.15. This represents the noise level that the laser actually produced at the output mirror. The laser output coupling efficiency η_c is given by

$$\eta_c = \frac{L^{-1}\ln\left(R_1^{-1/2}\right)}{\alpha_{in} + L^{-1}\ln\left[(R_1 R_2)^{-1/2}\right]}$$

Here, L = 250 μm represents the cavity length, $R_1 = 0.32$ and $R_2 = 0.6$ are the power reflectivities of the front and rear facets, respectively, and $\alpha_{in} = 22\,\text{cm}^{-1}$ is the internal absorption loss. If the laser output-coupling efficiency due to nonideal rear facet power reflectivity $\eta_m = 0.57$ and that due to internal loss $\eta_\alpha = 0.70$ are also corrected, the observed noise level corresponds to amplitude squeezing of −7 dB (80%) below the shot-noise limit. This represents the intrinsic noise level achievable if the rear facet power reflectivity is increased to 100% and internal absorption loss is eliminated. This optimistic projection could be fueled by the advancements in semiconductor laser cavity design and the ongoing integration of the entire system onto a low-loss single chip [40]. The achievement of 85% (−8.3 dB) amplitude squeezing through the face-to-face positioning of the detector and the laser, as demonstrated in previous research [47], provides compelling evidence for the feasibility of these improvements.

4.2.4 Limitation of Squeezing Bandwidth

Yamamoto et al. envisaged that a DFB semiconductor laser with a high internal quantum efficiency η could manifest a broad squeezing frequency bandwidth, typically exceeding 100 GHz, ultimately determined by the semiconductor laser photon lifetime (τ_p ~1 ps) [32]. However, merely relying on a constant-current source with reduced shot noise does not unambiguously dictate an anti-correlation between successive injection events into the active layer. This is due to the fact that each individual carrier injection constitutes a (Poisson) random point process, driven solely by the junction voltage and the temperature of the junction [63]. It should be underlined that the injection rate can be influenced by the Coulomb blockade effect, a term borrowed from mesoscopic junctions. For instance, the injection of a carrier results in a reduction of the junction voltage by e/C_{dep}, where C_{dep} represents the depletion layer capacitance. This voltage reduction accordingly leads to a decrease in the carrier injection rate, establishing a negative feedback mechanism that suppresses noise in the carrier injection process. This mechanism has been successfully demonstrated in the context of the single-photon turnstile device operating within the mesoscopic domain at low temperatures, as described in [64]. In this case, the effectiveness of the mechanism relies on the fact that when a single electron crosses the junction, it prevents additional electrons from crossing if the electron charging energy due to a single transit, e^2/C_{dep}, is significantly greater than the thermal energy $k_B T$. But in the macroscopic high-temperature regime, the junction voltage after each injection drops by e/C_{dep}, the depletion layer charging energy is much smaller than $k_B T$ and each microscopic event is unregulated and thus completely random. However, in the macroscopic high-temperature regime, where $N = (\frac{k_B T}{e})/(\frac{e}{C_{\text{dep}}}) = k_B T C_{\text{dep}}/e^2$, the presence of this number of electrons leads to a significant reduction in junction voltage, equivalent to the thermal voltage $k_B T/e$. This reduction results in a pronounced regulation of the carrier injection rate. In other words, it creates a enough high barrier against further injection.

Considering that electrons are provided by the current source at a rate of I/e, the time required for N electrons to be delivered is $\tau_{te} = k_B T C_{\text{dep}}/eI$. This leads to anti-bunched electrons on a time scale associated with τ_{te}. As the injection current is reduced, τ_{te} increases and may exceed the recombination lifetime, in which case there is negligible charge storage. Each injection event results in instantaneous recombination, and the statistics of the pump determine the sub-shot nature of the photon flux. For measurement times shorter than τ_{te}, the negative feedback mechanism is disrupted, and the sub-shot photon features can not be observed. Therefore, this collective behaviour of numerous electrons introduces an additional limitation to the squeezing bandwidth [63]. Combining the two effects of τ_{te} and τ_p, the squeezing bandwidth B:

$$B = \frac{1}{2\pi(\tau_{te} + \tau_p)} = \frac{1}{2\pi(\frac{k_B T C_{\text{dep}}}{eI} + \tau_p)} \tag{4.18}$$

The squeezing bandwidth is expected to be directly proportional to the current I, but inversely proportional to the temperature T and the capacitance C_{dep}.

4.3 Amplitude Squeezing in Interband Cascade Lasers

In this section, I will theoretically investigate a novel category of optoelectronic devices operating in the mid-infrared (MIR) spectrum, where photon-number fluctuations are suppressed below the shot-noise limit. To achieve this, one should focus on interband cascade lasers (ICLs) that emit in a broad wavelength range of 3–6 μm.

These lasers typically consist of an active region composed of multiple cascade stages with type-II quantum wells (QWs). ICLs leverage a combination of factors, including a relatively long upper-level recombination lifetime and an efficient cascading mechanism, to generate coherent light through interband transitions. As a result, their active medium differs from that of quantum cascade lasers (QCLs), where photons are emitted from intersubband transitions. Additionally, ICLs are known for their significantly lower power consumption, often by one or two orders of magnitude, compared to QCLs.

Early demonstrations of amplitude-squeezed light were realized by different types of semiconductor lasers, as detailed in Table 4.2. However, there have been no reports of such demonstrations in the MIR wavelength range [65]. At the end of this section, ICLs will be theoretically proven as suitable sources for generating amplitude-squeezed light. These findings hold promise for expediting the development of novel quantum hardware in the MIR range, an area currently deficient in such technology, yet ripe with prospects for valuable applications.

4.3.1 Semiclassical Langevin Rate Equation

The theoretical description of squeezing light in semiconductor lasers was initially proposed by Yamamoto et al. in the case of Fabry-Perot semiconductor lasers, using a fully quantum-mechanical-based model [32], which was consistent with experimental results [21]. Although quantum fluctuations of optical fields can be rigorously expounded within the framework of quantum electrodynamics (QED), the thorough quantum approach can become laborious and complex. As a result, several alternative semi-classical methods have been introduced to offer good descriptions of squeezing phenomena while avoiding the computational complexities of rigorous QED calculations. For instance, Marcenac and Caroll introduced a method based on the transmission line approach [66], which incorporates vacuum fluctuations to describe the generation of amplitude-squeezing light. Jacques Arnaud developed a classical corpuscular theory that doesn't require the quantization of the electromagnetic field, yet it coincides with the quantum counterparts [67]. In this work, the formalism is developed with certain assumptions [68]. It assumes that the main source of noise lies in the particle-like nature of radiation and the random arrival times of photons, which do not interact with each other. Shot noise, therefore, is considered an intrinsic property of radiation rather than a property of the interaction between light and matter. Thus, mirror partition noise from the laser facet is regarded as the source of shot noise, resulting from the particle-like behaviour of photons. It's important to note that this model remains semi-classical because it treats photons as classical particles without mutual interaction. Nevertheless, it provides a qualitative and straightforward description of amplitude-squeezed light in semiconductor lasers.

This study is an extension of a rate equation approach recently developed [69], further integrating Green's function method to facilitate numerical investigations of squeezed light generation. Notably, this work accounts for the negatively correlated partition noise coming from the laser's out-coupling process, known as the standard quantum limit or shot-noise limit (SNL). The result predicts that ICLs are particularly well-suited for this purpose due to their high quantum efficiency, enabling a substantial reduction in amplitude noise. The Green's function method was originally introduced by Charles H. Henry [70], especially in the context of semiconductor lasers with complex cavities [71]. A comprehensive investigation of a semiconductor laser necessitates a three-dimensional examination of the electromagnetic field and its propagation. However, in this particular study, several assumptions are made. Firstly, it is assumed that the cavity medium is isotropic, allowing the matrix propagation equation to be simplified to the Helmholtz scalar equation. Secondly, it is assumed that within the laser, optical waves predominantly propagate along the longitudinal z-axis, while the transverse and lateral axes are disregarded. Lastly, it is assumed that spontaneous emission is delta-correlated both in space and time. The complex Fourier component $E_\omega(z)$ is governed by the one-dimensional scalar wave equation:

$$\nabla_z^2 E_\omega(z) + k_0^2 \varepsilon E_\omega(z) = F_\omega(z) \tag{4.19}$$

where $\nabla_z^2 = \frac{\partial^2}{\partial z^2}$ is the Laplacian operator for the longitudinal coordinate z, ω is the lasing angular frequency, $k_0 = \frac{\omega}{c}$ is the wavenumber, c is the velocity of light in vacuum, ε is the complex dielectric constant, and $F_\omega(z)$ is the Langevin noise term representing distributed spontaneous emission, Eq. (4.19) can be solved using the Green's function method [70]:

$$(\nabla_z^2 + k_0^2 \varepsilon) G_\omega(z, z') = \delta(z - z')$$

where $\delta(z - z')$ is the Dirac delta function. Then the general solution is obtained through a spatial integration of Green's function weighted by the correspondent excitation source over the total cavity length L:

$$E_\omega(z) = \int_{(L)} dz' G_\omega(z, z') F_\omega(z') \tag{4.20}$$

The Green's function $G_\omega(z, z')$ has been discussed at length [72] and its one-dimensional formula is given by

$$G_\omega(z, z') = \frac{Z_R(z) Z_L(z') \Theta(z - z') + Z_R(z') Z_L(z) \Theta(z' - z)}{W(\omega, N)}$$

In the provided equation, $\Theta(z)$ is the Heaviside step function. The terms $Z_L(z)$ and $Z_R(z)$ are two independent solutions of the homogeneous wave equation (i.e., when $F_\omega = 0$). These solutions satisfy the boundary conditions at the left laser facet and the right facet, respectively. $W(\omega, N)$ is the explicitly z-independent Wronskian operator of these two solutions defined as [72]

$$W(\omega, N) = Z_L(z)\frac{\mathrm{d}}{\mathrm{d}z}Z_R(z) - Z_R(z)\frac{\mathrm{d}}{\mathrm{d}z}Z_L(z)$$

Certainly, the laser's longitudinal mode corresponds to a zero point of the Wronskian operator $W(\omega_0, N_{th}) = 0$. Both the lasing frequency ω_0 and the carrier density N_{th} at the threshold are determined from this relation. Given the assumption that the semiconductor laser operates exclusively in one longitudinal mode with a field distribution denoted as $Z_L(z) = Z_R(z) = Z_0(z)$, (4.20) can be reformulated as:

$$\frac{E_\omega(z)}{Z_0(z)} = \frac{1}{W(\omega, N)}\int_{(L)} Z_0(z')F_\omega(z')dz' \tag{4.21}$$

Since the Wronskian operator W depends on two independent variables, ω and N, its expansion around the operating point can be expressed as:

$$W = \left.\frac{\partial W}{\partial \omega}\right|_{\omega_0, N_{th}}(\omega - \omega_0) + \frac{1}{L}\int_{(L)}\left.\frac{\partial W}{\partial N}\right|_{\omega_0, N_{th}}(N - N_{th})dz \tag{4.22}$$

Using (4.21)–(4.22) and the inverse Fourier transform, the rate equation for the electrical field $\beta_0(t)$ is

$$\frac{\mathrm{d}\beta_0(t)}{\mathrm{d}t} = -\frac{j}{L}\int_{(L)} W_N(N - N_{th})dz \times \beta_0(t) + F(t)$$

where $W_N = (\frac{\partial W}{\partial N})/(\frac{\partial W}{\partial \omega})$ and $F(t)$ is the Langevin force associated with the complex amplitude in the time domain. In addition, $\beta_0(t)$ signifies the slowly varying envelope of the electric field within the laser cavity:

$$\beta_0(t) = \frac{1}{2\pi}\int_{-\infty}^{+\infty}\beta_\omega e^{j\omega t}d\omega$$

where $\beta_\omega = E_\omega(z)/Z_0(z)$. The resolution yields two generalized differential equations for the photon density S and the phase ϕ, which are defined through $\beta_0(t) = \sqrt{S(t)}e^{j\phi(t)}$:

$$\frac{\mathrm{d}S}{\mathrm{d}t} = \frac{2}{L}\int_{(L)}\Im\mathrm{m}(W_N)(N - N_{th})dz \times S(t) + F_{S'}(t)$$
$$\frac{\mathrm{d}\Phi}{\mathrm{d}t} = -\frac{1}{L}\int_{(L)}\Re\mathrm{e}(W_N)(N - N_{th})dz \times S(t) + F_{\phi'}(t)$$

Here "$\Re$e" signifies the real part of the complex force, while "$\Im$m" represents the imaginary part. These equations enable the analysis of both static and dynamic behaviours in various types of semiconductor lasers, including those constructed with complex structures like multisection devices [71]. Furthermore, the Petermann excess noise K factor can be generally defined as:

$$K = \frac{(\int_0^L |Z_0(z)|^2 dz)^2}{|\int_0^L Z_0^2(z)dz|^2} \tag{4.23}$$

From a broader perspective, the K-factor can be used to calculate the enhanced spontaneous emission rate in the lasing mode of semiconductor lasers [73]. More importantly, as discussed later, it also takes into account the spatial longitudinal inhomogeneities of the cavity, which can influence the amplitude-squeezing performance. A Fabry-Perot ICL cavity design with facet power reflectivities R_1 (rear) and R_2 (front) is going to be analyzed. The field intensity distribution is assumed nearly uniform inside the cavity, and thus, the spatial hole-burning effects are neglected. Consequently, the Wronskian operator is expressed as [74]:

$$W(\omega, N) = -2jke^{jkL}[1 - r_1 r_2 e^{-2jkL}]$$

Next, consider the partial derivative of the Wronskian [74]:

$$W_N = \frac{1}{2}(1 + j\alpha_H)G_0$$

Here, $G_0 = \Gamma_p v_g a_0/A$, where a_0 is the gain coefficient, A is the area of the active region, v_g is the group velocity, and Γ_p is the confinement factor. As a result, the comprehensive characterization of the ICL structure is now entirely encapsulated by the Wronskian operator. In this case, the photon and phase rate equations per gain stage can be transformed as:

$$\frac{dS}{dt} = G_0(N - N_{th})S + F_{S'}(t)$$
$$\frac{d\phi}{dt} = \frac{1}{2}\alpha_H G_0(N - N_{th}) + F_{\phi'}(t)$$

Taking into account the spontaneous emission and the relation between carrier number at transparency N_0 and carrier number at threshold N_{th}, the rate equations for the total photon number S and phase ϕ of all gain stages are given by:

$$\frac{dS}{dt} = mG_0(N - N_0)S - \frac{S}{\tau_p} + m\beta_{sp}\frac{N}{\tau_{sp}} + F_S(t) \tag{4.24}$$
$$\frac{d\phi}{dt} = \frac{1}{2}\alpha_H[mG_0(N - N_0) - \frac{1}{\tau_p}] + F_\phi(t) \tag{4.25}$$

with m the number of cascading gain stage, τ_p the photon lifetime and β_{sp} the spontaneous emission factor. Moreover, it's widely acknowledged that the dynamic evolution of the carrier density per gain stage is governed by the conventional relation [69]:

$$\frac{dN}{dt} = \eta\frac{I}{q} - \frac{N}{\tau_e} - G_0(N - N_0)S \tag{4.26}$$

where $N(t)$, τ_e, η and I are the carrier number within the active region, the carrier lifetime, the injection efficiency and the pump current, respectively. In the case of the ICL, the carrier lifetime τ_e is primarily influenced by two key factors: Auger recombination τ_{aug} and spontaneous emission τ_{sp}. Thus,

$$\frac{1}{\tau_e} = \frac{1}{\tau_{aug}} + \frac{1}{\tau_{sp}}$$

Given a Fabry-Perot structure, the Petermann excess noise K-factor is derived according to (4.23):

$$K = \left[\frac{(R_1 + R_2)(1 - R_1 R_2)}{2R_1 R_2 \ln(R_1 R_2)} \right]^2 \tag{4.27}$$

It's noteworthy that this formalism, which includes Eqs. (4.24), (4.25), and (4.26), is consistent with the recently developed rate equation approach [69], as well as the Eqs. (3.36) and (3.37) used in Sect. 3.2.3. Figure 4.16 provides an overview of the reservoir model employed to characterize the noise performance. Carriers are injected into the laser structure at a rate of I/q, and only a specific fraction ($\eta I/q$) of them actually reach the active region. Once within the carrier reservoir, some carriers recombine non-radiatively (Auger recombination τ_{aug}), while others undergo spontaneous recombination (τ_{sp}). It's noteworthy that a portion of photons β_{sp} from spontaneous emission contributes to the lasing mode. These photons, along with those generated from stimulated emission, experience further amplification through the cascade stages m inherent in the ICL structure. The photon reservoir comprehensively describes this interplay between stimulated and spontaneous emission processes. Subsequently, photons exiting the cavity after completing one round trip are

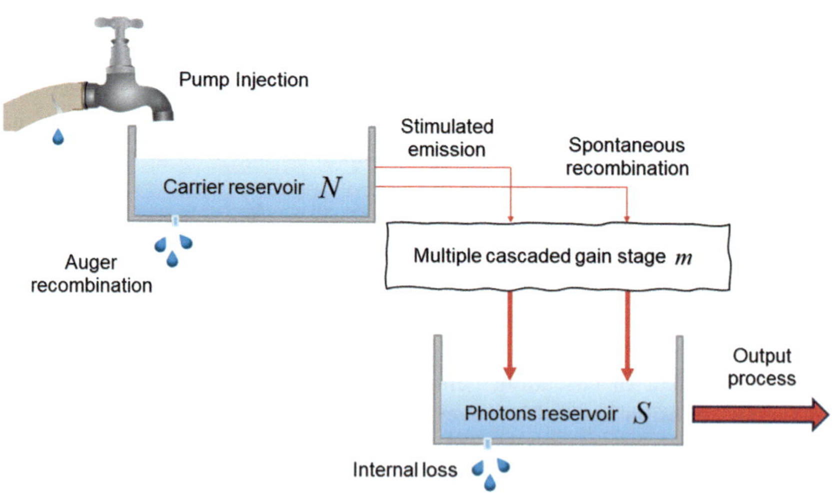

Fig. 4.16 Reservoir model used in the rate equation analysis of the ICL

controlled by the photon lifetime τ_p, eventually resulting in stable output power. The multiple carrier reservoirs given the existence of many cascaded stages in ICLs have not been taken into account by now and will be studied in the future.

Therefore, due to the incorporation of Langevin noise forces but neglecting the phase Eq. (4.25) in (3.37):

$$\begin{bmatrix} \gamma_{11} + j\omega & \gamma_{12} \\ -\gamma_{21} & \gamma_{22} + j\omega \end{bmatrix} \begin{bmatrix} \delta N(\omega) \\ \delta S(\omega) \end{bmatrix} = \begin{bmatrix} F_N(\omega) \\ F_S(\omega) \end{bmatrix}$$

It becomes evident that the intra-cavity relative intensity noise (RIN_{in}) through the application of Cramer's rule:

$$\begin{aligned} \text{RIN}_{\text{in}}(\omega) &= \frac{\langle \delta S(\omega) \delta S(\omega)^* \rangle}{S^2} = \frac{1}{S^2 |\Omega|^2} (\alpha + \beta\omega^2) \\ &= \frac{1}{S^2 |\Omega|^2} \left[(\gamma_{11}^2 + \omega^2) \langle F_S F_S^* \rangle + 2\gamma_{11}\gamma_{21} \langle F_S F_N^* \rangle + \gamma_{21}^2 \langle F_N F_N^* \rangle \right] \end{aligned} \quad (4.28)$$

where $\alpha = \gamma_{11}^2 \langle F_S F_S^* \rangle + 2\gamma_{11}\gamma_{21} \langle F_S F_N^* \rangle + \gamma_{21}^2 \langle F_N F_N^* \rangle$ and $\beta = \langle F_S F_S^* \rangle$. And $\omega_R^2 = \gamma_{12}\gamma_{21} + \gamma_{11}\gamma_{22}$ is the relaxation resonance frequency and $\gamma = \gamma_{11} + \gamma_{22}$ is the damping factor. Moreover, $H(\omega) = \omega_R^2 / \Omega$ is the modulation transfer function and $\Omega = \omega_R^2 - \omega^2 + j\omega\gamma$ is the determinant of the matrix. The diffusion coefficients can be found in Appendix B. Using the above theories, the RIN characteristics of an ICL are studied. The material and optical parameters of the ICL used in the simulations are all listed in Table A.3. Figure 4.17 highlights that incorporating more gain stages marginally enhances the intra-cavity RIN due to the presence of uncorrelated photon noise sources in each stage. However, this trend lacks sufficient evidence for the shot-noise limit scenario. Additionally, the influence of the facet power reflectivity on the intra-cavity RIN is also investigated. In Fig. 4.18, the low frequency RIN decreases from −163 dB/Hz for front facet $R_2 = 0.7$ to −168 dB/Hz for $R_2 = 0.1$. This behaviour can be understood by recognizing that facet power reflectivity directly affects the intra-cavity photon lifetime.

When accounting for quantum features, it is essential to exercise caution owing to the inherent quantization properties, such as the particle-like behaviour of photons. Here, it should be emphasized that the photons emitted as output due to the out-coupling process differ from the intra-cavity photons [32]. This distinction comes from the fact that each individual photon must either be transmitted through or reflected from the laser facet. While photons with a fixed average rate are reflected back into the laser cavity, this random division results in partition noise with both transmitted and reflected photons [75]. Hence, it is feasible to employ the Langevin method to describe the consequences of laser facet partition noise, treating output photons as another photon reservoir subject to its own Langevin noise force. In Refs. [32, 68], these can be regarded as vacuum fluctuations, which destructively interfere with photon fluctuations. By adapting (4.28), the fluctuations in output power can be expressed in relation to the intra-cavity RIN:

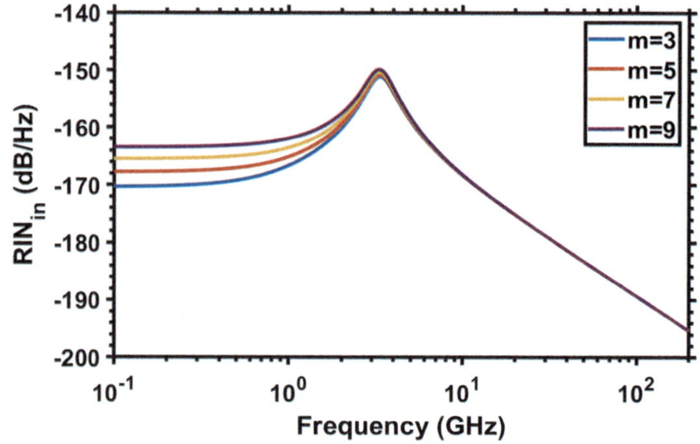

Fig. 4.17 Intra-cavity RIN spectra for different stage number m while the photon number is fixed at 4.2×10^6

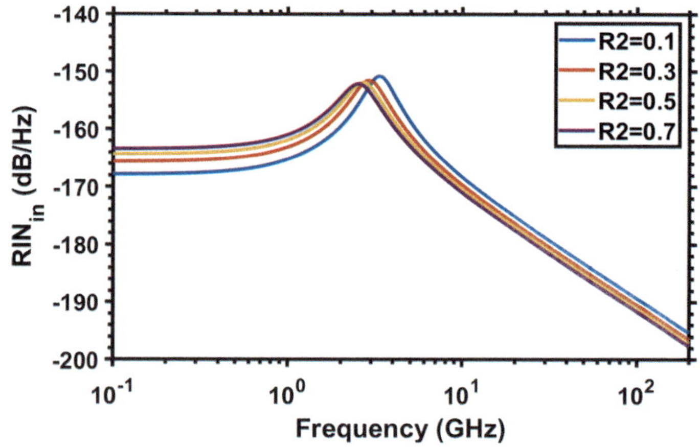

Fig. 4.18 Intra-cavity RIN spectra for the laser facet power reflectivity taking rear facet $R_1 = 1$ and front facet R_2 from 0.1 to 0.7

$$\delta S_{\text{out}}(\omega) = \lambda \times \delta S(\omega) + F_{\text{out}}(t)$$

$\lambda = \eta_0 h\nu/\tau_p$ is the fundamental conversion factor between output photons and output optical power. And η_0 is the quantum efficiency and $h\nu$ is the photon energy. Therefore,

$$\delta S_{\text{out}}(\omega) = \lambda^2 \delta S(\omega) + 2\Re\text{e}\left\{\lambda \left\langle \delta S F_{\text{out}} \right\rangle\right\} + \left\langle F_{\text{out}} F_{\text{out}} \right\rangle \tag{4.29}$$

The first term in this equation corresponds to what one would intuitively expect from the relationship. However, the presence of partition noise at the mirror facet introduces two additional noise contributions. Likewise, by applying Cramer's rule, one can determine:

$$\langle \delta S F_{\text{out}} \rangle = \frac{1}{\Omega} \left[(\gamma_{11} + j\omega) \left\langle F_S F_{\text{out}}^* \right\rangle + \gamma_{21} \left\langle F_N F_{\text{out}}^* \right\rangle \right] \tag{4.30}$$

where

$$\left\langle F_{\text{out}} F_{\text{out}}^* \right\rangle = h\nu S_0$$
$$\left\langle F_S F_{\text{out}}^* \right\rangle = -S_0$$

with $S_0 = \lambda S$ the output power of ICL. In addition, one has $\left\langle F_N F_{\text{out}}^* \right\rangle = 0$ because it is obvious that there is no correlation between the carrier noise and the photon partition noise. When correctly expanding (4.29) with the Langevin noise correlation strengths and the division of a non-zero complex number, the output RIN takes on the following form:

$$\begin{aligned} \text{RIN}_{\text{out}}(\omega) &= \frac{\langle \delta S_{\text{out}}(\omega) \delta S_{\text{out}}(\omega)^* \rangle}{S_0^2} = \frac{1}{S_0^2} \left[\lambda^2 \left\langle \delta S(\omega) \delta S(\omega)^* \right\rangle - \frac{2\lambda S_0}{|\Omega|^2} (\gamma_{11}\omega_R^2 + \gamma_{22}\omega^2) + h\nu S_0 \right] \\ &= \text{RIN}_{\text{in}} - \frac{2}{S\,|\Omega|^2} (\gamma_{11}\omega_R^2 + \gamma_{22}\omega^2) + \frac{h\nu}{S_0} = \frac{1}{S^2\,|\Omega|^2} (\alpha' + \beta'\omega^2) + \frac{h\nu}{S_0} \end{aligned} \tag{4.31}$$

with $\alpha' = \alpha - 2\gamma_{11}\omega_R^2$ and $\beta' = \beta - 2\gamma_{22}$ in (4.28).

By comparing $\text{RIN}_{\text{in}}(\omega)$ and $\text{RIN}_{\text{out}}(\omega)$, the presence of anti-correlation, marked by the negative sign in the second term of (4.31), signifies that when a photon is coupled out, the number of intra-cavity photons decreases, and the likelihood of other photons being coupled out also diminishes. This phenomenon reveals the anti-bunching mechanism governing the behaviour of the output photons. Additionally, it's worth noting that a constant term emerges in the third term of (4.31). This extra term ensures that the output RIN never falls below the minimum value of $h\nu/S_0$ across all frequencies. This quantum noise floor is usually referred to as the standard quantum limit or the shot-noise limit (SNL). Figure 4.19 depicts the characteristics of the output RIN for the investigated ICL. As evident, excess noise prevails at low frequencies and experiences significant enhancement in the vicinity of the relaxation resonance frequency, mirroring the behaviours observed in the intra-cavity RIN. However, ICLs tend to exhibit a reduction in noise towards the SNL at high frequencies beyond the resonance point. It's clear that increasing the pump current results in a reduction in the output RIN across the entire frequency spectrum, even reaching the SNL. Simulations presented in Fig. 4.20 demonstrate that the number of cascading stages does not influence the SNL of ICLs.

The Petermann excess noise K-factor can be calculated through (4.27). Figure 4.21 illustrates the K-factor for asymmetrical Fabry-Perot ICLs, where one of the laser rear facets has an anti-reflective coating ($R_1 = 1$) while the reflectivity of the front facet (R_2) varies from 0.1 to 1. For high values of R_2, the K-factor can be approximated to unity, but it becomes pronounced for low values of R_2. In a recent

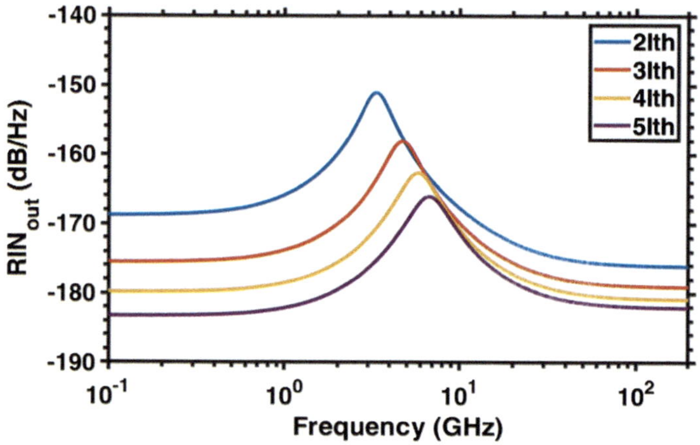

Fig. 4.19 Output RIN spectra for different pump currents while the stage number is fixed at $m = 5$

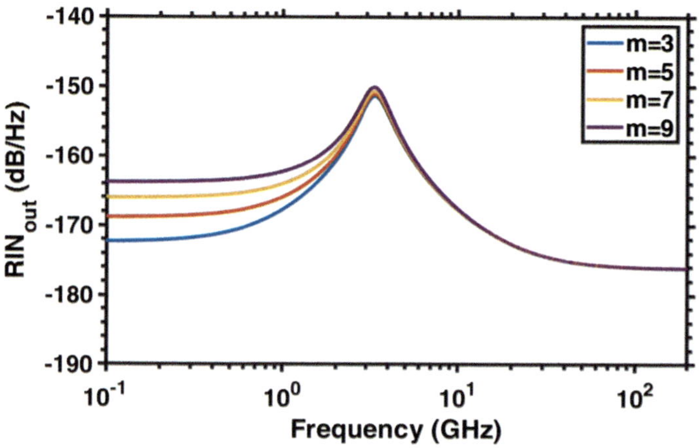

Fig. 4.20 Output RIN spectra for different pump currents while the stage number is fixed at $m = 5$

study [76], it was shown that a minor nonorthogonality of the laser cavity eigenmodes might explain the bottleneck observed in squeezing, implying that intensity noise can hardly be reduced below the SNL once the K-factor surpasses a critical value. This outcome underscores the importance of ICL cavity design for creating ultra-low noise light sources.

Subsequently, I shall explore how a noise-suppressed pump affects the squeezing performance of ICLs. In (4.31), the noise contribution of the current source should be evident in the α' term. This term can be minimized through the utilization of a "quiet" current source, as discussed in previous work [32]. Within (4.31), the term $\beta'\omega^2|\Omega|^2$ is responsible for the RIN peak at ω_R, while $h\nu/S_0$ shapes the RIN spectrum beyond

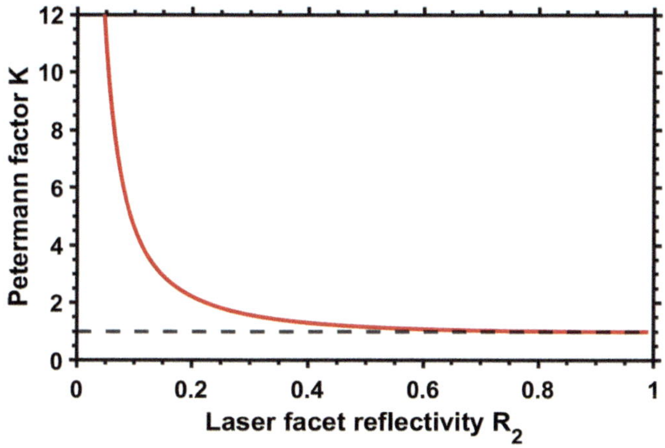

Fig. 4.21 The Petermann excess noise K factor varies as a function of the front facet power reflectivity R_2 while keeping the rear one R_1 fixed at 1. The black dashed line represents the limit of the K-factor value at 1 and serves as a visual guide to the eye in the plot

the peak at high frequencies. Consequently, $\alpha'|\Omega|^2$ is the dominant factor affecting the low-frequency behaviour away from the RIN peak. Now, let's focus on the low-frequency regime (neglecting the term $\beta'\omega^2|\Omega|^2$ so that $\langle F_S F_S^* \rangle = 0$) and assume that the laser output is highly efficient (e.g., $\eta_0 = 1$). In this case, for $\omega \ll \omega_R$, one can approximate $(\omega_R^2 - \omega^2)^{\approx}\omega_R^4$ without introducing significant errors in the calculation. Additionally, one can combine the "a'" term with the shot noise term "$h\nu/S_0$". With these modifications, the expression for the output RIN simplifies from (4.31) to:

$$
\begin{aligned}
\text{RIN}_{\text{out}}(\omega) &= \frac{h\nu}{S_0}\left[\frac{\lambda}{h\nu}\frac{\alpha'}{S\,|\Omega|^2} + 1\right] \approx \frac{h\nu}{S_0}\left[\frac{2\gamma_{11}\gamma_{21}\langle F_S F_N^* \rangle + \gamma_{21}^2 \langle F_N F_N^* \rangle - 2\gamma_{11}\omega_R^2}{\tau_p S(\omega_R^4 + \omega^2\gamma^2)} + 1\right] \\
&= \frac{h\nu}{S_0}\left[\frac{(2\gamma_{11}\gamma_{21}\langle F_S F_N^* \rangle + \gamma_{21}^2 \langle F_N F_N^* \rangle - 2\gamma_{11}\omega_R^2 + \tau_p S\omega_R^4)/(\tau_p S\omega_R^4) + \omega^2\tau_{\text{out}}^2}{1 + \omega^2\tau_{\text{out}}^2}\right]
\end{aligned}
\tag{4.32}
$$

with $\tau_{\text{out}} = \gamma/\omega_R^2$. The impact of the injected current on the output RIN spectra can be diminished by applying a current source with sub-shot noise characteristics. To accommodate such cases, one can generalize the first term in the numerator of (4.32) as S_I/qI, where S_I represents the double-sided spectral density of the injection current. Hence, for a perfectly noise-suppressed or "quiet" current source, $K_I \to 0$. In the case of a shot-noise-limited current source, $K_I = qI$.

$$
\text{RIN}_{\text{out}}(\omega) = \frac{h\nu}{S_0}\frac{K_I/qI + \omega^2\tau_{\text{out}}^2}{1 + \omega^2\tau_{\text{out}}^2} \tag{4.33}
$$

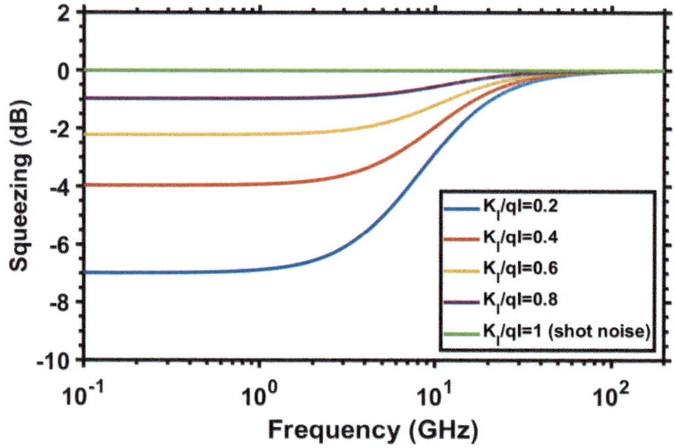

Fig. 4.22 Normalized output RIN varies as a function of pump current noise. Low-frequency noise can be reduced below the SNL if the pump current source is "quiet". The pump current is $4 \times I_{th}$

Figure 4.22 illustrates the previously mentioned amplitude-squeezing effect. When using a shot-noise-limited current source (e.g., $K_I/qI = 1$), the output RIN remains constant at the SNL. However, with a nearly noiseless pumping source (e.g., $K_I/qI = 0.2$), the output RIN can be significantly reduced 7 dB below the SNL at the low frequencies $f \ll 1/(2\pi\tau_{\text{out}})$. Amplitude-squeezed states can thus be qualitatively predicted, with the squeezing level that depends on the pump noise conditions.

As mentioned earlier, Yamamoto [61] demonstrated that the current noise in a semiconductor laser is dominated by the thermal noise of the series resistor, resulting in a double-sided spectral density of $K_I = 2k_BT/R_s$. To reduce the current noise below the SNL of qI, it is necessary for the series resistance to satisfy the condition $R_s \gg 2k_BT/qI$. For instance, with a 1 mA drive current at room temperature, this condition translates into $R_s \gg 50\ \Omega$. Furthermore, for such lasers to be successful, they must exhibit very high optical efficiency ($\eta_0 \to 1$) and minimal internal loss.

Regarding ICLs, the device's performance can be compromised by nonradiative multi-carrier Auger recombination, although it is suppressed compared to bulk materials, and internal losses caused by free carrier absorption. Nevertheless, several studies on ICLs emitting in the λ = 3–4 μm spectral range have reported internal efficiencies of approximately 80–90% [77]. Currently, it remains unclear why the internal efficiency should noticeably deviate from 100%. This is particularly puzzling because it is unlikely that carriers escape from the active quantum wells due to the large conduction and valence band offsets, and direct injection into the electron injector is obstructed by a very thick barrier. Furthermore, considering reasonable energy relaxation times, the impact of carrier heating, which reduces internal efficiency by necessitating a higher carrier density and, consequently, higher current to maintain the same optical gain, should also be relatively minor. Recent studies demonstrate that the causative underlying mechanism limiting efficiency is the resonant intersubband absorption in the valence band [78]. Experimentally, a direct correlation between this resonant absorption and the thickness of the $Ga_{1-x}In_xSb$ hole-quantum well is confirmed. These findings could be reflected in the improvement of the laser's characteristic temperature, threshold current density, slope efficiency, and output power. Another aspect of research is dedicated to estimating internal losses by analyzing the observed slope efficiencies when ICLs are operated in pulsed mode. The findings reveal that the internal loss exhibits a minimum value around λ = 3.3–3.8 μm, reaching as low as 2.9 cm^{-1} [77]. It is suggested that the optimal stage number is 5–7 for emitting wavelengths in the range of 3–4 μm, but for wavelengths exceeding 5 μm, it may be more preferable to employ at least 10 stages [77].

4.3.2 *Stochastic Gillespie Algorithm*

Laser rate equations represent a well-established phenomenological approach for modelling the dynamics of semiconductor lasers. These equations describe the evolution of both carrier density and photon density within the active region, allowing for the determination of static and dynamic characteristics, including properties related to relative intensity noise (RIN), spectral linewidth, and the laser's response to external optical feedback, among others [75]. However, when dealing with squeezed light, a vital issue emerges due to the inherent quantum granularity of photons and carriers. The limitations of continuous rate equations in accurately modelling quantum light have been widely acknowledged in the literature [79, 80]. In the previous Sect. 4.3.1, a semiclassical approach based on Green's function method was employed to conduct initial investigations into the squeezing properties of ICLs. This approach is reminiscent of previous descriptions [32, 68] that focus on assessing amplitude squeezing in terms of the RIN spectrum. This formalism assumes that the fundamental source of shot noise is associated with the particle-like nature of radiation, treating it as classical particles without mutual interaction. To this end, intensity noise fluctuations are assimilated to partition noise, which results from the coupling of the energy with the exterior of the cavity. While this model offers the advantage of providing a straightforward means to describe amplitude-squeezed states of light in semiconductor laser structures compared to more intricate quantum electrodynamics (QED) approaches, it merely introduces an alternative Langevin noise term instead of fully accounting for the granularity of photons and carriers. This limitation implies that it cannot provide insights into photon statistics and intrinsic quantum fluctuations. To address these issues, we have expanded upon a recent stochastic model [81, 82] to investigate the generation of amplitude-squeezed light in the mid-infrared range, particularly with regard to the band structure of ICLs. It's important to note the distinction between intracavity photons and external output photons, with a specific focus on the laser outcoupling process and the sub-Poissonian pumping process, both of which are necessary for the interpretation of sub-shot-noise-limited light emission.

In this part, a theoretical framework has been developed for the evaluation of squeezing performance, without relying on mathematical assumptions often associated with the derivation of a differential description [75]. Instead, the approach leverages the Gillespie stochastic simulation, based on the Monte Carlo algorithm, to predict the statistical dynamics of lasers. This framework intuitively highlights the discrete nature of photon and carrier interactions, as well as the inherent noise associated with many physical processes such as pumping, spontaneous emission, stimulated emission, and photon transmission through the laser facet. This method closely adheres to the principles of the rate equation description regarding the frequency spectrum. Nevertheless, it offers additional predictions and insights into the regime of photon distribution and other statistical features that the conventional method does not specify. As depicted in Fig. 4.23, when ICL is subjected to an external electric pump, holes and electrons migrate into each cascading stage via the tunnelling effect, where they undergo recombination through spontaneous emission, Auger recombination, and stimulated emission owing to population inversion. To replenish the carriers removed by the field, an equal number of additional electrons and holes must be continuously generated and transported to the next stage. The stochastic equations for carrier N and photon S in ICLs can be expressed as follows:

$$\begin{aligned} N_{t+1} &= N_t + R_{pump} - R_{st} - R_{sp} - R_{aug} \\ S_{t+1} &= S_t + mR_{st} - R_p + m\beta_{sp}R_{sp} \end{aligned} \tag{4.34}$$

Here the random variables are defined as follows: R_{pump} represents the pump rate into the active region, R_{st} and R_{sp} denote the rates of laser stimulated emission and spontaneous emission, R_{aug} represents the non-radiative Auger recombination rate, and $R_p = R_{int} + R_{out}$ accounts for the laser photon decay rate, which is further broken down into R_{int} (the photon internal decay rate due to internal losses) and R_{out} (the out-coupling rate through the laser facets). Additionally, m represents the cascade stage number, and β_{sp} stands for the spontaneous emission factor. The stochastic equations (4.34) capture the operational characteristics of ICLs, as each

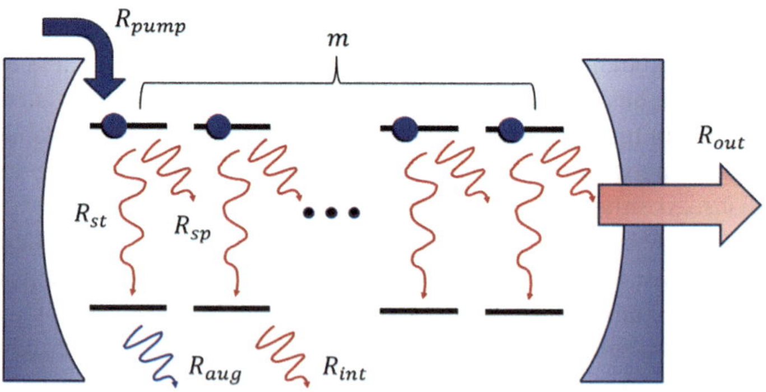

Fig. 4.23 Sketch of the m-cascaded ICL model

Table 4.3 Average rates of stochastic events occurring in the ICLs stochastic equations [83]

Event type	Symbols	Average rate
Stimulated emission	R_{st}	$\Gamma_p v_g g S$
Spontaneous emission	R_{sp}	N/τ_{sp}
Photon internal decay	R_{int}	$v_g \alpha_{in} S$
Auger recombination	R_{aug}	N/τ_{aug}
Pumping	R_{pump}	$\eta I/q$
Out-coupling process	R_{out}	$v_g \alpha_m S$

term corresponds to a random event occurring within the laser cavity, as summarized in Table 4.3. All the values used in the simulations are derived from the recently developed ICLs rate equations model and can be found in Table A.3. The term Γ_p is the optical confinement factor, v_g is the group velocity, g is the optical gain with a_0, N_0, and A is the differential gain, the carrier number at transparency and the area of the active region, respectively, whereas τ_{sp} is the spontaneous emission lifetime, α_{in} is the internal loss, τ_{aug} is the Auger lifetime, η is the injection efficiency, I is the pump current, q is the elementary charge and α_m is the laser facet loss with L, R_1, and R_2 being the laser cavity length and rear or front facet power reflectivity.

This stochastic simulation method is designed to model photons and carriers as discrete particles denoted as N and S, respectively. The numbers of these particles fluctuate due to different events, including particle loss (such as Auger recombination or photon decay), particle exchange (emission or out-coupling processes), and the addition of new particles (pumping). Unlike rate equation methods, we establish specific rules that capture the physical interactions between photons and carriers. For instance, when a pumping event occurs, we introduce a certain number of carriers N into the carrier reservoir. During spontaneous or stimulated emission, carriers N are transferred to intra-cavity photons S_{in}. Subsequently, we anticipate the expected number of distinct events occurring at each discrete time step in (4.34). For a sufficiently small time increment Δt (e.g., typically 1 picosecond to ensure convergence of noise features), the number of ongoing events can be approximated as a sequence of random integers drawn from a Poisson distribution, with the Poisson parameter λ equal to the product of the corresponding average rate calculated in Table 4.3 and the time increment Δt [81]. Our model basically assumes that the occurrence of one event does not affect the probability of other events [84]. The general idea behind our approach involves characterizing photon distribution by incorporating out-coupling as an additional process alongside the injection of sub-Poissonian pumped electrons. In traditional theory, only the internal field is considered the primary "system," while the output field outside the laser cavity is treated as a "heat bath" [75]. However, as we are interested in external photon statistics, we treat it as another "system" [32]. As a consequence, we divide photon decay into two events: out-coupling R_{out} related to laser facet losses α_m and internal decay R_{in} associated with all other cavity decay processes summarized as α_{in}. Regarding the pumping event description, it depends on how the laser is pumped. In the case of shot-noise-limited or normal pumping, we use a classical current source with the rate of $R_{pump} = R_{pump,conv} = \eta I/q$, following a Poisson distribution. In contrast, in the case of noise-suppressed or quiet pumping, the pumping term $R_{pump} = R_{pump,quiet}$ corresponds to the injection of pumped electrons at a constant rate, which is naturally linked to sub-Poissonian electron statistics.

It has been widely acknowledged that an ideal laser, when operating significantly above its threshold, produces coherent states of light [8]. In these coherent states, field fluctuations have equal amplitudes in two quadrature components, but minimize the product described by Heisenberg's uncertainty principle. This theory has found strong experimental support through the measurement of Poissonian photon statistics and shot-noise-limited photocurrent fluctuations. Such measurements are often conducted using an optical balanced homodyne detector [22]. Extensive studies have uncovered that lasers operating far above the threshold do not always generate coherent states. Instead, a range of other quantum states becomes possible, including squeezed states of light [15], which manifest reduced fluctuations in one quadrature component of the coherent state but at the cost of increased fluctuations in the other quadrature component. To investigate why ICLs naturally produce coherent states instead of amplitude-squeezed states, our analysis begins by examining the characteristics of intra-cavity photons in ICLs. It's important to note that photon statistics are typically described using the first two statistical moments, namely, the mean value $\langle S \rangle$ and the variance $\langle \Delta S^2 \rangle = \langle S^2 \rangle - \langle S \rangle^2$. Additionally, there's the second-order photon correlation function $g^{(2)}(0)$, which indicates the likelihood of two photons arriving simultaneously [16]:

$$g^{(2)}(0) = \frac{\langle S(S-1) \rangle}{\langle S \rangle^2} \tag{4.35}$$

As aforementioned, for chaotic (LED) light, $g^{(2)}(0) = 2$, and for coherent (laser) light, $g^{(2)}(0) = 1$ because photon statistics follow a Poissonian distribution. Although the photon number in ICLs fluctuates throughout the simulation, the mean value remains constant. For example, the intra-cavity photon number S_{in} maintains a stable mean value. In Fig. 4.24a, we present the time traces of the intra-cavity photon number S_{in} for three different pump rates within the vicinity of the threshold region. It is quite apparent that fluctuations in intra-cavity photon numbers decrease as ICLs transition from below the threshold to above the threshold. In Fig. 4.24b, we compare stochastic simulations (black markers) with analytical results obtained from rate equations (solid red line) for the mean internal photon number $\langle S_{in} \rangle$. This figure clearly demonstrates a good agreement between the two approaches. There are abrupt jumps in the mean internal photon number $\langle S_{in} \rangle$ as a function of pump rate. This phenomenon is frequently employed to identify micro- and meso-lasers with values of $10^{-6} \leq \beta_{sp} \leq 10^{-2}$ crossing the lasing threshold, and it is particularly noticeable at $\beta_{sp} = 10^{-4}$ for ICLs in this simulation. However, this feature becomes less pronounced in the case of nanolasers with $\beta_{sp} \approx 1$ [85, 86]. As depicted in Fig. 4.24c,

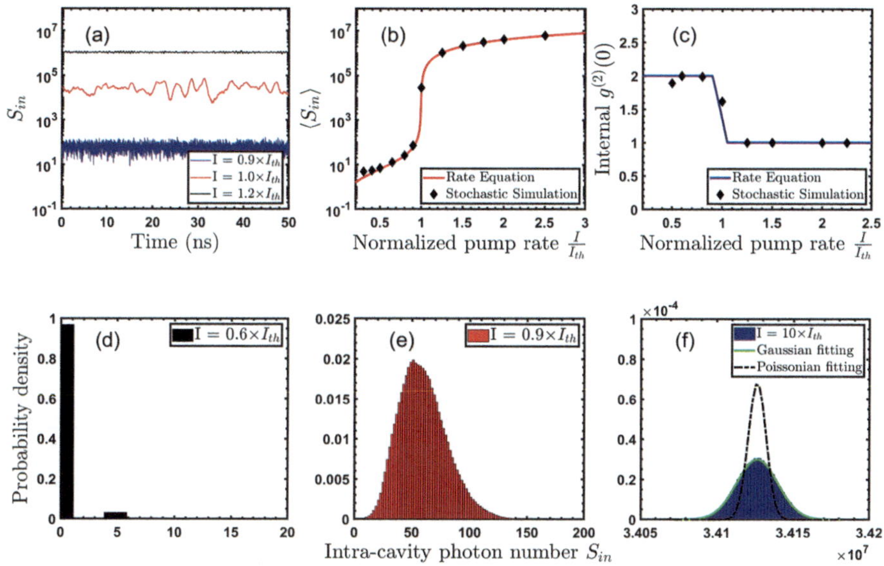

Fig. 4.24 **a** Time series of intra-cavity photon number S_{in} for three different values of pump rate near the threshold region, **b** Stochastic results (diamond markers) and analytical results from rate equations (solid lines) for mean photon number $\langle S_{in} \rangle$ and also **c** Internal photon correlation $g^{(2)}(0)$. Simulations for low pump rates are rejected because of statistical uncertainty. In **d** the intra-cavity photons show an exponential distribution. In **e** photons exhibit an intermediate distribution near the threshold. In **f** the dashed curve indicates a Poisson distribution with parameter given by the mean photon number $\langle S_{in} \rangle$ and the green curve indicates Gaussian distribution with mean and variance given by the mean photon number $\langle S_{in} \rangle$ and photon number variance $\langle \Delta S_{in}^2 \rangle$

the photon correlation function $g^{(2)}(0)$ also demonstrates agreement between the two different methods, with only a minor deviation observed around the ICLs' threshold. Additionally, there is a clear transition from chaotic behaviour or thermal statistics ($g^{(2)}(0) = 2$) below the threshold, through intermediate statistics ($1 < g^{(2)}(0) < 2$) near the threshold, to coherent behaviour with Poissonian statistics ($g^{(2)}(0) = 1$) well above the threshold [86]. To analyze the corresponding photon distribution, we select different pump rates near the threshold and create histograms of the intra-cavity photon number S_{in}. In Fig. 4.24d, we observe a Bose-Einstein or exponential distribution when ICLs operate below the lasing threshold. Conversely, in Fig. 4.24f, at high pump rates, the photon distribution follows a Gaussian distribution that is slightly broader than the Poissonian distribution with the same mean value. This deviation generally aligns with earlier theoretical and experimental findings [87].

The frequency histograms depicted in Fig. 4.25 illustrate the distribution of the filtered output photon numbers S_{out}, obtained with a pump current of $25 \times I_{th}$ for both pump configurations. In experiments, photon statistics are measured using photon counters, which, in the near-infrared range, employ sensitive detector devices such as avalanche photodiodes with user-defined detection filters of specific durations.

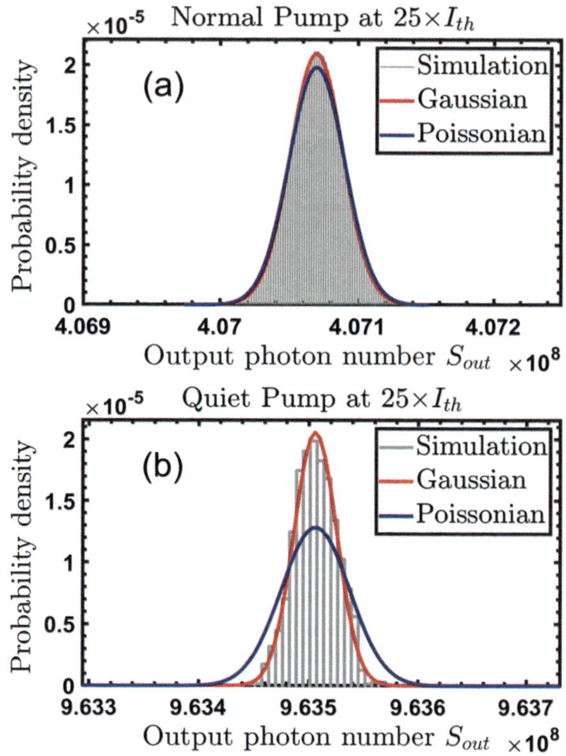

Fig. 4.25 Output photon number S_{out} distribution in 100 ps time slot for ICLs **a** Normal or shot-noise-limited pumping and **b** Quiet or noise-suppressed pumping. Red curves indicate Gaussian fitting and blue curves are Poissonian fittings with the same mean value. The pump current is set as $25 \times I_{th}$

Here, we set $T_d = 100$ ps to achieve a bit rate B of 10 Gb/s, in accordance with the relation $T_d = 1/B$ [82]. Figure 4.25 highlights the potential for non-classical photon statistics in ICLs, where conventional rate equation methods may fall short [88]. In the case of normal pumping, as shown in Fig. 4.25a, the output photon distribution is well-fitted by both the Gaussian and Poissonian probability density functions with the same mean value. In contrast, for the quiet pumping configuration, Fig. 4.25b reveals sub-Poissonian characteristics, featuring a narrower full width at half maximum (FWHM) compared to the Poissonian distribution. The maximum probability density in the Gaussian distribution is nearly twice as large as in the Poissonian distribution. Lastly, the slightly lower output photon number observed for normal pumping is attributed to the reduced electron injection rate, which does not significantly impact the conclusions.

The instability in the power level of semiconductor lasers is quantified using relative intensity noise (RIN) [75] and RIN can be determined by analyzing the spectrum of the photo-detected signal with the assistance of an electrical spectrum analyzer. However, to observe the phenomenon of laser squeezing, it is imperative to employ a balanced detection scheme because it allows for the cancellation of both excess noise and quantum noise associated with the local oscillator after the conversion of the photoelectric signal [22].

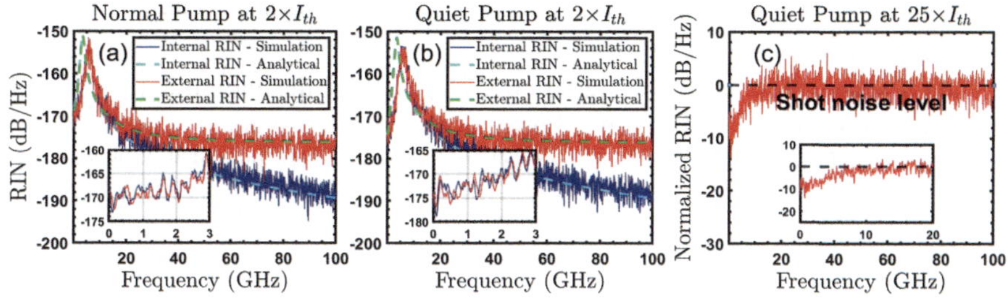

Fig. 4.26 Normal (**a**) and Quiet (**b**) pumping condition of ICLs for both internal (blue curves) and external (red curves) RIN spectra through Fourier transform. The green and cyan dashed lines describe the previous analytical rate equation description [88]. The insets are the detailed view of those two figures within the low-frequency region The pump current is fixed at $2 \times I_{th}$. **c** External RIN spectrum at $25 \times I_{th}$ but with $\alpha_{in} = 1\,\text{cm}^{-1}$ as an ideal condition. It should be noted that the black dashed line represents the shot noise level

In Fig. 4.26a, b, we present the external and internal RIN spectra for both normal and quiet pumping conditions, obtained through Fourier transform analysis. The parameters in Table A.3 are still considered and the pump current is set as $2 \times I_{th}$. Upon comparing these two RIN spectra, it becomes evident that the RIN of the outcoupled photon remains nearly the same as the intra-cavity RIN at low frequencies, well before the emergence of the relaxation oscillation peak, as recently observed in a previous study [89]. However, at higher frequencies, the external RIN approaches the standard quantum limit. In contrast, the internal RIN continues to decrease thanks to a filtering effect, which is defined as the response function in classical laser dynamics [75]. Furthermore, considering the inset of Fig. 4.26a, b, one can observe that the noiseless pumping condition is expected to reduce the intensity noise by approximately 5 dB when compared to the case with noise in the pump current source. This outcome implies the potential generation of amplitude-squeezed stats, but only when ICLs are pumped far from their threshold with a noise-free current source. Importantly, stochastic simulations align perfectly with the previous analytical results in the frequency spectra regime [88]. Figure 4.26c highlights the potential for sub-shot-noise-limited external RIN with the suppressed-pump-noise configuration. As mentioned earlier, using a high pump current ($25 \times I_{th}$) without pump fluctuations is desirable for generating amplitude-squeezed states. This is because ICLs' differential resistance becomes much smaller under strong forward-bias conditions [90], making the generation of amplitude-squeezed light more convenient. The external RIN in this scenario shows a strong squeezing level of a couple of dB and a broad squeezing bandwidth of nearly 10 GHz. It's worth noting that ICLs should have low internal losses for optimal squeezing performance. Recent n-type doping strategies can operate at ambient room temperature with internal losses (α_{in}) as low as $4\,\text{cm}^{-1}$ [77], compared to the typical value of $12\,\text{cm}^{-1}$ at the inception of ICL design. In this figure, we assume that the internal loss α_{in} approaches $1\,\text{cm}^{-1}$ to maximize the squeezing performance. Another noteworthy point is that the external RIN at the high pump rate exhibits less frequency dependence compared to the situation with low pump rates. This feature aligns with previous theoretical predictions [32] and experimental observations [46].

The Q-parameter is a valuable tool used to characterize the photon statistical properties by quantifying the divergence between coherent Poissonian states and other potential non-classical quantum states [43]. It is particularly useful for characterizing amplitude-squeezed states, where negative values of the Q-parameter indicate sub-Poissonian photon statistics. The Q-parameter is typically defined in (4.9):

$$Q = \frac{\left\langle \Delta S^2 \right\rangle - \langle S \rangle}{\langle S \rangle}$$

Additionally, another frequently employed quantum factor, the Fano factor, can be theoretically defined as the ratio between the variance and the mean of the output photons [10]. Experimentally, it is determined as the ratio between the spectral noise power of the laser field and the shot-noise level at the same direct current photocurrent of the detector. Consequently, when the Fano factor is smaller than unity, it indicates that the laser field demonstrates amplitude-squeezed quantum properties.

$$F = \frac{\left\langle \Delta S^2 \right\rangle}{\langle S \rangle} = 1 + Q$$

In macroscopic ICL systems with a large mean photon number, even a slight reduction of the external $g^{(2)}(0)$ below unity can result in highly sub-Poissonian statistics. This effect is evident through the expression derived from the second-order factorial moment:

$$g^{(2)}(0) = \frac{\langle S(S-1) \rangle}{\langle S \rangle^2} = 1 + \frac{\left\langle \Delta S^2 \right\rangle - \langle S \rangle}{\langle S \rangle^2} = 1 + \frac{Q}{\langle S \rangle}$$

This effect has been reported in a recent article [91], demonstrating that an increase in the mean photon number $\langle S \rangle$ can diminish the range and magnitude of anti-bunching (e.g., $g^{(2)}(0) < 1$). Nevertheless, it is still feasible to maintain a significant level of amplitude squeezing.

Figure 4.27 illustrates the importance of internal loss (α_{in}) in the amplitude squeezing performance. Using two distinct quantum criteria, namely the external photon correlation $g^{(2)}(0)$ and the Q-parameter, our simulations emphasize that the amplitude squeezing level is limited by the internal loss of the semiconductor materials. This limitation can be explained by considering that the output emitted photons face stiff competition from large internal loss, making the out-coupling process less likely. Therefore, the sub-Poissonian performance of the laser is diluted. As shown in Fig. 4.27a, b, ICLs display clear photon anti-bunching characteristics or amplitude-squeezed light when the internal loss is negligible ($g^{(2)}(0) < 1$ and $Q < 0$). On the contrary, when the internal loss is large enough, the squeezing properties can be significantly altered, with both $g^{(2)}(0)$ and the Q-parameter increasing and even exceeding values of 0, respectively. This trend is likely to be reflected in the Fano

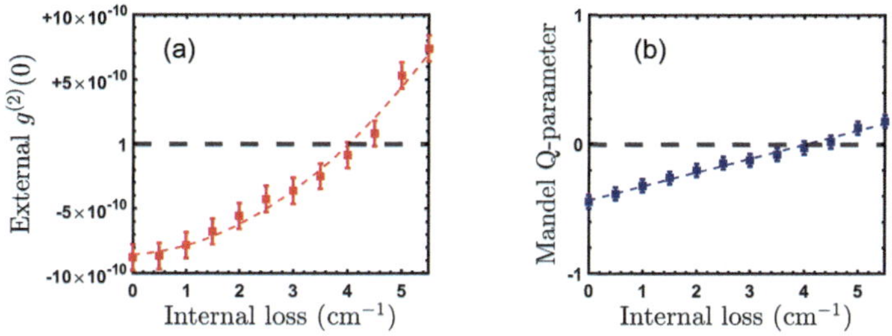

Fig. 4.27 **a** External $g^{(2)}(0)$ and **b** The Q-parameter of the external output photons as a function of internal loss coefficient α_{in}. The measurement duration is still 100 ps and other simulation parameters remain unchanged as Fig. 4.26c. In **a**, an extremely slight deviation is found. Hence, for the sake of visibility, ticks of the longitudinal axis represent effectively this deviation from 1, for instance, $\pm 5 \times 10^{-10}$ indicates $1 \pm 5 \times 10^{-10}$ in this case. In **b**, a value of -1 manifests the maximum squeezing level. The black dashed line in both figures represents the shot noise level, the red (blue) dashed lines are guides for the eye, and error bars are displayed as a result of stochastic uncertainty

factor as well. Hence, minimizing internal loss as much as possible is the first goal in future ICL quantum engineering designs.

The numerical results presented in this section provide compelling evidence that ICLs have the potential to achieve substantial intensity noise squeezing across a bandwidth of several gigahertz. These characteristics collectively establish a solid basis for realizing ultra-low noise laser sources at mid-infrared wavelengths. Furthermore, this result also paves the way for the implementation of mid-infrared continuous-variable quantum key distribution in free space, whose feasibility is further supported by the ongoing theoretical investigations [92], as well as the availability of mid-infrared homodyne detection techniques that utilize two commercial HgCdTe detectors, covering a spectral range from 2.5 μm to 5.0 μm [93]. In previous studies, research has been carried out using QCLs [94], where Alice transmits polarization-coded pseudo-single photons generated by a QCL at a wavelength of 4.6 μm to Bob. Bob employs a nonlinear crystal and a silicon avalanche photodiode for the detection process, which is achieved through frequency generation. Theoretical predictions, built on the foundation of a proof-of-principle experiment, demonstrate that, under specific foggy atmospheric conditions, the proposed system remains largely unaffected. In contrast, conventional systems operating at standard near-infrared wavelengths would be rendered non-functional under such circumstances. A recent article also presented a comprehensive characterization of 3.8 μm ring-ICLs operating at room temperature [95]. These lasers operate in close proximity to the standard quantum limit.

4.4 Amplitude Squeezing in Quantum Dot Lasers

In this section, I will theoretically and experimentally investigate the amplitude squeezing in electrically driven quantum dot (QD) lasers. Since the early 1980s, following the proposal of semiconductor quantum dots and their utilization in lasers to enhance the performance of semiconductor laser diodes [96], numerous research groups worldwide have made significant strides in both the development of fabrication techniques for quantum nanostructures [97] and the establishment of consistent theoretical models to elucidate their underlying physical properties [98, 99]. A semiconductor quantum dot typically comprises a small region of semiconductor material, approximately 10 nm in size, surrounded by another semiconductor with a larger bandgap. The surrounding material acts as a confining potential barrier, allowing electrons in the conduction band and holes in the valence band to be confined within the quantum dot [100]. This three-dimensional confinement leads to the electrons and holes within the quantum dot exhibiting a density of states (DoS) similar to a delta function [96]. When coupled with the fermionic nature of electrons, this density of states enables the spectral isolation of individual optical transitions involving single electrons. Thanks to their discrete energy levels, analogous to atomic energy levels, semiconductor quantum dots are often referred to as "artificial atoms." The exact energy of the emitted photons depends on several factors, including the bandgaps of the chosen material system, the size of the quantum dot, and induced strain effects [101].

In recent years, there has been a sudden surge in research focu sed on QD nanostructures [102]. Particular attention has been given to self-assembled QDs that are embedded into microcavities, which enables the generation of single photons with a high degree of purity and indistinguishability [101]. As a result, these sources play a crucial role in quantum computing [103] as well as discrete variables (DV) quantum key distribution (QKD) [104]. In sharp contrast to DV QKD, which relies on single-photon sources and detectors, continuous variable (CV) QKD employs lasers and balanced detection to continuously extract the light's quadrature components. This approach offers the advantage of using readily available equipment and seamless integration into existing optical telecommunications networks. Among the various CV QKD protocols, GG02 [105] is well-regarded for its security, largely due to the no-cloning theorem associated with coherent states [8]. However, recent research has delved into the use of squeezed states to enhance security and robustness even further [106]. This new approach strives to eliminate information leakage to potential eavesdroppers in a pure-loss channel and minimize it in a symmetric noisy channel. In this protocol, information is encoded exclusively using Gaussian modulation of amplitude-squeezed states [107]. These states display reduced fluctuations in photon number $\langle \Delta \hat{n}^2 \rangle < \langle \hat{n} \rangle$ compared to coherent states. However, they are subject to enhanced phase fluctuations, a consequence of the minimum uncertainty principle.

As aforementioned, squeezed states of light have been frequently generated using $\chi^{(2)}$ or $\chi^{(3)}$ nonlinear interactions through parametric down-conversion and four-wave mixing [17, 18]. A variety of nonlinear materials have been employed for these processes, including $LiNbO_3$ (PPLN) [35], $KTiOPO_4$ (PPKTP) [36], silicon [37], atomic vapour [108], disk resonators [109], and Si_3N_4 [38]. Recent advancements have also led to the development of more compact single-chip designs, moving away from traditional benchtop instruments [39, 110–113]. In contrast to these approaches, Yamamoto and colleagues [21, 32] initially proposed an alternative method to directly produce amplitude-squeezed states using off-the-shelf semiconductor lasers with a "quiet" pump, i.e., a constant-current source. A notable feature of semiconductor lasers is their ability to be pumped by injection current via an electrical circuit. Unlike optical pumping, electrical pumping is not inherently a Poisson point process due to the Coulomb interaction [1]. Additionally, the intensity noise of the laser output field, rather than the internal field, may be reduced below the shot-noise limit at high pump currents. The shot noise limit arises from two sources: pump fluctuations in the low-frequency region and reflected vacuum-field fluctuations in the high-frequency region [32]. This method could leverage the mature fabrication processes in the semiconductor industry, significantly enhancing its feasibility. It relies on the improved performance of recently developed semiconductor lasers, known for their compact size, ultra-low intensity noise, and narrow frequency linewidth. Subsequent experiments involving different types of laser diodes have generated significant interest in this field, such as commercial quantum well (QW) lasers [50, 52], vertical-cavity surface-emitting lasers [60], and semiconductor microcavity lasers [114]. However, the observed bandwidth in these experiments has remained relatively limited until now. The widest bandwidth achieved was 1.1 GHz with a QW transverse junction stripe laser operating at a cryogenic temperature of 77 K [46]. This limitation has imposed strict constraints on practical room-temperature implementations and impeded the realization of high-speed quantum communications. Although a previous study predicted the theoretical potential of achieving this phenomenon in interband cascade lasers [83], no experimental demonstration of producing broadband amplitude-squeezed states had been presented.

4.4.1 Experimental Evidence

Figure 4.28 provides an overview of the experimental setup, with individual components and details presented separately.

To eliminate pump current fluctuations, a high-impedance configuration is supposed to be implemented, ensuring uniform regulation of the pumping electrons [21]. To precisely characterize the current noise characteristics of both the quiet current source (ILX Lightwave LDX-3620) and the normal current source (Keithley 2400), one needs to conduct direct measurements, as depicted in Fig. 4.29.

In this setup, driven by these two current sources, the current noise can be evaluated by measuring voltage fluctuations across a 10 Ω resistor, passing them through a 100 nF capacitor, and then amplifying the signal using a low-noise amplifier with an input impedance of 50 Ω. Subsequently, the amplified signal was analyzed using an electrical spectrum analyzer. Figure 4.30 illustrates that the current noise power

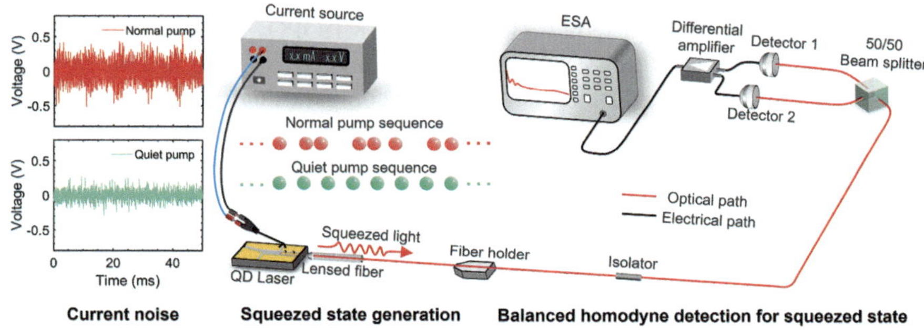

Fig. 4.28 Diagram of the key components of the apparatus used for generating amplitude-squeezed states through QD lasers. Left panel: Comparison of noise characteristics between two different current sources, namely Keithley 2400 (normal pump, red) and ILX Lightwave LDX-3620 (quiet pump, green). Right panel: Experimental setup. The quiet pump injects electrons at a constant rate. ESA is the electrical spectrum analyzer

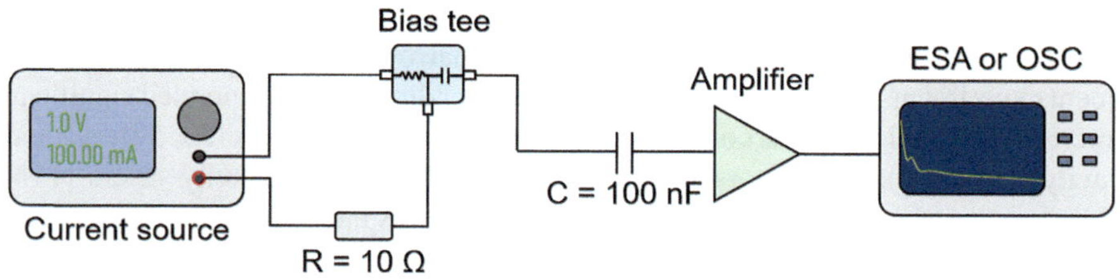

Fig. 4.29 Scheme of the experimental setup for the characterization of the intensity noise of the current noise

generated by the quiet current sources is significantly lower, approximately 20–30 dB lower, compared to that produced by the normal current sources at 50 MHz. At 6 GHz, while the difference is less pronounced, there is still a reduction of approximately 1 dB.

A distributed-feedback (DFB) single-mode QD laser emitting at a wavelength of 1.31 μm is used in this study. The laser has a highly reflective (HR) coating (95%) on the rear facet and an antireflection (AR) coating (3%) on the front facet. To maintain a constant operating temperature of 20 °C±0.005 °C, a thermoelectric temperature controller (ILX Lightwave LDT-5748) is used. The laser's threshold current I_{th} was measured to be approximately 9 mA. The optical losses degrade the regularity of the photon flux and decrease the squeezing level [46]. We measure a detection efficiency of 55% and a collection efficiency of 40% because of the losses during laser-fibre coupling. Consequently, the overall efficiency is estimated at 22%. However, this does not affect the internal quantum efficiency, which remains close to unity and pertains to the conversion from injected electrons to emitted photons. The laser exhibits stable single-longitudinal-mode operation over a wide range of pump currents, with a side mode suppression ratio (SMSR) of more than 50 dB at high pump currents to avoid side-mode competition interference [48].

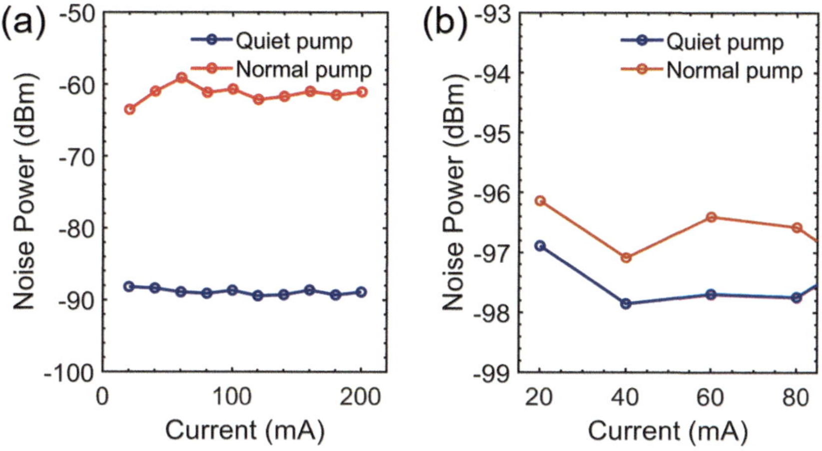

Fig. 4.30 The current noise power of different pumps taken at **a** 50 MHz and **b** 6 GHz

The epitaxial structure of the QD lasers closely follows designs previously reported in the literature [115]. The laser device has a mesa width of 2 μm and a cavity length of 750 μm. The active region of the device is formed by growing 8 layers of InAs QDs on a (100) GaAs substrate using molecular beam epitaxy. The QD density is approximately 5.9×10^{10} cm^{-2} per layer, and each dot layer is separated by a p-doped GaAs spacer, enhancing modal gain. The active region exhibits an exceptionally narrow luminescence width at half maximum, indicating minimal inhomogeneous broadening. This narrow luminescence width is because of the consistent size and shape distribution of the QDs. The uniformity of the QDs offers several advantages for laser performance, including enhanced thermal stability, reduced linewidth enhancement factor, increased insensitivity to back reflections, and the ability to achieve low-threshold, high-efficiency single-mode emission. As shown in Fig. 4.31, the QD DFB laser exhibits a side-mode suppression ratio exceeding 50 dB at 40 mA.

Amplitude-squeezed states demonstrate sub-Poissonian photon statistics, photon antibunching, and squeezing owing to the reduced uncertainty in photon number [116]. Therefore, we employ a standard balanced homodyne detection for the squeezing measurement [22]. In this measurement process, a DFB QD laser is coupled into a fibre holder and precisely driven by a low-noise current source (ILX Lightwave LDX-3620). The current leaking around the gain medium, rather than flowing into it, does not disrupt the regular flow of electrons from the pump current [47]. The laser's differential resistance is approximately 20 Ω when the injection current is set at 40 mA. The inclusion of a 30 dB optical isolator is crucial because even a small amount of back-reflected light can introduce excessive intensity noise. After passing through the optical isolator, the output beam is split into two separate paths using a tunable beamsplitter to ensure 50/50 optical balancing. These two paths are then detected by two identical photodiodes (Discovery Semiconductors DSC-R405ER

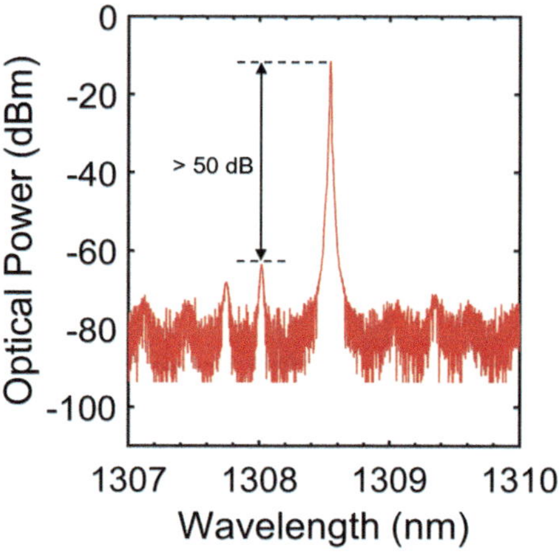

Fig. 4.31 The optical spectrum of a DFB QD laser at 40 mA and 20 °C

20 GHz Linear Balanced Photoreceivers). Care is taken to avoid detector saturation at high optical power levels. The difference between the photocurrent fluctuations is amplified by a homemade low-noise electronic amplifier and subsequently analyzed using an electrical spectrum analyzer (ESA) (Rohde & Schwarz, 43 GHz) to obtain the radio frequency (RF) noise power spectrum. Balanced detection substantially suppresses excessive intensity noise compared to single photodiode detection, with the common-mode rejection ratio (CMRR) consistently exceeding 30 dB across the entire frequency spectrum up to 18 GHz.

Accurate calibration of the shot noise level, often referred to as the standard quantum limit (SQL), is of paramount importance in this experiment. The SQL serves as a normalization parameter crucial for quantifying squeezing. Here, we not only calibrate the SQL using either a filtered incandescent lamp or a semiconductor optical amplifier operating at the same wavelength [46], but also adhere to the original setup while replacing the low-noise current source with the normal pump source (Keithley 2400) [82]. In both measurements, the RF frequency spectra of the subtracted photocurrent remain consistent, regardless of alterations in the pump current. This coherence unequivocally demonstrates the high correspondence between these two calibration methods, thereby providing an authentic representation of the SQL.

Figure 4.32 presents the measured RF noise power spectral density for the QD laser biased at 40 mA, illustrating amplitude squeezing (green curve) in comparison to the calibrated SQL (red dashed curve). As seen, the RF spectrum in the frequency range of 3 GHz to 12 GHz exhibits a significant reduction below the SQL. The most substantial noise reduction occurs around 8 GHz, with a power reduction of 0.9 ± 0.1 dB (6.7%) below the SQL. It's important to note that the measurement frequency range discussed here is significantly higher than the cutoff frequency of Flicker (1/f) noise. Additionally, at such a high pump rate, the relaxation oscillation resonance

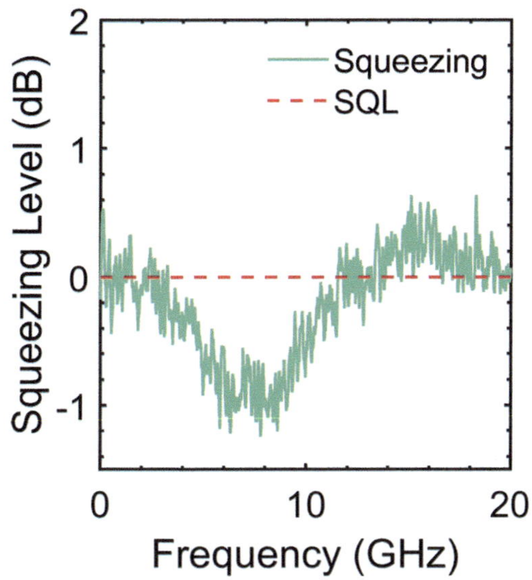

Fig. 4.32 The measured RF spectra at 40 mA are represented by the green line (amplitude squeezing). The spectrum analyzer was set to a resolution bandwidth (RBW) of 200 kHz and a video bandwidth (VBW) of 500 Hz. All traces have been corrected for thermal background noise subtraction

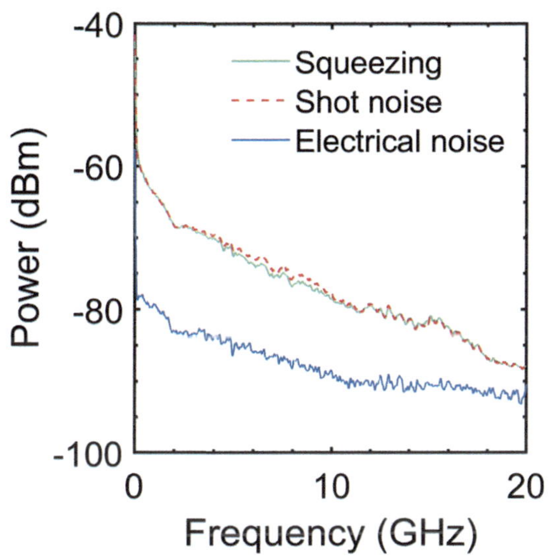

Fig. 4.33 The original RF spectrum of electrical noise, shot noise level, and squeezing light. The shot-noise clearance (SNC) of our homodyne detection is equal to 20 dB, which is the difference between calibrated shot noise and electrical noise

is effectively suppressed. The original RF spectrum of the shot noise level, anti-squeezing, and squeezing are displayed in Fig. 4.33. Then, we redirect our attention to photon statistics to provide further evidence of the amplitude squeezing.

The normalized second-order correlation function, denoted as $g^{(2)}(\tau)$, is used to assess whether a light source exhibits antibunching, bunching, or behaves as a coherent source. For the photon-number squeezed state, if the normalized second-order correlation function at zero delay ($g^{(2)}(0)$) is less than 1, it would provide

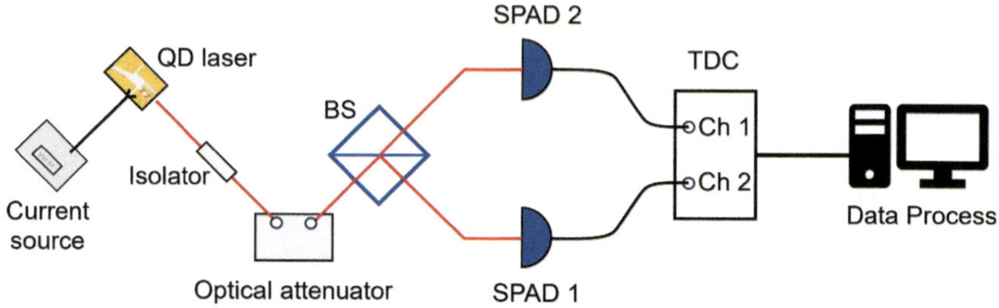

Fig. 4.34 The HBT setup for $g^{(2)}(0)$ with QD laser under different current sources

additional confirmation of the squeezed state light source. Figure 4.34 shows the Hanbury Brown and Twiss (HBT) setup for measuring $g^{(2)}(0)$.

In this setup, the QD laser is driven by two different types of current sources. The laser light then passes through two optical attenuators, with a maximum attenuation of 110 dB, to ensure that the optical power entering the single-photon detectors remains within a safe range. Subsequently, an exact 50/50 optical beamsplitter is used to equally divide the laser light into two paths, ensuring optical power balance between the two single-photon avalanche detectors (SPAD, ID Quantique ID200). The photocurrent signals generated by these detectors are then routed into a time-to-digital converter (TDC, ID Quantique ID800) to determine the coherence between the signals. During the measurement process, the coincidence time interval δ is set to be 50 ns, and the total testing time T is 2 s. Over this duration, the count rates R_1 and R_2 for the two channels are continuously recorded. Furthermore, the coincidence events n between the two channels are also collected. Using this dataset, the normalized second-order correlation function $g^{(2)}(0)$ can be computed using the following equation and analysis process [117]:

$$g^{(2)}(0) = \frac{n}{\delta T R_1 R_2} \tag{4.36}$$

4.4.2 Theoretical Investigation

In the literature, many methods have been historically used to analyze quantum fluctuations and quantum-statistical properties in semiconductor lasers. Here, I will briefly describe three of these methods, along with a more recent approach:

1 Langevin Forces in Rate Equations: This approach incorporates Langevin forces into semiclassical rate equations, supplemented with vacuum fluctuation terms. It partially accounts for quantum-mechanical corrections to the diffusion coefficients and provides a simple description of shot noise [68, 88, 118]. However, it can be considered a mean-field theory from a statistical perspective. While it doesn't explicitly treat fluctuations in a detailed manner, it is designed to be

consistent with the full quantum theory under specific conditions known as the "thermodynamic limit" [119].

2 Quantum-Mechanical Langevin Operator Equation: This approach is based on the quantum-mechanical Langevin operator equation or Fokker-Planck equations [32]. It closely follows the Heisenberg picture and provides power spectra through Fourier transformation. However, it may have challenges in obtaining photon-number statistics.
3 Density Matrix Master Equation: This approach involves the density matrix master equation [120, 121], essentially resembles the Schrödinger picture. This approach enables the comprehensive gathering of information on quantum state statistics, but it does entail higher computational complexity when compared to other approaches.
4 Stochastic Simulation: More recently, an efficient stochastic simulation method has been introduced, particularly for nanolasers and interband cascade lasers [82, 83]. This approach translates deterministic ordinary differential equations into a probabilistic framework, implicitly capturing quantum fluctuations during the simulation. It faithfully represents the stochastic and discrete nature of physical processes, offering a valuable tool for studying semiconductor lasers realistically.

Each of these approaches has its advantages and limitations, making them suitable for different aspects of semiconductor laser analysis, depending on the level of detail and computational resources available. Therefore, to facilitate a rigorous quantitative comparison between the experimental findings and theoretical predictions, we have meticulously refined a standard multi-level model for QD lasers [99, 122, 123]. This model encompasses three key energy states: a reservoir state (RS), a four-fold degenerate excited state (ES) with $\sigma_{ES} = 4$, and a two-fold degenerate ground state (GS) with $\sigma_{GS} = 2$. In this model, the QDs are assumed to be consistently neutral, and the system is composed of excitonic energy states, treating electrons and holes as electron-hole pairs. According to the QD laser parameters, we do not consider dot size dispersion, which leads to the inhomogeneous broadening of the gain profile [124]. To further simplify the QD laser model, the dynamics of the microscopic interband polarization can be eliminated adiabatically. This elimination is justified because the polarization dephases relatively quickly within about 100 fs at room temperature. Given that charge-carrier scattering times and the photon lifetime are typically a few picoseconds, this de-coherence process allows one to disregard the extra quantum-mechanical correlations.

Figure 4.35 offers a detailed depiction of the diverse processes occurring within the QD laser. Under forward bias conditions, carriers injected from an external source directly populate the two-dimensional RS (N_{RS}). Some of these carriers are subsequently captured into the ES (N_{ES}) through intraband carrier-carrier and carrier-phonon collisions within a characteristic capture time τ_{ES}^{RS}, while others undergo spontaneous recombination with a characteristic spontaneous emission time τ_{spon}^{RS}. Once inside the ES, carriers can relax to the GS (N_{GS}) within a relaxation time τ_{GS}^{ES} or recombine spontaneously. Furthermore, carriers can thermally reemit from the ES

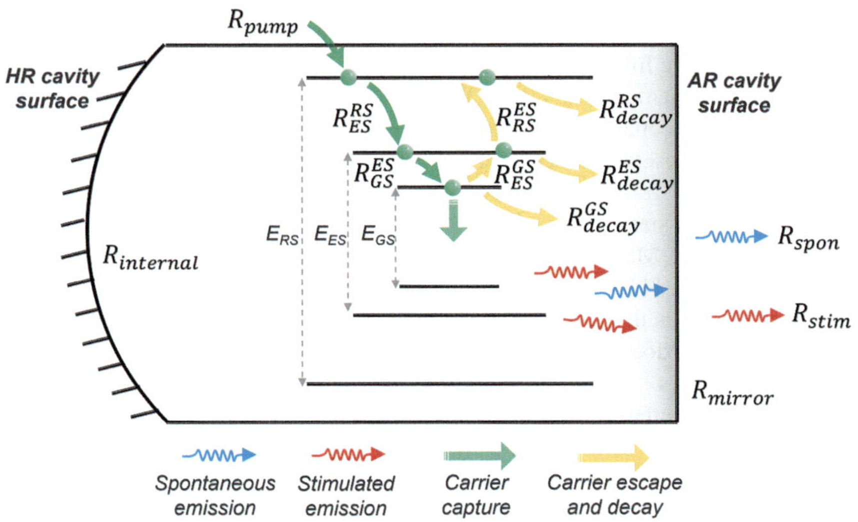

Fig. 4.35 Schematic diagram of carrier transport process in a QD laser. See Table A.5 for definitions and values

back to the RS and from the GS back to the ES. These processes are governed by the Fermi distribution, adhering to the detailed balance principle, and occur in the absence of external excitation [125]. The relevant timescales for these processes τ_{RS}^{ES} and τ_{ES}^{GS}, are expressed as follows:

$$\tau_{RS}^{ES} = \tau_{ES}^{RS}\frac{\sigma_{GS}}{\sigma_{ES}}e^{(E_{ES}-E_{GS})/(k_B T)}$$

$$\tau_{ES}^{GS} = \tau_{GS}^{ES}\frac{\sigma_{ES}N_B}{D_{RS}}e^{(E_{RS}-E_{ES})/(k_B T)}$$

Here, $D_{RS} = \frac{m_{RS}^*}{\pi\hbar^2}k_B T$ represents the effective density of states in the RS, with m_{RS}^* denoting the effective mass of an electron in the RS. A fundamental aspect of a QD laser system pumped with a noise-suppressed current source is its inherent capability to directly convert sub-Poissonian statistics of electrons into nonclassical photon statistics. We employ a set of coupled stochastic differential equations. These equations are adeptly tailored to the three-energy-level QD model [99]:

$$
\begin{aligned}
\frac{dN^{RS}}{dt} &= R_{\text{pump}} + R_{ES\to RS} - R_{RS\to ES} - R^{RS}_{\text{decay}} \\
\frac{dN^{ES}}{dt} &= R_{RS\to ES} + R_{GS\to ES} - R_{ES\to RS} - R_{ES\to GS} - R^{ES}_{\text{decay}} \\
\frac{dN^{GS}}{dt} &= R_{ES\to GS} - R_{GS\to ES} - R_{\text{stim}} - R^{GS}_{\text{decay}} \\
\frac{dS^{GS}}{dt} &= R_{stim} - R_{\text{internal}} - R_{\text{mirror}} + R_{\text{spon}}
\end{aligned}
\tag{4.37}
$$

where N^i (i = RS, ES, GS) represents the carrier density in the RS, ES, and GS, respectively. S^{GS} denotes the photon density in the GS, R^i_{decay} (i = RS, ES, GS) represents the spontaneous emission rate that reduces the carrier density in each energy state. $R_{ES\to RS}$ and $R_{RS\to ES}$ describe the carrier scattering rates between RS and ES, while $R_{ES\to GS}$ and $R_{GS\to ES}$ depict the carrier scattering rates between ES and GS. R_{stim} accounts for the stimulated emission rate solely on GS, and R_{spon} represents the fraction of the spontaneous emission coupled into the lasing mode. Furthermore, the sum of R_{internal} and R_{mirror} comprises the photon decay rate, with R_{mirror} being the outcoupling rate via facet mirrors into the output channel and R_{internal} containing other internal photon losses. More detailed device information can be found in Table A.5 in Appendix A.

The stochastic simulation emulates a series of physical events occurring randomly at discrete time intervals, representing a sequence of potential occurrences [80]. To ensure an accurate representation of all processes, we select a time increment Δt that is significantly smaller than the fastest timescale involved, typically set to 0.1 ps in our specific case. At each time increment, the number of ongoing events is anticipated through a series of non-negative integer random variables drawn from the Poisson distribution ($\mathbb{P}$). The first arguments $\mathscr{P}$ of the Poisson distribution are directly proportional to the corresponding terms in the previously described QD model, as demonstrated in Table 4.4. Among these rates, the term R_{pump} represents the pumping rate induced by the external pump current. The precise definition of this term can vary depending on the method used for laser pumping. In conventional pumping, a current source with classically Poisson-distributed shot noise is employed, resulting in a variable rate. Conversely, in the case of quiet pumping, the pump is defined by a periodic injection of electrons at a constant rate, leading to a constant value. This results in sub-Poissonian electron statistics. Additionally, the mirror loss is calculated using the formula $\alpha_m = \frac{1}{2L}\ln\left(\frac{1}{R_1R_2}\right)$, where the facet power reflectivities are given as $R_1 = 0.03$ (AR) and $R_2 = 0.95$ (HR), resulting in a mirror loss value of α_m equal to $23.72\,\text{cm}^{-1}$. However, it's crucial to note that the internal loss α_{in} is about $0.5\,\text{cm}^{-1}$, a deliberate choice made to maximize the display of amplitude squeezing.

The stimulated RF spectrum is presented in Fig. 4.36a. In this approach, we independently derive the SQL and amplitude squeezing for both normal pumping and quiet pumping conditions by applying Fourier transforms to the time traces. While a large squeezing level can be theoretically expected at low frequencies, it is often overshadowed by technical noise in practical settings [46]. Nevertheless, both in

Table 4.4 Synopsis of the underlying physical processes considered in the stochastic simulation. Each event (column 3) is determined based on the Poissonian probability distribution (column 4), with the first arguments $\mathscr{P}$ specified in column 5

Process	Physical event	Symbol	Expression	$\mathscr{P}$
Excitation	Normal pumping	R_{pump}	$\mathbb{P}(\mathscr{P}_{\text{pump}}\Delta t)$	$\eta I/q$
	Quiet pumping	R_{pump}	$\mathscr{P}_{\text{pump}}\Delta t^1$	$\eta I/q$
Disexcitation	RS decay	R^{RS}_{decay}	$\mathbb{P}(\mathscr{P}^{RS}_{\text{decay}}\Delta t)$	$N_{RS}/\tau^{\text{spon}}_{RS}$
	ES decay	R^{ES}_{decay}	$\mathbb{P}(\mathscr{P}^{ES}_{\text{decay}}\Delta t)$	$N_{ES}/\tau^{\text{spon}}_{ES}$
	GS decay	R^{GS}_{decay}	$\mathbb{P}(\mathscr{P}^{GS}_{\text{decay}}\Delta t)$	$N_{GS}/\tau^{\text{spon}}_{GS}$
Relaxation	RS→ES	$R_{RS\rightarrow ES}$	$\mathbb{P}(\mathscr{P}_{RS\rightarrow ES}\Delta t)$	$N_{RS}(1-\rho_{ES})/\tau^{RS2}_{ES}$
	ES→RS	$R_{ES\rightarrow RS}$	$\mathbb{P}(\mathscr{P}_{ES\rightarrow RS}\Delta t)$	N_{ES}/τ^{ES}_{RS}
	ES→GS	$R_{ES\rightarrow GS}$	$\mathbb{P}(\mathscr{P}_{ES\rightarrow GS}\Delta t)$	$N_{ES}(1-\rho_{GS})/\tau^{ES3}_{GS}$
	GS→ES	$R_{GS\rightarrow ES}$	$\mathbb{P}(\mathscr{P}_{GS\rightarrow ES}\Delta t)$	$N_{GS}(1-\rho_{ES})/\tau^{GS}_{ES}$
Emission	Spontaneous	R_{spon}	$\mathbb{P}(\mathscr{P}_{\text{spon}}\Delta t)$	$\beta_{sp}N_{GS}/\tau^{\text{spon}}_{GS}$
	Stimulated	R_{stim}	$\mathbb{P}(\mathscr{P}_{\text{stim}}\Delta t)$	$\Gamma_p v_g g_{GS} S_{GS}{}^4$
Photon Losses	Internal loss	R_{internal}	$\mathbb{P}(\mathscr{P}_{\text{internal}}\Delta t)$	$v_g\alpha_{in}S_{GS}$
Out-Coupling	Mirror loss	R_{mirror}	$\mathbb{P}(\mathscr{P}_{\text{mirror}}\Delta t)$	$v_g\alpha_m S_{GS}$

[1] This term is a constant value instead of a random variable
[2] $\rho_{ES} = N_{ES}/4N_B$, the carrier occupation probabilities in the ES
[3] $\rho_{GS} = N_{GS}/2N_B$, the carrier occupation probabilities in the GS
[4] $g_{GS} = \frac{a_{GS}}{1+\xi S_{GS}}\frac{N_B}{V_B}(2\rho_{GS}-1)$, the GS material gain and $v_g = \frac{c}{n_r}$, the group velocity

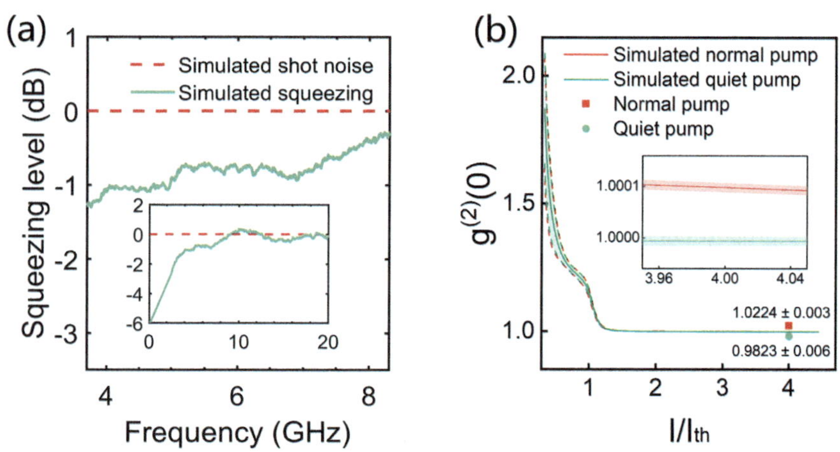

Fig. 4.36 **a** The stimulated RF spectra are shown in green, while the stimulated SQL is represented in the red dashed line. **b** The stimulated normalized second-order correlation function at zero delay $g^{(2)}(0)$ as a function of pump current. The green and grep areas show the standard deviation from the average obtained from 100 simulation runs. Two experimental data points are also presented

our experiments and simulations, we observe a discernible signature of amplitude squeezing at relatively high frequencies. Furthermore, we computed the normalized second-order correlation at zero delay $g^{(2)}(0)$, as illustrated in Fig. 4.36b, using the simulation data. Following the relationship $\langle \Delta S^2 \rangle = [g^{(2)}(0) - 1] \langle S \rangle^2 + \langle S \rangle$ [82], with S representing the external photon number, we observe that stimulated $g^{(2)}(0)$ rapidly converges to approximately 1 shortly after surpassing the laser threshold I_{th}. However, under normal pump conditions, $g^{(2)}(0)$ slightly exceeds 1 at $4 \times I_{th}$, whereas with quiet pumping, $g^{(2)}(0)$ dips just below unity at $4 \times I_{th}$ due to the relatively high average photon number. Moreover, we conducted corresponding $g^{(2)}(0)$ measurements at 40 mA, and the experimental results are represented by two symbols within the same figure, corroborating our simulation findings. A recent theoretical study has affirmed that a single-mode Gaussian state can exhibit a $g^{(2)}(0)$ value below zero [126]. The spectral distribution of the field is observed to be centred around 1308.5 nm in Fig. 4.31. Additionally, the contribution of thermal photons is found to be negligible at room temperature, and hence we obtain a first signature of a displaced squeezed state. This observation will be unambiguously verified through further investigations, involving the full reconstruction of the Wigner function using quantum tomography [26, 127].

Figure 4.37 gives a comparison between measured and simulated photon statistics. The photon number distributions are extracted from experimental data recorded by the oscilloscope and from stochastic simulations. In both cases, the histogram for shot noise confirms a Poissonian distribution. However, under quiet pumping conditions, the distribution becomes narrower, displaying sub-Poissonian characteristics that are supported by the measurement of $g^2(0)$ below 1.

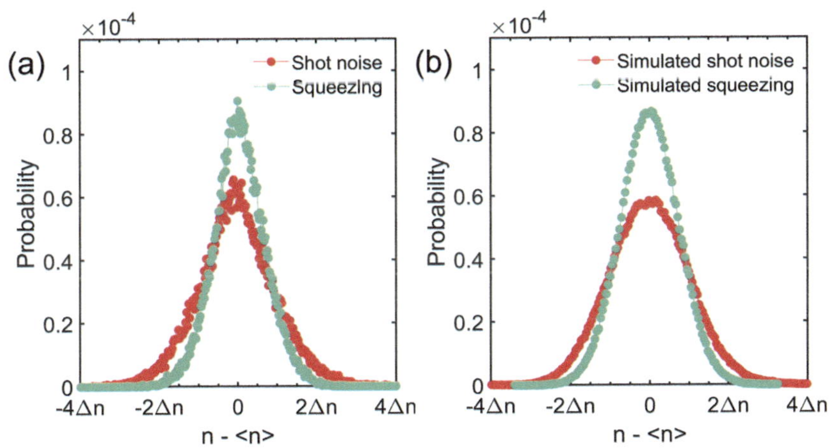

Fig. 4.37 **a** Histogram calculated from the raw data acquired by the oscilloscope. Each time trace is averaged over 5 frames and the sampling rate is 20 Gbps. **b** Histogram computed from the simulation data, where each time trace has 2×10^6 data points, spanning a time period of 1 μs

4.4.3 Data Correction

Amplitude-squeezed states that occur in practical experiments necessarily suffer from losses present in sources, transmission channels and detectors. To calculate the overall efficiency, we conducted careful measurements and estimations of diverse optical attenuation factors in our experimental setup, as summarized in Table 4.5. Consequently, the estimated overall efficiency stands at 22%. Subsequently, we conducted the following calculations, where a lossy optical element with a transmission coefficient (T) is substituted by a beam splitter [128]. A vacuum state is assumed to be present at the second input port of the beam splitter, as shown in Fig. 4.38. Hence, the interference between the variance $\langle \Delta X_a^2 \rangle$ of signal mode $\hat{a}$ and the variance $\langle \Delta X_v^2 \rangle$ of vacuum mode $\hat{v}$ will yield the variance $\langle \Delta X_{\text{out}}^2 \rangle$ in the beam splitter output:

$$\langle \Delta X_{\text{out}}^2 \rangle = T \langle \Delta X_a^2 \rangle + (1 - T) \langle \Delta X_v^2 \rangle = T \langle \Delta X_a^2 \rangle + (1 - T)/2 \tag{4.38}$$

Since $\langle \Delta X_v^2 \rangle = 1/2$. Considering the total losses, $T = 0.4 \times 0.794 \times 0.933 \times 0.933 \times 0.891 \times 0.9 \times 0.99 = 0.219$. With $s_{\text{out}}^{dB} = 0.9$ dB, we can calculate $s_{\text{out}} = 10^{(s_{\text{out}}^{dB}/10)} = 1.23$, which is equal to $\langle \Delta X_{\text{out}}^2 \rangle / \langle \Delta X_v^2 \rangle = 2 \langle \Delta X_{\text{out}}^2 \rangle$. Solving for $\langle \Delta X_{\text{out}}^2 \rangle$, we find it to be 0.6150. Using Eq. 4.38, we can determine $\langle \Delta X_a^2 \rangle = 1.0251$. Then $s_a = 2 \langle \Delta X_a^2 \rangle = 2.0502$, and converting this to s_a^{dB} gives $s_a^{dB} = 10 \times \log_{10}(s_a) = 3.12$ dB.

We estimate, by tracking photon losses within our experimental setup, that utilizing direct light collection, fused fibres, and pigtailing could bridge the gap between the observed squeezing level at the spectrum analyzer and the corrected squeezing level at the laser output facet.

Table 4.5 Measured and estimated optical attenuation factors

Source of loss	Efficiency (%)	Reference
Fibre coupling	40.0	$*^{a}$
Isolator	79.4	$*$
1st Beamsplitter	93.3	$*$
2nd Beamsplitter	93.3	$*$
Fibre+connector	89.1	$*$
Detector	90.0	$\dagger^{b}$
Electrical noise	99.0	$\ddagger^{c}$

[a] The asterisk ($*$) notation denotes that the data originates from our experimental measurements

[b] $\eta_{PD} = S_{PD} \times \hbar\omega/e$, with $S_{PD} = 0.95$ A/W being the typical photosensitivity of our InGaAs photodiode at 1300 nm, $\hbar$ being the reduced Planck constant, ω being the laser angular frequency and e being the elementary charge

[c] $\eta_{elec} = 1 - 1/\text{SNC}$, with SNC= 20 dB being the shot-noise clearance of our homodyne detector, which is the difference between calibrated shot noise and electrical noise. It is shown that the effect of the electronic noise on the detector performances is equivalent to that of optical losses [129]

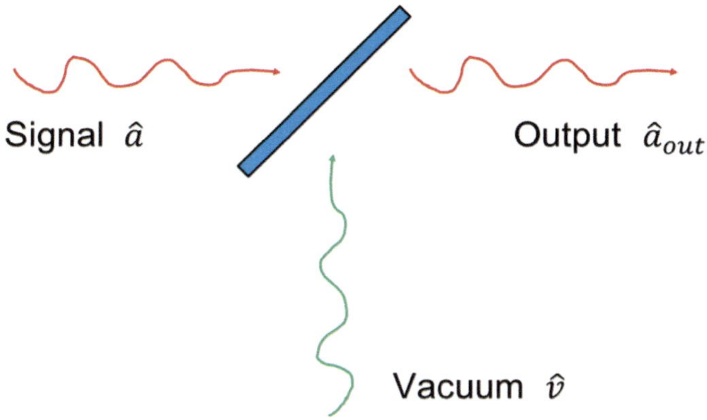

Fig. 4.38 The beam splitter model of loss [128]

Conclusion

Why do QD lasers exhibit such a broad squeezing bandwidth? Examining the specific numerical parameters detailed in Ref. [47], we find that the depletion layer capacitance of the QW laser is approximately 280 pF. According to (4.18), the estimated squeezing bandwidth at 66 K is approximately 1 GHz. On the contrary, QD lasers typically exhibit significantly lower values of C_{dep}, sometimes as low as 3.5 pF [130]. This discrepancy leads to a situation where $\tau_{te} \ll \tau_p$, resulting in a calculated squeezing bandwidth of several tens of gigahertz at room temperature, considering that τ_p is on the order of a few picoseconds. This demonstration provides, for the first time, access to a regime where the squeezing electrical bandwidth is dictated by the photon lifetime of the semiconductor laser.

In summary, this work has achieved a great milestone by successfully generating broadband amplitude-squeezed states at room temperature using constant-current-driven QD DFB lasers. We observed a reduction of up to 0.9 dB (3.1 dB after the correction) in the RF spectrum through homodyne detection across a wide frequency range from 3 GHz to 12 GHz when compared to the shot noise level. Achieving higher squeezing levels, however, would necessitate significant engineering advancements in laser cavity design [131], external stabilization techniques [50, 52], and the ambitious ongoing integration of the entire system onto a single chip, which, nonetheless, appears realistic [39, 40]. In our upcoming research, we plan to increase the squeezing level beyond 10 dB using quantum dot technology, focusing on related applications in quantum sensing. Moreover, in practical scenarios, 3–6 dB of squeezing has proven sufficient to outperform coherent states in quantum communication [106]. This squeezing level is particularly effective in mitigating the impact of channel noise, such as that induced by atmospheric turbulence in free-space transmission, which could otherwise compromise the rate of secret key exchange [132]. Our results carry direct implications for various quantum information applications, especially in the CV QKD over free-space channels. Bright squeezed semiconductor lasers, as proposed in Ref.[92], offer excellent light sources in this context. We firmly believe that this technology represents a significant stride toward the imminent on-chip implementation of CV QKD protocols.

References

1. Yamamoto Y (1991) Coherence, amplification, and quantum effects in semiconductor lasers. Wiley series in pure and applied optics. Wiley
2. Einstein A (1909) Zum gegenwartigen Stand des Strahlungsproblesm. Phys Zeits 10:185–193
3. Davidovich L (1996) Sub-Poissonian processes in quantum optics. Rev Mod Phys 68(1): 127
4. Maiman T (1960) Stimulated optical radiation in ruby. Nature 187(4736): 493–494. URL: https://doi.org/10.1038/187493a0
5. Arecchi FT (1965) Measurement of the statistical distribution of Gaussian and laser sources. Phys Rev Lett 15(24): Acirc; 912
6. Brown RH, Twiss RQ (1956) Correlation between photons in two coherent beams of light. Nature 177(4497):27–29
7. Glauber RJ (1963) The quantum theory of optical coherence. Phys Rev 130(6):2529
8. Glauber RJ (1963) Coherent and incoherent states of the radiation field. Phys Rev 131(6):2766
9. Sudarshan E (1963) Equivalence of semiclassical and quantum mechanical descriptions of statistical light beams. Phys Rev Lett 10(7):277
10. Gerry CC, Knight PL (2005) Introductory quantum optics. Cambridge University Press
11. Furusawa A (2015) Quantum states of light. Springer
12. Bachor H-A, Ralph TC (2004) A guide to experiments in quantum optics. Wiley
13. Yuen H, Shapiro J (1978) Optical communication with two-photon coherent states–Part I: quantum-state propagation and quantum-noise. IEEE Trans Inf Theory 24(6):657–668
14. Lawrie BJ, Lett PD, Marino AM, Pooser RC (2019) Quantum sensing with squeezed light. Acs Photon 6(6):307–1318
15. Yuen HP (1976) Two-photon coherent states of the radiation field. Phys Rev A 13(6): 2226
16. Walls DF (1983) Squeezed states of light. Nature 306(5939): Acirc; 141–146
17. Slusher R, Hollberg LW, Yurke B, Mertz JC, Valley JF (1985) Observation of squeezed states generated by four-wave mixing in an optical cavity. Phys Rev Lett 55(22):2409
18. Wu LA, Kimble HJ, Hall JL, Wu H (1986) Generation of squeezed states by parametric down conversion. Phys Rev Lett 57(20):2520
19. Carruthers P, Nieto MM (1968) Phase and angle variables in quantum mechanics. Rev Mod Phys 40(2):411
20. Yamamoto Y, Imoto N, Machida S (1986) Amplitude squeezing in a semiconductor laser using quantum nondemolition measurement and negative feedback. Phys Rev A 33(5):3243
21. Machida S, Yamamoto Y, Itaya Y (1987) Observation of amplitude squeezing in a constant-current–driven semiconductor laser. Phys Rev Lett 58(10):1000
22. Yuen HP, Chan VW (1983) Noise in homodyne and heterodyne detection. Optics Lett 8(3):177–179
23. Glauber RJ (1963) Photon correlations. Phys Rev Lett 10(3): Acirc; 84
24. Scully MO, Zubairy MS (1999) Quantum optics
25. Mandel L (1982) Squeezed states and sub-Poissonian photon statistics. Phys Rev Lett 49(2):136
26. Smithey DT, Beck M, Raymer MG, Faridani A (1993) Measurement of the Wigner distribution and the density matrix of a light mode using optical homodyne tomography: application to squeezed states and the vacuum. Phys Rev Lett 70(9):1244
27. Johnson JB (1928) Thermal agitation of electricity in conductors. Phys Rev 32(1):97

28. Nyquist H (1928) Thermal agitation of electric charge in conductors. Phys Rev 32(1):110
29. Van Der Ziel A (1955) Theory of shot noise in junction diodes and junction transistors. Proc IRE 43(11):1639–1646
30. Callen HB, Welton TA (1951) Irreversibility and generalized noise. Phys Rev 83(1):34
31. Yamamoto Y (2012) Lecture notes: fundamentals of noise processes. https://www.nii.ac.jp/qis/firstquantum/e/forStudents/lecture/index.html
32. Yamamoto Y, Machida S, Nilsson O (1986) Amplitude squeezing in a pump-noise-suppressed laser oscillator. Phys Rev A 34(5):4025
33. Shelby RM, Levenson MD, Perlmutter SH, DeVoe RG, Walls DF (1986) Broad-band parametric deamplification of quantum noise in an optical fiber. Phys Rev Lett 57(6):691
34. Andersen UL, Gehring T, Marquardt C, Leuchs G (2016) 30 years of squeezed light generation. Phys Scripta 91(5):053001
35. Kashiwazaki T, Yamashima T, Enbutsu K, Kazama T, Inoue A, Fukui K, Endo M, Umeki T, Furusawa A (2023) Over-8-dB squeezed light generation by a broadband waveguide optical parametric amplifier toward fault-tolerant ultra-fast quantum computers. Appl Phys Lett 122(23):1
36. Vahlbruch H, Mehmet M, Danzmann K, Schnabel R (2016) Detection of 15 dB squeezed states of light and their application for the absolute calibration of photoelectric quantum efficiency. Phys Rev Lett 117(11):110801
37. Safavi-Naeini AH, Gröblacher S, Hill JT, Chan J, Aspelmeyer M, Painter O (2013) Squeezed light from a silicon micromechanical resonator. Nature 500(7461):185–189
38. Zhao Y, Okawachi Y, Jang JK, Ji X, Lipson M, Gaeta AL (2020) Near-degenerate quadrature-squeezed vacuum generation on a silicon-nitride chip. Phys Rev Lett 124(19):193601
39. Moody G, Chang L, Steiner TJ, Bowers JE (2020) Chip-scale nonlinear photonics for quantum light generation. AVS Quantum Sci 2(4). https://doi.org/10.1116/5.0020684
40. Tasker JF, Frazer J, Ferranti G, Allen EJ, Brunel LF, Tanzilli S, D'Auria V, Matthews JC (2021) Silicon photonics interfaced with integrated electronics for 9 GHz measurement of squeezed light. Nat Photon 15(1):11–15
41. Teich MC, Saleh BE (1985) Observation of sub-Poisson Franck-Hertz light at 253.7 nm. JOSA B 2(2):275–282
42. Tapster P, Rarity J, Satchell J (1987) Generation of sub-Poissonian light by high-efficiency light-emitting diodes. Europhys Lett 4(3):293
43. Short R, Mandel L (1983) Observation of sub-Poissonian photon statistics. Phys Rev Lett 51(5):384
44. Machida S, Yamamoto Y (1986) Observation of sub-Poissonian photoelectron statistics in a negative feedback semiconductor laser. Optics Commun 57(4):290–296
45. Golubev YM, Sokolov I (1984) Photon antibunching in a coherent light source and suppression of the photorecording noise. Sov Phys JETP 60(2):234–234
46. Machida S, Yamamoto Y (1988) Ultrabroadband amplitude squeezing in a semiconductor laser. Phys Rev Lett 60(9):792
47. Richardson W, Machida S, Yamamoto Y (1991) Squeezed photon-number noise and sub-Poissonian electrical partition noise in a semiconductor laser. Phys Rev Lett 66(22):2867
48. Marin F, Bramati A, Giacobino E, Zhang T-C, Poizat J-P, Roch J-F, Grangier P (1995) Squeezing and intermode correlations in laser diodes. Phys Rev Lett 75(25):4606
49. Inoue S, Machida S, Yamamoto Y, Ohzu H (1993) Squeezing in an injection-locked semiconductor laser. Phys Rev A 48(3):.&Acirc2230
50. Wang H, Freeman MJ, Steel DG (1993) Squeezed light from injection-locked quantum well lasers. Phys Rev Lett 71(24):Â 3951
51. Freeman M, Wang H, Steel DG, Craig R, Scifres D (1993) Wavelength-tunable amplitude-squeezed light from a room-temperature quantum-well laser. Optics Lett 18(24):2141–2143
52. Kitching J, Yariv A, Shevy Y (1995) Room temperature generation of amplitude squeezed light from a semiconductor laser with weak optical feedback. Phys Rev Lett 74(17):3372
53. Li Y-Q, Xiao M (1998) Generation and applications of amplitude-squeezed states of light from semiconductor diode lasers. Optics Express 2(3):110–117

54. Jérémie F, Chabran C, Gallion P (1999) Room-temperature generation of amplitude-squeezed light from 1550-nm distributed-feedback semiconductor lasers. JOSA B 16(3):460–464
55. Richardson W, Shelby R (1990) Nonclassical light from a semiconductor laser operating at 4 K. Phys Rev Lett 64(4):400
56. Machida S, Yamamoto Y (1989) Observation of amplitude squeezing from semiconductor lasers by balanced direct detectors with a delay line. Optics Lett 14(19):1045–1047
57. Richardson W, Yamamoto Y (1991) Quantum correlation between the junction-voltage fluctuation and the photon-number fluctuation in a semiconductor laser. Phys Rev Lett 66(15):1963
58. Freeman MJ, Wang H, Steel DG, Craig R, Scifres DR (1993) Amplitude-squeezed light from quantum-well lasers. Optics Lett 18(5):Â 379–381
59. Kilper DC, Roos PA, Carlsten JL, Lear KL (1997) Squeezed light generated by a microcavity laser. Phys Rev A 55(5):R3323
60. Kaiser J, Degen C, Elsässer W (2001) Amplitude-squeezed emission from a transverse single-mode vertical-cavity surface-emitting laser with weakly anticorrelated polarization modes. Optics Lett 26(21):1720–1722
61. Yamamoto Y, Machida S (1987) High-impedance suppression of pump fluctuation and amplitude squeezing in semiconductor lasers. Phys Rev A 35(12):5114
62. Kim J, Yamamoto Y (1997) Theory of noise in pn junction light emitters. Phys Rev B 55(15):9949
63. Imamog A, Yamamoto Y et al (1993) Noise suppression in semiconductor p-i-n junctions: transition from macroscopic squeezing to mesoscopic coulomb blockade of electron emission processes. Phys Rev Lett 70(21):3327
64. Imamog A, Yamamoto Y, et al (1994) Turnstile device for heralded single photons: Coulomb blockade of electron and hole tunneling in quantum confined p-i-n heterojunctions. Phys Rev Lett 72(2):Â 210
65. Borri S, Siciliani de Cumis M, Viciani S, D'Amato F, De Natale P (2020) Unveiling quantum-limited operation of interband cascade lasers. APL Photon 5(3):036101
66. Marcenac D, Carroll J (1994) Modeling of intensity noise including squeezing in DFB and Fabry-Perot semiconductor laser diodes. IEEE J Quantum Electron 30(9):2064–2072
67. Arnaud J (1995) Classical theory of laser noise. Opt Quant Electron 27:63–89
68. Vey JV, Gallion P (1997) Semiclassical model of semiconductor laser noise and amplitude noise squeezing. I. Description and application to Fabry-Perot laser. IEEE J Quantum Electron 33(1):1Â 2097–2104
69. Deng Y, Wang C (2020) Rate equation modeling of interband cascade lasers on modulation and noise dynamics. IEEE J Quantum Electron 56(2):1–9
70. Henry C (1986) Theory of spontaneous emission noise in open resonators and its application to lasers and optical amplifiers. J Lightwave Technol 4(3):288–297
71. Grillot F, Thedrez B, Duan G-H (2004) Feedback sensitivity and coherence collapse threshold of semiconductor DFB lasers with complex structures. IEEE J Quantum Electron 40(3):231–240
72. Morse PM, Feshbach H (1954) Methods of theoretical physics. Am J Phys 22(6):410–413
73. Petermann K (1991) Laser diode modulation and noise, vol 3. Springer Science & Business Media, New York
74. Tromborg B, Olesen H, Pan X (1991) Theory of linewidth for multielectrode laser diodes with spatially distributed noise sources. IEEE J Quantum Electron 27(2):178–192
75. Coldren LA, Corzine SW, Mashanovitch ML (2012) Diode lasers and photonic integrated circuits. Wiley, New York
76. Van der Lee A, Van Druten N, Van Exter M, Woerdman J, Poizat J-P, Grangier P (2000) Critical Petermann K factor for intensity noise squeezing. Phys Rev Lett 85(22):4711
77. Meyer JR, Bewley WW, Canedy CL, Kim CS, Kim M, Merritt CD, Vurgaftman I (2020) The interband cascade laser. Photonics 7(3):75. https://doi.org/10.3390/photonics7030075
78. Knötig H, Nauschütz J, Opacak N, Höfling S, Koeth J, Weih R, Schwarz B (2022) Mitigating valence intersubband absorption in interband cascade lasers. Laser Photon Rev 16(9):Â 2200156

79. Lebreton A, Abram I, Takemura N, Kuwata-Gonokami M, Robert-Philip I, Beveratos A (2013) Stochastically sustained population oscillations in high-β nanolasers. New J Phys 15(3):033039. https://doi.org/10.1088/1367-2630/15/3/033039
80. Puccioni G, Lippi G (2015) Stochastic Simulator for modeling the transition to lasing. Optics Express 23(3):2369–2374. https://doi.org/10.1364/OE.23.002369
81. Mørk J, Lippi G (2018) Rate equation description of quantum noise in nanolasers with few emitters. Appl Phys Lett 112(14)
82. Mørk J, Yvind K (2020) Squeezing of intensity noise in nanolasers and nanoLEDs with extreme dielectric confinement. Optica 7(11):Â 1641–1644
83. Zhao S, Grillot F (2022) Stochastic model of sub-Poissonian quantum light in an interband cascade laser. Phys Rev Appl 18(6):064027
84. André EC, Mørk J, Wubs M (2020) Efficient stochastic simulation of rate equations and photon statistics of nanolasers. Optics Express 28(22):32632–32646
85. Ulrich SM, Gies C, Ates S, Wiersig J, Reitzenstein S, Hofmann C, Löffler A, Forchel FA, Jahnke F, Michler P (2007) Photon statistics of semiconductor microcavity lasers. Phys Rev Lett 98(4):Â 043906
86. Chow WW, Reitzenstein S (2018) Quantum-optical influences in optoelectronics–an introduction. Appl Phys Rev 5(4)
87. Lien Y, de Vries S, van Druten N, van Exter M, Woerdman J (2001) Photon statistics of a laser with slow inversion. Phys Rev Lett 86(13):2786
88. Zhao S, Grillot F (2022) Modeling of amplitude squeezing in a pump-noise-suppressed interband cascade laser. IEEE Photon J 14(3):1–8
89. Didier P, Spitz O, Cerutti L, Diaz-Thomas D, Baranov A, Carras M, Grillot F (2021) Relative intensity noise and intrinsic properties of RF mounted interband cascade laser. Appl Phys Lett 119(17):171107
90. Lin Y, Li L, Huang W, Yang RQ, Gupta JA, Zheng W (2020) Quasi-Fermi level pinning in interband cascade lasers. IEEE J Quantum Electron 56(4):1–10
91. Carroll MA, D'alessandro G, Lippi GL, Oppo GL, Papoff F (2021) Thermal, quantum antibunching and lasing thresholds from single emitters to macroscopic devices. Phys Rev Lett 126(6):Â 063902
92. Hosseinidehaj N, Winnel MS, Ralph TC (2022) Simple and loss-tolerant free-space quantum key distribution using a squeezed laser. Phys Rev A 105(3):032602
93. Gabbrielli T, Cappelli F, Bruno N, Corrias N, Borri S, De Natale P, Zavatta A (2021) Mid-infrared homodyne balanced detector for quantum light characterization. Opt Express 29(10):Â 14536–14547
94. Aellen TH, Giovannini MA, Faist JE, Von der Weid JP (2008) Feasibility study of free-space quantum key distribution in the mid-infrared. Quantum Inf Comput 8(1&2):Â 0001–0011
95. Marschick G, Pelini J, Gabbrielli T, Cappelli F, Weih R, KnoÌˆtig H, Koeth J, HoÌˆfling S, De Natale P, Strasser G et al (2023) Mid-infrared ring interband cascade laser: operation at the standard quantum limit. http://arxiv.org/abs/2306.11628arXiv:2306.11628arXiv:2306.11628
96. Arakawa Y, Sakaki H (1982) Multidimensional quantum well laser and temperature dependence of its threshold current. Appl Phys Lett 40(11):939–941. https://doi.org/10.1063/1.92959
97. Zhou Z, Ou X, Fang Y, Alkhazraji E, Xu R, Wan Y, Bowers JE (2023) Prospects and applications of on-chip lasers. Elight 3(1):Â 1–25. https://doi.org/10.1186/s43593-022-00027-x
98. Chow WW, Koch SW, Sargent MI (2012) Semiconductor-laser physics. Springer Science & Business Media
99. Wang C, Osiński M, Even J, Grillot F (2014) Phase-amplitude coupling characteristics in directly modulated quantum dot lasers. Appl Phys Lett 105(22):221114
100. Grillot F, Duan J, Dong B, Huang H (2021) Uncovering recent progress in nanostructured light-emitters for information and communication technologies. Light: Sci Appl 10(1):Â 156. https://doi.org/10.1038/s41377-021-00598-3

101. Arakawa Y, Holmes MJ (2020) Progress in quantum-dot single photon sources for quantum information technologies: a broad spectrum overview. Appl Phys Rev 7(2):1
102. Michler P (2017) Quantum dots for quantum information technologies, vol 237. Springer
103. Zhong H-S, Deng Y-H, Qin J, Wang H, Chen M-C, Peng L-C, Luo Y-H, Wu D, Gong S-Q, Su H et al (2021) Phase-programmable Gaussian boson sampling using stimulated squeezed light. Phys Rev Lett 127(18):180502
104. Vajner DA, Rickert L, Gao T, Kaymazlar K, Heindel T (2022) Quantum communication using semiconductor quantum dots. Adv Quantum Technol 5(7):2100116
105. Grosshans F, Grangier P (2002) Continuous variable quantum cryptography using coherent states. Phys Rev Lett 88(5):Â 057902
106. Jacobsen CS, Madsen LS, Usenko VC, Filip R, Andersen UL (2018) Complete elimination of information leakage in continuous-variable quantum communication channels. npj Quantum Inf 4(1):Â 32
107. Laudenbach F, Pacher C, Fung CH, Poppe A, Peev M, Schrenk B, Hentschel M, Walther P, Hübel H (2018) Continuous-variable quantum key distribution with Gaussian modulation—the theory of practical implementations. Adv Quantum Technol 1(1):Â 1800011
108. McCormick C, Boyer V, Arimondo E, Lett P (2007) Strong relative intensity squeezing by four-wave mixing in rubidium vapour. Optics Lett 32(2):178–180
109. Fürst J, Strekalov D, Elser D, Aiello A, Andersen UL, Marquardt C, Leuchs G (2011) Quantum light from a whispering-gallery-mode disk resonator. Phys Rev Lett 106(11):113901
110. Dutt A, Luke K, Manipatruni S, Gaeta AL, Nussenzveig P, Lipson M (2015) On-chip optical squeezing. Phys Rev Appl 3(4):Â 044005
111. Mondain F, Lunghi T, Zavatta A, Gouzien E, Doutre F, De Micheli M, Tanzilli S, D'Auria V (2019) Chip-based squeezing at a telecom wavelength. Photon Res 7(7):A36–A39
112. Yang Z, Jahanbozorgi M, Jeong D, Sun S, Pfister O, Lee H, Yi X (2021) A squeezed quantum microcomb on a chip. Nat Commun 12(1):4781
113. Nehra R, Sekine R, Ledezma L, Guo Q, Gray RM, Roy A (2022) Few-cycle vacuum squeezing in nanophotonics. Science 377(6612):Â 1333–1337
114. Choi W, Lee J-H, An K, Fang-Yen C, Dasari R, Feld M (2006) Observation of sub-Poisson photon statistics in the cavity-QED microlaser. Phys Rev Lett 96(9):093603
115. Nishi K, Kageyama T, Yamaguchi M, Maeda Y, Takemasa K, Yamamoto T, Sugawara M, Arakawa Y (2013) Molecular beam epitaxial growths of high-optical-gain InAs quantum dots on GaAs for long-wavelength emission. J Cryst Growth 378:459–462
116. Kitagawa M, Yamamoto Y (1986) Number-phase minimum-uncertainty state with reduced number uncertainty in a Kerr nonlinear interferometer. Phys Rev A 34(5):3974
117. McKeever J, Boca A, Boozer AD, Buck JR, Kimble HJ (2003) Experimental realization of a one-atom laser in the regime of strong coupling. Nature 425(6955):268–271
118. Tromborg B, Lassen HE, Olesen H (1994) Traveling wave analysis of semiconductor lasers: modulation responses, mode stability and quantum mechanical treatment of noise spectra. IEEE J Quantum Electron 30(4):939–956
119. Rice PR, Carmichael H (1994) Photon statistics of a cavity-QED laser: a comment on the laser–phase-transition analogy. Phys Revi A 50(5):4318. https://link.aps.org/doi/10.1103/PhysRevA.50.4318
120. Scully MO, Lamb Jr WE (1967) Quantum theory of an optical maser. I. General theory. Phys Rev 159(2):208
121. Lax M, Louisell W (1969) Quantum noise. XII. Density-operator treatment of field and population fluctuations. Phys Rev 185(2):Â 568
122. Sugawara M, Mukai K, Nakata Y, Ishikawa H, Sakamoto A (2000) Effect of homogeneous broadening of optical gain on lasing spectra in self-assembled In x Ga 1- x A s/G a A s quantum dot lasers. Phys Rev B 61(11):7595
123. Chow WW, Koch SW (2005) Theory of semiconductor quantum-dot laser dynamics. IEEE J Quantum Electron 41(4):495–505
124. Chow WW, Jahnke F (2013) On the physics of semiconductor quantum dots for applications in lasers and quantum optics. Prog Quantum Electron 37(3):109–184. https://doi.org/10.1016/j.pquantelec.2013.04.001

125. Grillot F, Veselinov K, Gioannini M, Montrosset I, Even J, Piron R, Homeyer E, Loualiche S (2009) Spectral Analysis of 1.55-μ m InAs–InP (113) B quantum-dot lasers based on a multipopulation rate equations model. IEEE J Quantum Electron 45(7):Â 872–878
126. Olivares S, Cialdi S, Paris MG (2018) Homodyning the g (2)(0) of Gaussian states. Optics Commun 426:547–552
127. Zavatta A, Marin F, Giacomelli G (2002) Quantum-state reconstruction of a squeezed laser field by self-homodyne tomography. Phys Rev A 66(4):043805
128. Lvovsky AI (2015) Squeezed light. Photonics. Wiley. Chap.Â 5, pp. 121–163. ISBN: 9781119009719
129. Appel J, Hoffman D, Figueroa E, Lvovsky AI (2007) Electronic noise in optical homodyne tomography. Phys Rev A 75(3):Â 035802
130. Inoue D, Jung D, Norman J, Wan Y, Nishiyama N, Arai S, Gossard AC, Bowers JE (2018) Directly modulated 1.3 μm quantum dot lasers epitaxially grown on silicon. Optics Express 26(6):Â 7022–7033
131. Ostrowski LA, Baker TJ, Saadatmand SN, Wiseman HM (2023) No tradeoff between coherence and sub-poissonianity for Heisenberg-Limited lasers. Phys Rev Lett 130(18):183602
132. Derkach I, Usenko VC, Filip R (2020) Squeezing-enhanced quantum key distribution over atmospheric channels. New J Phys 22(5):053006

Chapter 5
Conclusion and Perspectives

The fields of semiconductor physics and quantum optics are undergoing a transition from laboratory research to real-world applications. In order to facilitate the progress and development of innovative quantum-confined devices like quantum dot (QD) lasers and interband cascade lasers (ICL), device engineers will require simulation tools that combine classical device physics with the principles of quantum mechanics. This thesis primarily focuses on the device modelling, proof-of-concept experiments, and numerical simulation of electrically driven coherent and quantum light sources. As a result, it encompasses a diverse range of topics, all interconnected with the overarching goal of methodically advancing the state-of-the-art in this field. These topics include classical noise, nonlinear dynamics, and the generation of squeezed light in QD lasers and ICLs. Above all, this thesis commences with the fundamentals of quantum-confined lasers and provides an introduction to their electronic structure and carrier scattering processes. Subsequently, it presents three different modelling approaches, namely, semiclassical Langevin rate equations, the stochastic Gillespie algorithm, and a comprehensive quantum-mechanical description.

In the first part of the thesis, a theoretical investigation of various aspects of classical noise, including the relative intensity noise (RIN), frequency noise (FN), spectral linewidth, and the linewidth enhancement factor (α_H) of quantum-confined lasers, has been undertaken. These theoretical simulations rely on a semiclassical multi-level rate equation model that incorporates Langevin noise terms. This is achieved through a meticulous small-signal analysis that takes into consideration contributions from both gain theory and carrier effects. The calculations are carefully elaborated, enabling a precise description of laser noise characteristics that align with corresponding experimental findings. In particular, to assess the influence of the silicon substrate on QD lasers, a non-radiative recombination carrier lifetime τ_{SRH} via Shockley–Read–Hall recombination (SRH) is added to the original QD rate equations. Accounting for epitaxial defects induced by threading dislocations, the analysis demonstrates that the α_H factor decreases as τ_{SRH} values decrease. In the context of modulation dynamics, there is an observation of a narrower 3-dB bandwidth but a

S. Zhao, *Noise, Dynamics and Squeezed Light in Quantum Dot and Interband Cascade Lasers*, Springer Theses, https://doi.org/10.1007/978-3-031-66818-0_5

higher damping factor in QD laser systems with faster nonradiative recombination lifetimes. This change in the SRH recombination time effectively alters the timescale ratio between carrier and photon lifetimes, affecting the damping of the turn-on dynamics of the solitary laser. Subsequently, a comprehensive investigation of optical feedback and optoelectronic feedback dynamics in QD lasers on silicon is conducted. Thanks to the near-zero α_H factor, QD lasers on silicon exhibit a high degree of tolerance to external optical feedback. Numerical simulations based on the Lang–Kobayashi method also indicate that as τ_{SRH} values decrease, the chaotic region contracts and the first Hopf bifurcation point shifts to higher feedback strength. This result offers valuable insights into the design of feedback-resistant lasers for future photonic integrated circuits, operating without the need for an optical isolator. On the contrary, in terms of optoelectronic feedback, with increasing feedback strength, the QD laser undergoes several dynamical regimes, including steady-state, square waves, and a mixed waveform characterized by fast and slow oscillations. Unlike QW lasers, QD lasers exhibit greater sensitivity to optoelectronic feedback, which promotes dynamic generation. This phenomenon is substantiated through the integral-differential delay model (Ikeda-like model). Finally, in the context of ICLs, the anticipation of broadband dynamics and chaos generation is supported not only by numerical simulations but also by experimental evidence. In these experiments, an unprecedented broad and flat chaos in ICLs is reported, extending beyond 1 GHz at mid-infrared wavelength. The chaos in ICLs has the potential to be harnessed for the development of secure free-space optical communication links and remote chaotic Lidar systems.

In the second part of the thesis, a theoretical examination of quantum fluctuations and amplitude squeezing has been conducted. This section starts with a general introduction to numerous aspects, including coherent states of light, squeezed states of light, methods for measuring quantum states of light, quantum noise from electronics, and the generation of quantum light. Then, a detailed explanation is provided regarding how a semiconductor laser, when subjected to a noise-suppressed current pump, can generate sub-shot-noise light. The limitations that exist with respect to the achievable squeezing level and bandwidth in this context have been also demonstrated. Afterwards, two different methods, namely, semiclassical Langevin rate equations and Gillespie stochastic simulation, are employed to establish the eligibility of ICLs for generating amplitude-squeezed light in the mid-infrared range. Both of these methods are in agreement and support this prediction through the analysis of sub-Poissonian photon distribution, sub-shot-noise-limited RF spectrum, and a negative Mandel Q parameter. The results reveal that ICLs possess the potential to achieve substantial amplitude squeezing across a wide bandwidth, extending into the gigahertz range. Finally, both experimental and theoretical investigations of amplitude squeezing are carried out in QD lasers. A major achievement of this work is the successful generation of broadband amplitude-squeezed states using single-mode QD lasers operated at room temperature. In homodyne detection, the RF spectrum shows a reduction of up to 0.9 dB across a wide frequency range from 3 to 12 GHz when compared to the shot noise level. These experimental results are further validated through a stochastic simulation, specifically tailored to the three-energy-level QD model. All these outcomes hold direct relevance for many quantum information applications, particularly in the domain of continuous-variable quantum key distribution (CV QKD). Bright squeezed semiconductor lasers, as proposed in Ref. [1], emerge as ideal light sources for this purpose.

The novel results presented in this thesis carry profound implications for the design of future high-speed and non-classical quantum sources, regardless of whether they are built upon QD lasers or mid-infrared ICLs.

Why laser squeezing is so interesting? Laser squeezing distinguishes itself significantly from conventional squeezing methods. Take parametric amplification, for instance, where the reduction in phase fluctuations is counterbalanced by an equal amplification of amplitude fluctuations, thus preserving their inherent uncertainty relationship. In contrast, laser squeezing achieves a reduction in amplitude fluctuations without a corresponding increase in phase fluctuations [2]. This distinct characteristic arises because the laser output exhibits excess phase fluctuations. In essence, laser squeezing takes advantage of the fact that the laser output is not in a minimum uncertainty state to suppress amplitude noise. Consequently, laser squeezing always manifests as amplitude squeezing. An intriguing question would be the transformation of amplitude squeezing into phase squeezing through the α_H factor in laser and the determination of the fundamental limit for laser linewidth [3]. Future phase noise measurements or the construction of Wigner functions can provide valuable insights into validating this hypothesis [4].

Why there is a need to enhance squeezing performance? Figure 5.1 gives an overview of historical experiments related to amplitude squeezing from semiconductor lasers. Within these experiments lies a huge untapped potential, particularly

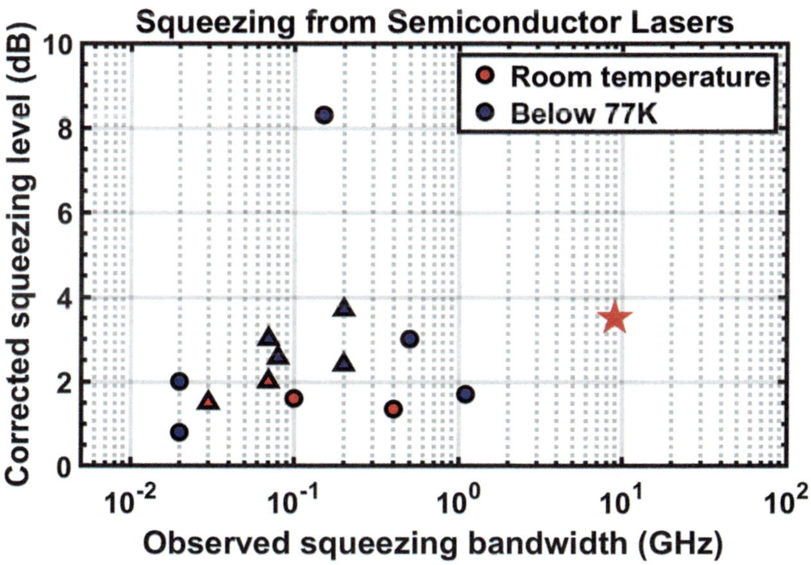

Fig. 5.1 A summary of the amplitude squeezing from semiconductor lasers. All the data points come from Table 4.2

in achieving higher squeezing levels and broader squeezing bandwidths. In terms of practical requirements, it is generally acknowledged that achieving a more important squeezing level holds immense significance for quantum sensors [5] and gravitational wave detection [6].

- In sensing and metrology platforms using optical readout fields, the minimum resolvable signal faces growing constraints imposed by the standard quantum limit. This limit is mainly determined by the sum of photon shot noise and back-action noise [5]. The advancement of the next generation of sensors, spanning applications from high-energy physics to biochemistry, and the development of microscopy platforms capable of unveiling material properties previously concealed by quantum noise, will hinge upon a synergistic approach that combines back-action and shot noise reduction techniques.
- A global network of Earth-based gravitational wave observatories is presently dedicated to the direct detection of this subtle radiation using precision laser interferometry [6]. The quantum nature of light introduces photon shot noise, which sets a fundamental limit on the attometre-level sensitivity required for the kilometre-scale Michelson interferometers. In this context, the introduction of squeezed states has emerged as a method to enhance the performance of one of the detectors within the Laser Interferometer Gravitational-Wave Observatory (LIGO). Recent advancements have shown the generation of 15 dB squeezing light for this purpose [7]. The incorporation of squeezed states has empowered this particular LIGO detector to achieve unparalleled broadband sensitivity and unprecedented precision to gravitational waves.

Simultaneously, the progress in photonics-based quantum technologies is poised to usher in a transformative era in data processing and communication protocols, with the quantum emitter playing a pivotal role in this landscape. A semiconductor laser with a broader squeezing bandwidth is critically desirable for applications such as quantum cryptography [9], high-speed continuous-variable quantum key distribution (CV QKD) [8, 10, 11], and quantum computing [12–14].

- CV QKD has emerged as one of the foremost practical applications of quantum information theory. It furnishes trusted parties with protocols that enable the provably secure distribution of secret cryptographic keys, as detailed in Fig. 5.2. In this protocol, quantum signals are typically encoded in light through Gaussian modulation of its electromagnetic fields. Detection in CV QKD relies on homodyne receivers, which not only provide fast and efficient detection capabilities but also seamlessly integrate with existing optical networks. While weak coherent states are utilized in certain scenarios [15], the distinctive characteristics of squeezed states make them a preferred choice, particularly when the objective is to enhance resistance to eavesdropping and ensure robust immunity to channel noise [10]. In real-world applications, a squeezing level in the range of 3–6 dB has demonstrated its effectiveness in outperforming coherent states [16]. Such a squeezing level is especially valuable in mitigating channel noise arising from atmospheric turbulence during free-space transmission. Without this mitigation, such noise could

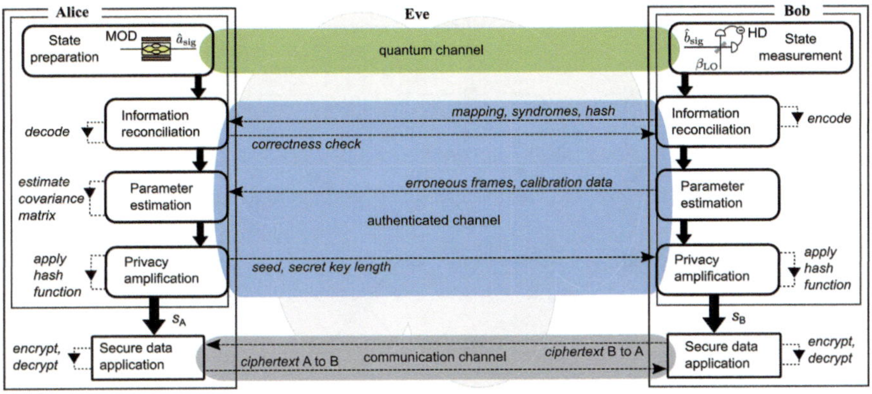

Fig. 5.2 A diagram of the CV QKD protocol. Alice and Bob establish quantum correlations through the use of modulation (MOD) and local oscillator (LO) in homodyne/heterodyne detection (HD). They employ these techniques to prepare and measure optical quantum states. Subsequently, after completing the remaining steps of the protocol, which involve an authenticated channel, they obtain correlated bitstreams, denoted as s_A and s_B respectively. For the application to guarantee composable security, it is imperative that certain criteria related to the correctness, robustness, and secrecy of the protocol are met. (*Source* In Ref. [8])

potentially undermine the secret key rate and compromise the overall security of communication systems. Current squeezing technologies face limitations basically in terms of their bandwidth and the necessity for low operating temperatures. These constraints present huge challenges when aiming for practical implementations under room-temperature conditions and impede the realization of high-speed quantum communications. The broadest bandwidth achieved in all methods thus far has reached dozens of gigahertz, as highlighted in [17]. The potential for significant improvement in this regard can be realized through an amplitude-squeezed semiconductor laser.

- Quantum computers promise to deliver extraordinary computational power and solve problems that are beyond the capabilities of conventional classical machines. The development of a programmable and scalable photonic circuit has turned out to be a leader in the race toward realizing practical quantum computing [13], as shown in Fig. 5.3. It has become evident that a relatively simple optical circuit, relying exclusively on squeezed light, beam splitters, and photon counters, can execute a sampling algorithm at a speed that surpasses the capabilities of classical computers. Furthermore, this algorithm has revealed numerous practical applications. For instance, it proves valuable in simulating transitions between molecular states and identifying matching configurations between two molecules, a process commonly referred to as molecular docking [18]. Another intriguing application is the coherent Ising machine [12]. A coherent Ising machine constitutes a network of optical parametric oscillators where the "strongest" collective mode of oscillation, occurring well above the threshold, corresponds to an optimum solution of a given NP problem. However, as the pump rate or network coupling rate is increased

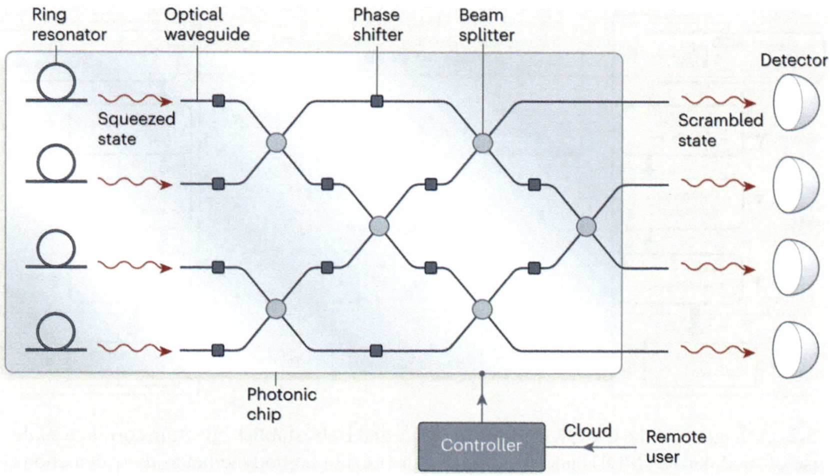

Fig. 5.3 Quantum algorithms implemented on a photonic chip. Ring resonators were employed to generate quantum states of light. These states were launched into an optical network consisting of optical waveguides and beam splitters. As the squeezed states encountered one another within the beam splitters, they became mixed together, leading to a complete scrambling of all states as they exited the network. Subsequently, highly sensitive detectors were used to count the number of photons within each scrambled state. The chip was employed to execute quantum algorithms, with the input squeezed states representing input variables and the count of photons within each scrambled state signifying the output resulting from the algorithm's processing of those variables. The chip's functionality could be reconfigured to execute different algorithms by manipulating the beam splitters and employing phase shifters. This control over the chip's behaviour could be accessed remotely by a user through cloud-based means. (*Source* In Ref. [13])

from below to above the threshold, the eigenvectors associated with the smallest eigenvalues of the Ising coupling matrix appear near the threshold, hindering the machine's relaxation to true ground states. To overcome this challenge, the proposed approach leverages the squeezed or anti-squeezed vacuum noise generated by optical parametric oscillators operating below the threshold. This utilization of quantum noise correlation facilitates coherent exploration across numerous local minima. Ultimately, this enables the machine to access either true ground states or excited states whose eigenenergies are sufficiently close to those of the ground states occurring above the threshold. The concept of an Ising machine can also be extended to include semiconductor lasers subjected to optoelectronic feedback [19], as shown in Fig. 5.4. The performance of such systems can be improved by implementing amplitude squeezing techniques in semiconductor lasers.

How to improve the squeezing performance?

- **Cryogenic temperature**: As illustrated in Fig. 5.1, the blue data points correspond to previous squeezing experiments conducted at extremely low temperatures, typically below 77 K. These experiments tend to exhibit superior squeezing

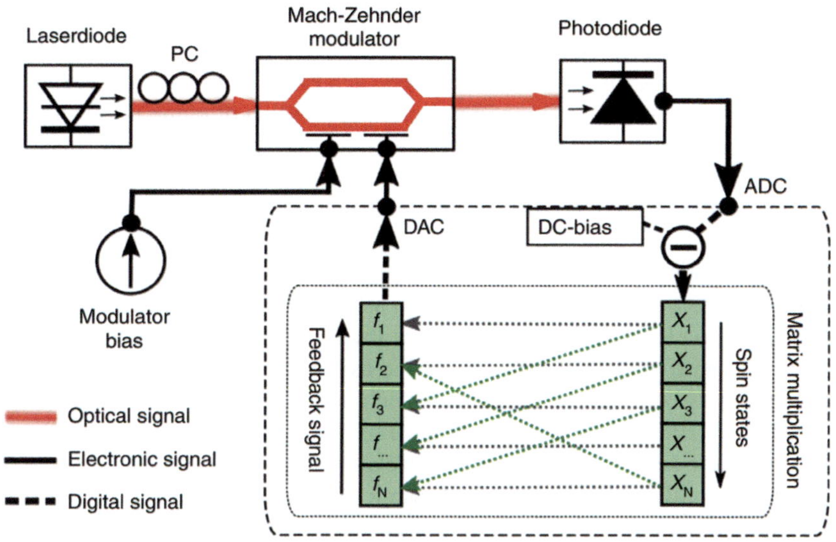

Fig. 5.4 Experimental schematic of an optoeletronic-feedback-based Ising machine. PC, polarization controller, ADC, analogue-digital converter, DAC, digital-analogue converter. (*Source* In Ref. [19])

performance compared to the red data points, which represent experiments conducted at room temperature. This observation provides a valuable insight. The underlying physics can be understood through the interplay of two crucial factors: the junction capacitance (C_{dep}) and the operating temperature (T). This transition from the macroscopic to the mesoscopic regime hinges on the ratio $r = e^2/k_B T C_{\text{dep}}$, which quantifies the single electron charging energy relative to the characteristic energy associated with thermal fluctuations [20]. For an in-depth exploration of this concept, please refer to Sect. 4.2.4. The value of the ratio r significantly influences the extent of correlations among the injected electrons. In the macroscopic regime and at high temperatures ($r \ll 1$), individual photon emission events have minimal impact on the expected injection time of the subsequent electron. However, when a substantial number of emission events occur collectively, their combined effect maintains the sub-Poissonian characteristics. Conversely, in the mesoscopic regime and at low temperatures ($r \gg 1$), each electron is acutely aware of the previous one, primarily due to the considerable change in the junction voltage brought about by the photon emission event. This phenomenon is known as the Coulomb blockade regime and can even lead to single-photon emission events at the end [21]. Furthermore, cooling both the current source and the laser to cryogenic temperatures can lead to a huge reduction in the thermal noise within the current source, as indicated by the expression $S_{I_n}(\omega) = 4k_B T/R_s$ in (4.11). In Ref. [22], employing this approach, a constant-current-driven semiconductor laser generated a field with amplitude noise reduced to levels below the standard quantum limit by −1.7 dB (32%) across a broad frequency range, spanning from near direct current to 1.1 GHz.

- **External feedback**: Recently, researchers achieved amplitude squeezing by employing a combination of high impedance suppression of pump noise and line-narrowing techniques. Through injection-locking, they attained a 3 dB squeezing level (corrected for detection efficiency, it was 4.3 dB) from AlGaAs diode lasers operating at 10 K [23]. Currently, the state-of-the-art intensity noise reduction of a collimated beam from a single-mode semiconductor laser at its front facet is 2–3 dB below shot noise at room temperature and 5–9 dB at low temperatures. These results are achieved through methods such as feedback from an external grating [24] or injection locking [23, 25]. Previous studies [26] have explored the impact of these line-narrowing techniques on the intensity and phase noise of diode lasers. It's well-understood that these techniques help reduce the noisy longitudinal side modes, which, although small, contribute enormously to noise. The key to noise reduction lies in achieving perfect anti-correlation between the side modes and the main mode, similar to the antiphase dynamics observed in classical fluctuations in multimode lasers. Imperfect noise anti-correlation was identified as the source of excess noise in laser intensity. Furthermore, anti-correlation was also observed between the main mode and a small mode with orthogonal polarization [27]. Full noise reduction was only achieved when the total intensity of both modes was detected [28]. Another way to generate amplitude squeezing is to use electro-optic feedback from one of the balanced detectors [29]. This approach can indeed produce an in-loop photocurrent with exceptionally low noise. However, it's important to clarify that this phenomenon is not considered evidence of "genuine" squeezing because squeezed light cannot be extracted from the feedback loop using a linear beam splitter but can be extracted through a quantum non-demolition measurement [30]. Electro-optical feedback from the balanced homodyne photocurrent is often regarded as a more effective and viable solution [31, 32]. Another more recent method involves the use of an electrically controllable beam splitter [33].
- **Miniature**: Integrated quantum photonics draws upon classical integrated photonic technologies and devices to address quantum applications [34–38]. Similar to classical photonics, achieving chip-scale integration is critical for advancing from laboratory demonstrations to applications. In integrated quantum photonics, a prime objective is to minimize internal loss while ensuring scalability, stability, and cost-effectiveness. The successful demonstration of 85% (−8.3 dB) amplitude squeezing, achieved through the face-to-face positioning of the detector and the laser [39], provides compelling evidence for the feasibility of these advancements. In the near future, the realization of an entire quantum system on a single chip is poised to become a reality.

What are the future applications? For over two decades, the go-to method for generating squeezed states has relied on the interaction between lasers and nonlinear materials like $\chi^{(2)}$ or $\chi^{(3)}$. However, these approaches often necessitate the inclusion of an additional nonlinear material known as a "squeezer". Additionally, they frequently face limitations in wavelength tunability due to phase-matching constraints, making them less versatile. Maintaining a high level of photon conversion efficiency can also pose problems within these setups. In a recent article [1], researchers conducted a comprehensive finite-size security analysis of a CV QKD protocol that used a bright amplitude-squeezed semiconductor laser. This study showcased the potential by achieving zero information leakage to potential eavesdroppers and effectively mitigating channel fluctuation noise [10]. These findings represent a promising advancement in the field of quantum communication.

To the best of our knowledge, there is currently no quantum hardware available in the mid-infrared range. Squeezed ICLs have the potential to expedite the development of this innovative technology and, ultimately, enable its implementation in this wavelength range. For instance, in the laser-based free-space secure communication systems, the atmospheric transmission spectrum features two transparency windows: one between 3 and 5 μm and another between 8 and 14 μm. Mid-infrared optical sources are ideal candidates for transmission applications, particularly in conditions like fog or drizzle, where they outperform near-infrared and visible-light sources significantly. Longer wavelengths in the mid-infrared range offer advantages such as reduced molecular (Rayleigh) and particulate (Mie) scattering, as well as decreased signal distortion caused by atmospheric conditions. Moreover, atmospheric turbulence along the propagation path can severely degrade optical signals, leading to phenomena like beam spreading, beam wandering, scintillation, and loss of spatial coherence. Numerical studies involving simulated scintillation data and turbulence strength have consistently shown the superiority of mid-infrared wavelengths over near-infrared ones for free-space transmissions [40]. From this perspective, there is a growing interest in the development of free-space data transfer, particularly using CV QKD protocols. This is because existing free-space communication systems are typically unidirectional and susceptible to eavesdropping. Recent theoretical analyses have extended CV QKD protocols, initially designed for fibre optic systems, to the context of free-space communications [41]. However, there are scenarios where dedicated fibre-optical infrastructure may be unavailable, such as in the case of mobile stations, the need for rapid channel deployment, or hostile environments. Additionally, long-distance intercontinental quantum communication via satellites relies on free-space channels [42]. Atmospheric turbulence introduces random variations in channel transmissivity over time, which can add extra noise to CV QKD systems, reducing the secret key rate and potentially compromising communication security in the presence of strong turbulence [43]. Given these challenges, the use of amplitude-squeezed states generated by mid-infrared ICLs becomes even more favourable, as depicted in Fig. 5.5. Mid-infrared ICLs present a persuasive choice for secure quantum communication under these demanding conditions, owing to their longer optical wavelengths, higher optical power, enhanced squeezing levels, and broader squeezing bandwidths. Specifically, they exhibit several times faster key rates, improved robustness, and tolerance to channel losses of up to dozens of decibels. This makes them well-suited for downlink communication channels from low-earth-orbit satellites in the future.

The field of silicon-based photonics integrated circuits has witnessed numberless advancements, particularly in QKD technology. These advancements encompass the

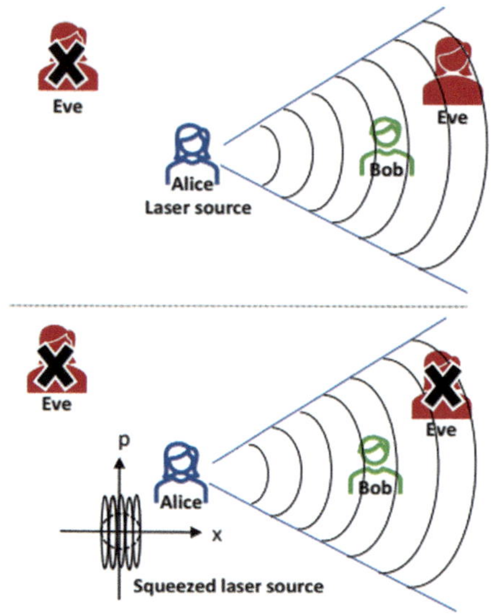

Fig. 5.5 A schematic representation of communication scenarios between Alice and Bob in the presence of an eavesdropper, Eve: **a** communication using laser-comm and **b** the QKD protocol using squeezed ICLs, which can completely eliminate information leakage to Eve. (*Source* In Ref. [1])

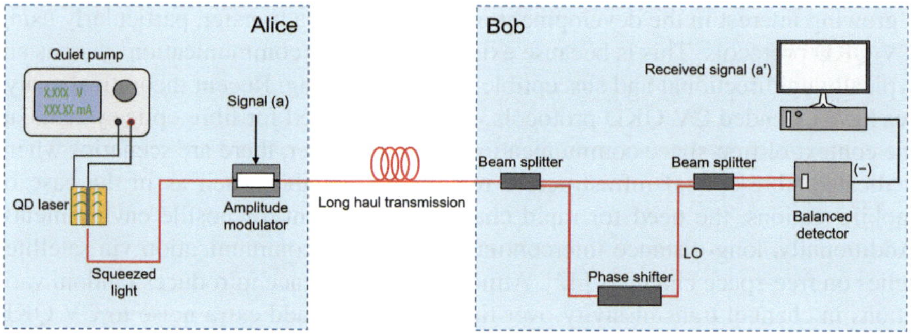

Fig. 5.6 Scheme of the optical implementation of the CV QKD protocol based on an amplitude-squeezed semiconductor laser

development of QKD transmitter chips, QKD receiver chips, and quantum random number generator chips. Our work is directed towards establishing a pathway for implementing the proposed protocol. The incorporation of amplitude-squeezed QD lasers offers several noteworthy advantages. Firstly, our approach offers a simplified setup, as illustrated in Fig. 5.6. Unlike conventional CV QKD protocols, where quantum states of light require additional nonlinear "squeezer" components, the QD laser itself can generate these states. In conventional CV QKD, information is encoded in both the amplitude and phase quadratures of light, and detection typically involves homodyne or heterodyne receivers, accompanied by a separate Local Oscillator (LO). In contrast, our proposed CV QKD protocol, which relies on intensity modulation of a squeezed QD laser, can be directly detected using a self-homodyne approach, thereby eliminating the need for a separate LO. Secondly, our approach offers compatibility with current technology. QD lasers possess many advantages, such as high output power and a wide range of wavelength tunability, which can leverage advancements in the well-established semiconductor industry. Additionally, QD lasers predominantly operate within the near-infrared spectrum, aligning with the peak quantum efficiency of InGaAs photodiodes. These photodiodes provide high detection bandwidth (tens of gigahertz) and nearly 100% detection efficiency. All of these devices seamlessly integrate with existing fibre-optical communication infrastructure, rendering them highly practical for many applications. Thirdly, recent breakthroughs have led to the development of stable and cost-effective systems specifically designed for Gaussian-modulated coherent state CV QKD protocols. These systems have achieved the integration of optical components onto a silicon chip, although it should be noted that the laser source itself remains an external component [44]. Furthermore, significant progress has been made in the integration of QD lasers on silicon substrates [45]. Collectively, these advancements hold substantial promise for the ongoing endeavours to miniaturize on-chip quantum communication systems. They represent a crucial step towards more efficient and compact quantum technologies in the foreseeable future.

References

1. Hosseinidehaj N, Winnel MS, Ralph TC (2022) Simple and loss-tolerant free-space quantum key distribution using a squeezed laser. Phys Rev A 105(3):032602
2. Ostrowski LA, Baker TJ, Saadatmand SN, Wiseman HM (2023) No Tradeoff between Coherence and Sub-Poissonianity for Heisenberg-Limited Lasers. Phys Rev Lett 130(18):183602
3. Baker TJ, Saadatmand SN, Berry DW, Wiseman HM (2021) The Heisenberg limit for laser coherence. Nat Phys 17(2):179–183
4. Smithey D, Beck M, Raymer MG, Faridani A (1993) Measurement of the Wigner distribution and the density matrix of a light mode using optical homodyne tomography: application to squeezed states and the vacuum. Phys Rev Lett 70(9):1244
5. Lawrie BJ, Lett PD, Marino AM, Pooser RC (2019) Quantum sensing with squeezed light. Acs Photonics 6(6):1307–1318
6. Aasi J, Abadie J, Abbott B, Abbott R, Abbott T, Abernathy M, Adams C, Adams T, Addesso P, Adhikari R et al (2013) Enhanced sensitivity of the LIGO gravitational wave detector by using squeezed states of light. Nat Photonics 7(8):613–619
7. Vahlbruch H, Mehmet M, Danzmann K, Schnabel R (2016) Detection of 15 dB squeezed states of light and their application for the absolute calibration of photoelectric quantum efficiency. Phys Rev Lett 117(11):110801
8. Jain N, Chin H-M, Mani H, Lupo C, Nikolic DS, Kordts A, Pirandola S, Pedersen TB, Kolb M, Ömer B et al (2022) Practical continuous-variable quantum key distribution with composable security. Nat Commun 13(1):4740
9. Pirandola S, Andersen UL, Banchi L, Berta M, Bunandar D, Colbeck R, Englund D, Gehring T, Lupo C, Ottaviani C et al (2020) Advances in quantum cryptography. Adv Optics Photonics 12(4):1012–1236

10. Jacobsen CS, Madsen LS, Usenko VC, Filip R, Andersen UL (2018) Complete elimination of information leakage in continuous-variable quantum communication channels. npj Quantum Inf 4(1):32
11. Zhang Y, Chen Z, Pirandola S, Wang X, Zhou C, Chu B, Zhao Y, Xu B, Yu S, Guo H (2020) Long-distance continuous-variable quantum key distribution over 202.81 km of fiber. Phys Rev Lett 125(1):010502
12. McMahon PL, Marandi A, Haribara Y, Hamerly R, Langrock C, Tamate S, Inagaki T, Takesue H, Utsunomiya S, Aihara K et al (2016) A fully programmable 100-spin coherent Ising machine with all-to-all connections. Science 354(6312):614–617
13. Arrazola JM, Bergholm V, Brádler K, Bromley TR, Collins MJ, Dhand I, Fumagalli A, Gerrits T, Goussev A, Helt LG et al (2021) Quantum circuits with many photons on a programmable nanophotonic chip. Nature 591(7848):54–60
14. Zhong H-S, Deng Y-H, Qin J, Wang H, Chen M-C, Peng L-C, Luo Y-H, Wu D, Gong S-Q, Su H et al (2021) Phase-programmable Gaussian boson sampling using stimulated squeezed light. Phys Rev Lett 127(18):180502
15. Grosshans F, Grangier P (2002) Continuous variable quantum cryptography using coherent states. Phys Rev Lett 88(5):057902
16. Derkach I, Usenko VC, Filip R (2020) Squeezing-enhanced quantum key distribution over atmospheric channels. J Phys 22(5):053006
17. Inoue A, Kashiwazaki T, Yamashima T, Takanashi N, Kazama T, Enbutsu K, Watanabe K, Umeki T, Endo M, Furusawa A (2022) 43-GHz bandwidth real-time amplitude measurement of 5-dB squeezed light using modularized optical parametric amplifier with 5G technology. http://arxiv.org/abs/2205.14061arXiv:2205.14061arXiv:2205.14061
18. Banchi L, Fingerhuth M, Babej T, Ing C, Arrazola JM (2020) Molecular docking with Gaussian boson sampling. Sci Adv 6(23):eaax1950
19. Böhm F, Verschaffelt G, Van der Sande G (2019) A poor man's coherent Ising machine based on opto-electronic feedback systems for solving optimization problems. Nat Commun 10(1):3538
20. Imamog A, Yamamoto Y et al (1993) Noise suppression in semiconductor p-i-n junctions: Transition from macroscopic squeezing to mesoscopic Coulomb blockade of electron emission processes. Phys Rev Lett 70(21):3327
21. Kim J, Benson O, Kan H, Yamamoto Y (1999) A single-photon turnstile device. Nature 397(6719):500–503
22. Machida S, Yamamoto Y (1988) Ultrabroadband amplitude squeezing in a semiconductor laser. Phys Rev Lett 60(9):792
23. Freeman M, Wang H, Steel DG, Craig R, Scifres D (1993) Wavelength-tunable amplitude-squeezed light from a room-temperature quantum-well laser. Optics Lett 18(24):2141–2143
24. Kitching J, Yariv A, Shevy Y (1995) Room temperature generation of amplitude squeezed light from a semiconductor laser with weak optical feedback. Phys Rev Lett 74(17):3372
25. Inoue S, Machida S, Yamamoto Y, Ohzu H (1993) Squeezing in an injection-locked semiconductor laser. Phys Rev A 48(3):2230
26. Marin F, Bramati A, Giacobino E, Zhang T-C, Poizat J-P, Roch J-F, Grangier P (1995) Squeezing and intermode correlations in laser diodes. Phys Rev Lett 75(25):4606
27. Van der Lee A, Van Druten N, Van Exter M, Woerdman J, Poizat J-P, Grangier P (2000) Critical Petermann K factor for intensity noise squeezing. Phys Rev Lett 85(22):4711
28. Bramati A, Jost V, Marin F, Giacobino E (1997) Quantum noise models for semiconductor lasers: is there a missing noise source? J Modern Optics 44(10):1929–1935
29. Yamamoto Y, Imoto N, Machida S (1986) Amplitude squeezing in a semiconductor laser using quantum nondemolition measurement and negative feedback. Phys Rev A 33(5):3243
30. Wiseman H (1998) In-loop squeezing is like real squeezing to an in-loop atom. Phys Rev Lett 81(18):3840
31. Yamamoto Y, Nilsson O, Saito S (1985) Theory of a negative frequency feedback semiconductor laser. IEEE J Quantum Electron 21(12):1919–1928
32. Wiseman HM, Milburn GJ (1993) Quantum theory of optical feedback via homodyne detection. Phys Rev Lett 70(5):548

33. Vashukevich E, Lebedev V, Ilichev I, Agruzov P, Shamrai A, Petrov V, Golubeva TY (2022) Broadband chip-based source of quantum noise with electrically controllable beam splitter. Phys Rev Appl 17(6):064039
34. Moody G, Chang L, Steiner TJ, Bowers JE (2020) Chip-scale nonlinear photonics for quantum light generation. AVS Quantum Sci 2(4). https://doi.org/10.1116/5.0020684
35. Moody G, Sorger VJ, Blumenthal DJ, Juodawlkis PW, Loh W, Sorace-Agaskar C, Jones AE, Balram KC, Matthews JC, Laing A et al (2022) 2022 Roadmap on integrated quantum photonics. J Phys: Photonics 4(1):012501
36. Bruynsteen C, Vanhoecke M, Bauwelinck J, Yin X (2021) Integrated balanced homodyne photonic-electronic detector for beyond 20 GHz shot-noise-limited measurements. Optica 8(9):1146–1152
37. Pelucchi E, Fagas G, Aharonovich I, Englund D, Figueroa E, Gong Q, Hannes H, Liu J, Lu C-Y, Matsuda N et al (2022) The potential and global outlook of integrated photonics for quantum technologies. Nat Rev Phys 4(3):194–208
38. Luo W, Cao L, Shi Y, Wan L, Zhang H, Li S, Chen G, Li Y, Li S, Wang Y et al (2023) Recent progress in quantum photonic chips for quantum communication and internet. Light: Sci & Appl 12(1):175
39. Richardson W, Machida S, Yamamoto Y (1991) Squeezed photon-number noise and sub-Poissonian electrical partition noise in a semiconductor laser. Phys Rev Lett 66(22):2867
40. Spitz O, Didier P, Durupt L, Díaz-Thomas DA, Baranov AN, Cerutti L, Grillot F (2021) Free-space communication with directly modulated mid-infrared quantum cascade devices. IEEE J Sel Topics Quantum Electron 28(1):1–9
41. Peuntinger C, Heim B, Müller CR, Gabriel C, Marquardt C, Leuchs G (2014) Distribution of squeezed states through an atmospheric channel. Phys Rev Lett 113(6):060502
42. Pirandola S (2021) Satellite quantum communications: fundamental bounds and practical security. Phys Rev Res 3(2):023130
43. Pirandola S (2021) Limits and security of free-space quantum communications. Phys Rev Res 3(1):013279
44. Zhang G, Haw JY, Cai H, Xu F, Assad S, Fitzsimons JF, Zhou X, Zhang Y, Yu S, Wu J et al (2019) An integrated silicon photonic chip platform for continuous-variable quantum key distribution. Nat Photonics 13(12):839–842. https://www.nature.com/articles/s41566-019-0504-5
45. Shang C, Feng K, Hughes ET, Clark A, Debnath M, Koscica R, Leake G, Herman J, Harame D, Ludewig P et al (2022) Electrically pumped quantum-dot lasers grown on 300 mm patterned Si photonic wafers. Light: Sci & Appl 11(1):299

Appendix A
Simulation Parameters

See Tables A.1, A.2, A.3, A.4 and A.5.

Table A.1 Material and optical parameters of the InAs/GaAs QD laser used in the simulation [1]

Symbol	Description	Value
E_{RS}	RS energy level	1.25 eV
E_{ES}	ES energy level	1.03 eV
E_{GS}	GS energy level	0.95 eV
T	Temperature	295 K
τ_{ES}^{RS}	RS to ES capture time	6.3 ps
τ_{GS}^{ES}	ES to GS relaxation time	2.9 ps
τ_{RS}^{spon}	RS spontaneous emission lifetime	0.5 ns
τ_{ES}^{spon}	ES spontaneous emission lifetime	0.5 ns
τ_{GS}^{spon}	GS spontaneous emission lifetime	1.2 ns
L	Cavity length	750 μm
W	Cavity width	2 μm
h_{RS}	RS barrier layer thickness	40 nm
h_{QD}	QD layer thickness	2 nm
α_i	Internal loss	6 cm^{-2}
R_1, R_2	Facet reflectivity	0.03 & 0.95
τ_p	Photon lifetime	3.9 ps
η	Injection efficiency	0.21
T_D	Polarization dephasing time	0.1 ps

(continued)

S. Zhao, *Noise, Dynamics and Squeezed Light in Quantum Dot and Interband Cascade Lasers*, Springer Theses, https://doi.org/10.1007/978-3-031-66818-0

Table A.1 (continued)

Symbol	Description	Value
β_{sp}	Spontaneous emission factor	1×10^{-4}
ξ	Gain compression factor	2×10^{-16} cm^3
Γ_p	Optical confinement factor	0.06
a_{GS}	GS differential gain	5×10^{-15} cm^2
a_{ES}	ES differential gain	10×10^{-15} cm^2
a_{RS}	RS differential gain	2.5×10^{-15} cm^2
α_H^{GS}	GS contribution to α_H-factor	0.5
N_D	QD density per layer	5.9×10^{10} cm^{-2}
n_{QD}	The number of QD layer	8
N_B	Total QD number	7.08×10^6
D_{RS}	Total RS state number	3.62×10^6
V_B	The volume of QD region	2.4×10^{-11} cm^3
V_{RS}	The volume of RS region	6×10^{-11} cm^3

Table A.2 Material and optical parameters of the InAs/InP QD laser on Silicon used in the simulation [2]

Symbol	Description	Value
E_{RS}	RS energy level	0.97 eV
E_{ES}	ES energy level	0.87 eV
E_{GS}	GS energy level	0.82 eV
T	Temperature	295 K
τ_{ES}^{RS}	RS to ES capture time	6.3 ps
τ_{GS}^{ES}	ES to GS relaxation time	2.9 ps
τ_{RS}^{spon}	RS spontaneous emission lifetime	0.5 ns
τ_{ES}^{spon}	ES spontaneous emission lifetime	0.5 ns
τ_{GS}^{spon}	GS spontaneous emission lifetime	1.2 ns
L	Cavity length	500 μm
W	Cavity width	4 μm
h_{RS}	RS barrier layer thickness	5 nm
h_{QD}	QD layer thickness	5 nm
α_i	Internal loss	6 cm^{-2}
R_1, R_2	Facet reflectivity	0.32 & 0.32
τ_p	Photon lifetime	4.1 ps
η	Injection efficiency	0.12
T_D	Polarization dephasing time	0.1 ps

(continued)

Table A.2 (continued)

Symbol	Description	Value
τ_{SRH}	Nonradiative recombination lifetime	0.1–10 ns (variable)
β_{sp}	Spontaneous emission factor	1×10^{-4}
ξ	Gain compression factor	2×10^{-16} cm^3
Γ_p	Optical confinement factor	0.06
a_{GS}	GS differential gain	5×10^{-15} cm^2
a_{ES}	ES differential gain	10×10^{-15} cm^2
a_{RS}	RS differential gain	2.5×10^{-15} cm^2
α_H^{GS}	GS contribution to α_H-factor	0.5
N_D	QD density per layer	10×10^{10} cm^{-2}
n_{QD}	The number of QD layer	5
N_B	Total QD number	1×10^7
D_{RS}	Total RS state number	4.8×10^6
V_B	The volume of QD region	5×10^{-11} cm^3
V_{RS}	The volume of RS region	1×10^{-11} cm^3

Table A.3 Material and optical parameters of the $InAs/In_xGa_{1-x}Sb$ ICL used in the simulation [3, 4]

Symbol	Description	Value
T	Temperature	300 K
L	Cavity length	2 mm
W	Cavity width	4.4 μm
n_{bg}	Refractive index	3.58
m	Stage number	5
α_i	Internal loss	6 cm^{-2}
R_1, R_2	Facet reflectivity	0.32 & 0.32
τ_p	Photon lifetime	10.5 ps
τ_{aug}	Auger lifetime	1.08 ns
η	Injection efficiency	0.64
N_0	Transparent carrier number	6.2×10^7
β_{sp}	Spontaneous emission factor	1×10^{-4}
Γ_p	Optical confinement factor	0.04
a_0	Differential gain	2.8×10^{-10} cm^2
α_H	Linewidth enhancement factor	2.2

Table A.4 Material and optical parameters of the $InAs/In_xGa_{1-x}Sb$ ICL used in the simulation [5]

Symbol	Description	Value
T	Temperature	300 K
L	Cavity length	3 mm
W	Cavity width	10 μm
n_{bg}	Refractive index	3.5
m	Stage number	7
α_i	Internal loss	5 cm^{-2}
R_1, R_2	Facet reflectivity	0.32 & 0.32
τ_p	Photon lifetime	10.5 ps
τ_{aug}	Auger lifetime	1.08 ns
τ_{sp}	Spontaneous emission time	15 ns
η	Injection efficiency	0.64
N_0	Transparent carrier number	6.2×10^7
η	Injection efficiency	0.64
β_{sp}	Spontaneous emission factor	1×10^{-4}
Γ_p	Optical confinement factor	0.15
a_0	Differential gain	2.8×10^{-10} cm^2
α_H	Linewidth enhancement factor	3

Table A.5 Material and optical parameters of the InAs/GaAs QD laser used in the simulation

Symbol	Description	Value	Reference
E_{RS}	RS energy level	1.25 eV	[6]
E_{ES}	ES energy level	1.03 eV	[7]
E_{GS}	GS energy level	0.95 eV	[7]
T	Temperature	293 K	*
n_r	Refractive index	3.5	[8]
L	Cavity length	750 μm	[8]
W	Cavity width	2 μm	[8]
n_{layer}	QD layer number	8	[8]
h_{QD}	QD layer height	5 nm	[8]
τ_{ES}^{RS}	RS to ES capture time	6.3 ps	[9]
τ_{GS}^{ES}	ES to GS relaxation time	2.9 ps	[9]

(continued)

Table A.5 (continued)

Symbol	Description	Value	Reference
τ_{RS}^{spon}	RS spontaneous emission lifetime	0.5 ns	[10]
τ_{ES}^{spon}	ES spontaneous emission lifetime	0.5 ns	[10]
τ_{GS}^{spon}	GS spontaneous emission lifetime	1.2 ns	[10]
τ_p	Photon lifetime	4.9 ps	†
η	Injection efficiency	0.18	*
β_{sp}	Spontaneous emission factor	1×10^{-4}	[6]
ξ	Gain compression factor	2×10^{-16} cm^3	[9]
Γ_p	Optical confinement factor	0.06	[6]
a_{GS}	GS differential gain	5×10^{-15} cm^2	[9]
N_B	Total QD number	7.08×10^6	[8]
V_B	Active region volume	6×10^{-11} cm^3	‡

$*$ The asterisk ($*$) notation denotes that the data originates from our experimental observations

† $\tau_p = (\frac{c}{n_r})\left[\alpha_{in} + \frac{1}{2L}\ln(\frac{1}{R_1 R_2})\right]$ with c being the velocity of light in vacuum

‡ $V_B = L \times W \times h_{QD} \times n_{\text{layer}}$

References

1. Dong B, Duan J, Huang H, Norman JC, Nishi K, Takemasa K, Sugawara M, Bowers JE, Grillot F (2021) Dynamic performance and reflection sensitivity of quantum dot distributed feedback lasers with large optical mismatch. Photonics Res 9(8):1550–1558
2. Zhao S, Grillot F (2021) Effect of Shockley-Read-Hall recombination on the static and dynamical characteristics of epitaxial quantum-dot lasers on silicon. Phys Rev A 103(6):063521
3. Deng Y, Wang C (2020) Rate equation modeling of interband cascade lasers on modulation and noise dynamics. IEEE J Quantum Electron 56(2):1–9
4. Zhao S, Grillot F (2022) Modeling of amplitude squeezing in a pump-noise-suppressed interband cascade laser. IEEE Photonics J 14(3):1–8
5. Didier P, Spitz O, Cerutti L, Diaz-Thomas D, Baranov A, Carras M, Grillot F (2021) Relative intensity noise and intrinsic properties of RF mounted interband cascade laser. Appl Phys Lett 119(17):171107
6. Sugawara M, Mukai K, Nakata Y, Ishikawa H, Sakamoto A (2000) Effect of homogeneous broadening of optical gain on lasing spectra in self-assembled In x Ga 1- x A s/G a A s quantum dot lasers. Phys Rev B 61(11):7595
7. Huang H, Lin L-C, Chen C-Y, Arsenijević D, Bimberg D, Lin F-Y, Grillot F (2018) Multimode optical feedback dynamics in InAs/GaAs quantum dot lasers emitting exclusively on ground or excited states: transition from short-to long-delay regimes. Optics Express 26(2):1743–1751
8. Nishi K, Kageyama T, Yamaguchi M, Maeda Y, Takemasa K, Yamamoto T, Sugawara M, Arakawa Y (2013) Molecular beam epitaxial growths of high-optical-gain InAs quantum dots on GaAs for long-wavelength emission. J Crystal Growth 378:459–462
9. Wang C, Osiński M, Even J, Grillot F (2014) Phase-amplitude coupling characteristics in directly modulated quantum dot lasers. Appl Phys Lett 105(22): 221114
10. Grillot F, Veselinov K, Gioannini M, Montrosset I, Even J, Piron R, Homeyer E, Loualiche S (2009) Spectral analysis of 1.55-μ m InAs–InP (113) B quantum-dot lasers based on a multipopulation rate equations model. IEEE J Quantum Electron 45(7):872–878

Appendix B
Small-Signal Analysis of Quantum-Confined Lasers

Quantum Dot Lasers

By following the approach developed in Refs. [1, 2, 3] and utilizing the steady-state solutions from (3.16) to (3.20), the diffusion coefficients can be expressed as follows:

$$
\begin{aligned}
\langle F_{RS}F_{RS}^{*}\rangle &= 2D_{RSRS} = \frac{\eta I}{q} + \frac{N_{ES}}{\tau_{RS}^{ES}} + \frac{N_{RS}}{\tau_{ES}^{RS}}(1-\rho_{ES}) + \frac{N_{RS}}{\tau_{RS}^{spon}} \\
\langle F_{ES}F_{ES}^{*}\rangle &= 2D_{ESES} = \left(\frac{N_{RS}}{\tau_{ES}^{RS}} + \frac{N_{GS}}{\tau_{ES}^{GS}}\right)(1-\rho_{ES}) + \frac{N_{ES}}{\tau_{RS}^{ES}} + \frac{N_{ES}}{\tau_{GS}^{ES}}(1-\rho_{GS}) + \frac{N_{ES}}{\tau_{ES}^{spon}} \\
\langle F_{GS}F_{GS}^{*}\rangle &= 2D_{GSGS} = \frac{N_{ES}}{\tau_{GS}^{ES}}(1-\rho_{GS}) + \frac{N_{GS}}{\tau_{ES}^{GS}}(1-\rho_{ES}) + \Gamma_{p}v_{g}g_{GS}S_{GS} + \frac{N_{GS}}{\tau_{GS}^{spon}} \\
\langle F_{S}F_{S}^{*}\rangle &= 2D_{SS} = \Gamma_{p}v_{g}g_{GS}S_{GS} + \frac{S_{GS}}{\tau_{p}} + \beta_{sp}\frac{N_{GS}}{\tau_{GS}^{spon}} \\
\left\langle F_{\phi}F_{\phi}^{*}\right\rangle &= 2D_{\phi\phi} = \frac{2D_{SS}}{4S_{GS}^{2}} \\
\langle F_{RS}F_{ES}^{*}\rangle &= 2D_{RSES} = -\left[\frac{N_{RS}}{\tau_{ES}^{RS}}(1-\rho_{ES}) + \frac{N_{ES}}{\tau_{RS}^{ES}}\right] \\
\langle F_{ES}F_{GS}^{*}\rangle &= 2D_{ESGS} = -\left[\frac{N_{GS}}{\tau_{ES}^{GS}}(1-\rho_{ES}) + \frac{N_{ES}}{\tau_{GS}^{ES}}(1-\rho_{GS})\right] \\
\langle F_{GS}F_{S}^{*}\rangle &= 2D_{GSS} = -\left[\beta_{sp}\frac{N_{GS}}{\tau_{GS}^{spon}} + \Gamma_{p}v_{g}g_{GS}S_{GS}\right] \\
2D_{RS\phi} &= 2D_{ES\phi} = 2D_{GS\phi} = 2D_{S\phi} = 0
\end{aligned}
\tag{B.1}
$$

The QD laser noise features are investigated via a small-signal analysis or linearization with the rate Eqs. (3.16)–(3.20). Therefore, the corresponding matrix can be derived as

S. Zhao, *Noise, Dynamics and Squeezed Light in Quantum Dot and Interband Cascade Lasers*, Springer Theses, https://doi.org/10.1007/978-3-031-66818-0

$$\begin{bmatrix} \gamma_{11}+j\omega & -\gamma_{12} & 0 & 0 & 0 \\ -\gamma_{21} & \gamma_{22}+j\omega & -\gamma_{23} & 0 & 0 \\ 0 & -\gamma_{32} & \gamma_{33}+j\omega & -\gamma_{34} & 0 \\ 0 & 0 & -\gamma_{43} & \gamma_{44}+j\omega & 0 \\ -\gamma_{51} & -\gamma_{52} & -\gamma_{53} & -\gamma_{54} & j\omega \end{bmatrix} \begin{bmatrix} \delta N_{RS} \\ \delta N_{ES} \\ \delta N_{GS} \\ \delta S_{GS} \\ \delta\phi \end{bmatrix} = \begin{bmatrix} F_{RS} \\ F_{ES} \\ F_{GS} \\ F_S \\ F_\phi \end{bmatrix} \tag{B.2}$$

with

$$\gamma_{11} = \frac{1-\rho_{ES}}{\tau_{ES}^{RS}} + \frac{1}{\tau_{RS}^{spon}}$$

$$\gamma_{12} = \frac{1}{\tau_{RS}^{ES}} + \frac{1}{4N_B}\frac{N_{RS}}{\tau_{ES}^{RS}}$$

$$\gamma_{21} = \frac{1-\rho_{ES}}{\tau_{ES}^{RS}}$$

$$\gamma_{22} = \frac{1-\rho_{GS}}{\tau_{GS}^{ES}} + \frac{1}{\tau_{RS}^{ES}} + \frac{1}{\tau_{ES}^{spon}} + \frac{1}{4N_B}\left(\frac{N_{RS}}{\tau_{ES}^{RS}} + \frac{N_{GS}}{\tau_{ES}^{GS}}\right)$$

$$\gamma_{23} = \frac{1-\rho_{ES}}{\tau_{ES}^{GS}} + \frac{1}{2N_B}\frac{N_{ES}}{\tau_{GS}^{ES}}$$

$$\gamma_{32} = \frac{1-\rho_{GS}}{\tau_{GS}^{ES}} + \frac{1}{4N_B}\frac{N_{GS}}{\tau_{ES}^{GS}}$$

$$\gamma_{33} = \frac{1-\rho_{ES}}{\tau_{ES}^{GS}} + \frac{1}{2N_B}\frac{N_{ES}}{\tau_{GS}^{ES}} + \frac{1}{\tau_{GS}^{spon}} + \Gamma_p v_g a S_{GS}$$

$$\gamma_{34} = -\Gamma_p v_g g_{GS} + \Gamma_p v_g a_p S_{GS}$$

$$\gamma_{43} = \Gamma_p v_g a S_{GS} + \frac{\beta_{sp}}{\tau_{GS}^{spon}}$$

$$\gamma_{44} = \frac{1}{\tau_p} - \Gamma_p v_g g_{GS} + \Gamma_p v_g a_p S_{GS}$$

$$\gamma_{51} = \Gamma_p v_g a'_{RS}\alpha_H^{RS}$$

$$\gamma_{52} = \frac{1}{4}\Gamma_p v_g a'_{ES}\alpha_H^{ES}$$

$$\gamma_{53} = \frac{1}{2}\Gamma_p v_g a\alpha_H^{GS}$$

$$\gamma_{54} = -\frac{1}{2}\Gamma_p v_g a_p\alpha_H^{GS}$$

where a and a_p take into account the gain compression and also a'_{ES}, a'_{RS} at high photon numbers for differential gain such as $dg_{GS} = a dN_{GS} - a_p dS_{GS}$, $dg_{ES} = a'_{ES} dN_{GS}$ and $dg_{RS} = a'_{RS} dN_{RS}$.

$$a = \frac{\partial g_{GS}}{\partial N_{GS}} = \frac{a_{GS}}{1+\xi S_{GS}}\frac{1}{V_B}$$

$$a_p = -\frac{\partial g_{GS}}{\partial S_{GS}} = \frac{\xi g_{GS}}{1+\xi S_{GS}}$$

$$a'_{ES} = \frac{a_{ES}}{V_B}$$

$$a'_{RS} = \frac{a_{RS}}{V_{RS}}$$

The RIN of the QD laser emitting on the GS transition is then expressed as follows [4]:

$$\mathrm{RIN}(\omega) = \frac{|\delta S_{GS}(\omega)|^2}{S_{GS}^2} \tag{B.3}$$

with S_{GS} being the average photon number and $\delta S_{GS}(\omega)$ being the photon number variation in the frequency domain. The determinant of the full matrix in (B.2) is

$$\Delta = \det \begin{bmatrix} \gamma_{11} + j\omega & -\gamma_{12} & 0 & 0 & 0 \\ -\gamma_{21} & \gamma_{22} + j\omega & -\gamma_{23} & 0 & 0 \\ 0 & -\gamma_{32} & \gamma_{33} + j\omega & -\gamma_{34} & 0 \\ 0 & 0 & -\gamma_{43} & \gamma_{44} + j\omega & 0 \\ -\gamma_{51} & -\gamma_{52} & -\gamma_{53} & -\gamma_{54} & j\omega \end{bmatrix}$$

By applying Cramer's rule and substituting the fourth column, which represents the effects of all five equations on the photon number S_{GS}, one obtain:

$$\Delta_S = \det \begin{bmatrix} \gamma_{11} + j\omega & -\gamma_{12} & 0 & F_{RS} & 0 \\ -\gamma_{21} & \gamma_{22} + j\omega & -\gamma_{23} & F_{ES} & 0 \\ 0 & -\gamma_{32} & \gamma_{33} + j\omega & F_{GS} & 0 \\ 0 & 0 & -\gamma_{43} & F_S & 0 \\ -\gamma_{51} & -\gamma_{52} & -\gamma_{53} & F_\phi & j\omega \end{bmatrix}$$

To continue the analytical approach, the Laplace expansion[1] is used, which translates Δ_S to:

$$\Delta_S = -F_{RS}M_{RS-S} + F_{ES}M_{ES-S} - F_{GS}M_{GS-S} + F_S M_{S-S} - F_\phi M_{\phi-S}$$

where the five minors M are defined as follows

$$M_{\mathrm{RS}-\phi} = \det \begin{bmatrix} -\gamma_{21} & \gamma_{22} + j\omega & -\gamma_{23} & 0 \\ 0 & -\gamma_{32} & \gamma_{33} + j\omega & 0 \\ 0 & 0 & -\gamma_{43} & 0 \\ -\gamma_{51} & -\gamma_{52} & -\gamma_{53} & j\omega \end{bmatrix}$$

$$M_{\mathrm{ES-S}} - \det \begin{bmatrix} \gamma_{11} + j\omega & -\gamma_{12} & 0 & 0 \\ 0 & -\gamma_{32} & \gamma_{33} + j\omega & 0 \\ 0 & 0 & -\gamma_{43} & 0 \\ -\gamma_{51} & -\gamma_{52} & -\gamma_{53} & j\omega \end{bmatrix}$$

$$M_{\mathrm{GS-S}} = \det \begin{bmatrix} \gamma_{11} + j\omega & -\gamma_{12} & 0 & 0 \\ -\gamma_{21} & \gamma_{22} + j\omega & -\gamma_{23} & 0 \\ 0 & 0 & -\gamma_{43} & 0 \\ -\gamma_{51} & -\gamma_{52} & -\gamma_{53} & j\omega \end{bmatrix}$$

[1] $|A| = \det(A) = \sum_n^{j=1} a_{ij}(-1)^{i+j}\det(A_{ij})$.

$$M_{S-S} = \det\begin{bmatrix} \gamma_{11}+j\omega & -\gamma_{12} & 0 & 0 \\ -\gamma_{21} & \gamma_{22}+j\omega & -\gamma_{23} & 0 \\ 0 & -\gamma_{32} & \gamma_{33}+j\omega & 0 \\ -\gamma_{51} & -\gamma_{52} & -\gamma_{53} & j\omega \end{bmatrix}$$

$$M_{\phi-S} = \det\begin{bmatrix} \gamma_{11}+j\omega & -\gamma_{12} & 0 & 0 \\ -\gamma_{21} & \gamma_{22}+j\omega & -\gamma_{23} & 0 \\ 0 & -\gamma_{32} & \gamma_{33}+j\omega & 0 \\ 0 & 0 & -\gamma_{43} & 0 \end{bmatrix}$$

Let's now apply the determinants and correlation strengths discussed above to the definition of RIN in (B.3). The variation in laser photon number, denoted as δS_{GS}, can be obtained as $\delta S_{GS} = \Delta_S/\Delta$, so that

$$\begin{aligned}
\text{RIN}(\omega) =& \frac{|\delta S_{GS}(\omega)|^2}{S_{GS}^2} = \frac{|\Delta_S|^2}{S_{GS}^2|\Delta|^2} \\
=& \frac{1}{S_{GS}^2|\Delta|^2}\left|-F_{\text{RS}}M_{\text{RS}-\text{S}} + F_{\text{ES}}M_{\text{ES}-\text{S}} - F_{\text{GS}}M_{\text{GS}-\text{S}} + F_S M_{\text{S}-\text{S}} - F_\phi M_{\phi-S}\right|^2 \\
=& \frac{1}{S_{GS}^2|\Delta|^2}\{\langle F_{\text{RS}}F_{\text{RS}}^*\rangle M_{\text{RS}-\text{S}}M_{\text{RS}-\text{S}}^* - \langle F_{\text{RS}}F_{\text{ES}}^*\rangle M_{\text{RS}-\text{S}}M_{\text{ES}-\text{S}}^* \\
&+ \langle F_{\text{RS}}F_{\text{GS}}^*\rangle M_{\text{RS}-\text{S}}M_{\text{GS}-\text{S}}^* - \langle F_{\text{RS}}F_S^*\rangle M_{\text{RS}-\text{S}}M_{\text{S}-\text{S}}^* + \left\langle F_{\text{RS}}F_\phi^*\right\rangle M_{\text{RS}-\text{S}}M_{\phi-S}^* \\
&+ \langle F_{\text{ES}}F_{\text{ES}}^*\rangle M_{\text{ES}-\text{S}}M_{\text{ES}-\text{S}}^* - \langle F_{\text{ES}}F_{\text{RS}}^*\rangle M_{\text{ES}-\text{S}}M_{\text{RS}-\text{S}}^* - \langle F_{\text{ES}}F_{\text{GS}}^*\rangle M_{\text{ES}-\text{S}}M_{\text{GS}-\text{S}}^* \\
&+ \langle F_{\text{ES}}F_S^*\rangle M_{\text{ES}-\text{S}}M_{\text{S}-\text{S}}^* - \left\langle F_{\text{ES}}F_\phi^*\right\rangle M_{\text{ES}-\text{S}}M_{\phi-S}^* + \langle F_{\text{GS}}F_{\text{GS}}^*\rangle M_{\text{GS}-\text{S}}M_{\text{GS}-\text{S}}^* \\
&+ \langle F_{\text{GS}}F_{\text{RS}}^*\rangle M_{\text{GS}-\text{S}}M_{\text{RS}-\text{S}}^* - \langle F_{\text{GS}}F_{\text{ES}}^*\rangle M_{\text{GS}-\text{S}}M_{\text{ES}-\text{S}}^* - \langle F_{\text{GS}}F_S^*\rangle M_{\text{GS}-\text{S}}M_{\text{S}-\text{S}}^* \\
&+ \left\langle F_{\text{GS}}F_\phi^*\right\rangle M_{\text{GS}-\text{S}}M_{\phi-S}^* + \langle F_S F_S^*\rangle M_{\text{S}-\text{S}}M_{\text{S}-\text{S}}^* - \langle F_S F_{\text{RS}}^*\rangle M_{\text{S}-\text{S}}M_{\text{RS}-\text{S}}^* \\
&+ \langle F_S F_{\text{ES}}^*\rangle M_{\text{S}-\text{S}}M_{\text{ES}-\text{S}}^* - \langle F_S F_{\text{GS}}^*\rangle M_{\text{S}-\text{S}}M_{\text{GS}-\text{S}}^* - \left\langle F_S F_\phi^*\right\rangle M_{\text{S}-\text{S}}M_{\phi-S}^* \\
&+ \left\langle F_\phi F_\phi^*\right\rangle M_{\phi-S}M_{\phi-S}^* + \langle F_\phi F_{\text{RS}}^*\rangle M_{\phi-S}M_{\text{RS}-\text{S}}^* - \langle F_\phi F_{\text{ES}}^*\rangle M_{\phi-S}M_{\text{ES}-\text{S}}^* \\
&+ \langle F_\phi F_{\text{GS}}^*\rangle M_{\phi-S}M_{\text{GS}-\text{S}}^* - \langle F_\phi F_S^*\rangle M_{\phi-S}M_{\text{S}-\text{S}}^*\}
\end{aligned}$$

Due to the presence of zero correlation strengths, the expression for RIN simplifies significantly.

$$\langle F_{RS}F_{GS}\rangle = \langle F_{RS}F_S\rangle = \left\langle F_{RS}F_\phi\right\rangle = \langle F_{ES}F_S\rangle = \left\langle F_{ES}F_\phi\right\rangle = \left\langle F_{GS}F_\phi\right\rangle = 0$$

Furthermore, a real part is introduced due to the property that the sum of an imaginary number z and its complex conjugate z^* is equal to $a + jb + a - jb = 2a = 2\Re\{z + z^*\}$. This property applies to the complex-conjugated minor determinants M^*, which can be expressed as follows:

$$\langle F_{\text{RS}}F_{\text{ES}}^*\rangle M_{\text{RS}-\text{S}}M_{\text{ES}-\text{S}}^* + \langle F_{\text{ES}}F_{\text{RS}}^*\rangle M_{\text{ES}-\text{S}}M_{\text{RS}-\text{S}}^* = 2\langle F_{\text{RS}}F_{\text{ES}}^*\rangle \Re\{M_{\text{RS}-\text{S}}M_{\text{ES}-\text{S}}^*\}$$

By applying the same principle as RIN and in (3.9), the frequency noise (FN) of the QD laser can be expressed as follows:

$$\mathrm{FN}(\omega) = |\delta\omega_S(\omega)|^2 \tag{B.4}$$

The angular variation in laser frequency is given by $\delta\omega_S(\omega) = j\omega\delta\phi = j\omega\Delta_\phi/\Delta$, where $\omega_S = 2\pi f$ represents the optical frequency of interest measured in radians per second. To apply Cramer's rule once more, one can define Δ_ϕ in a similar manner to Δ_S, expressed as:

$$\Delta_\phi = \det\begin{bmatrix} \gamma_{11}+j\omega & -\gamma_{12} & 0 & 0 & F_{RS} \\ -\gamma_{21} & \gamma_{22}+j\omega & -\gamma_{23} & 0 & F_{ES} \\ 0 & -\gamma_{32} & \gamma_{33}+j\omega & -\gamma_{34} & F_{GS} \\ 0 & 0 & -\gamma_{43} & \gamma_{44}+j\omega & F_S \\ -\gamma_{51} & -\gamma_{52} & -\gamma_{53} & -\gamma_{54} & F_\phi \end{bmatrix}$$

Therefore again,

$$\Delta_\phi = F_{RS}M_{\mathrm{RS}-\phi} - F_{ES}M_{\mathrm{ES}-\phi} + F_{GS}M_{\mathrm{GS}-\phi} - F_S M_{S-\phi} + F_\phi M_{\phi-\phi}$$

where the five minors M are defined as follows

$$M_{\mathrm{RS}-\phi} = \det\begin{bmatrix} -\gamma_{21} & \gamma_{22}+j\omega & -\gamma_{23} & 0 \\ 0 & -\gamma_{32} & \gamma_{33}+j\omega & -\gamma_{34} \\ 0 & 0 & -\gamma_{43} & \gamma_{44}+j\omega \\ -\gamma_{51} & -\gamma_{52} & -\gamma_{53} & -\gamma_{54} \end{bmatrix}$$

$$M_{\mathrm{ES}-\phi} = \det\begin{bmatrix} \gamma_{11}+j\omega & -\gamma_{12} & 0 & 0 \\ 0 & -\gamma_{32} & \gamma_{33}+j\omega & -\gamma_{34} \\ 0 & 0 & -\gamma_{43} & \gamma_{44}+j\omega \\ -\gamma_{51} & -\gamma_{52} & -\gamma_{53} & -\gamma_{54} \end{bmatrix}$$

$$M_{\mathrm{GS}-\phi} = \det\begin{bmatrix} \gamma_{11}+j\omega & -\gamma_{12} & 0 & 0 \\ -\gamma_{21} & \gamma_{22}+j\omega & -\gamma_{23} & 0 \\ 0 & 0 & -\gamma_{43} & \gamma_{44}+j\omega \\ -\gamma_{51} & -\gamma_{52} & -\gamma_{53} & -\gamma_{54} \end{bmatrix}$$

$$M_{S-\phi} = \det\begin{bmatrix} \gamma_{11}+j\omega & -\gamma_{12} & 0 & 0 \\ -\gamma_{21} & \gamma_{22}+j\omega & -\gamma_{23} & 0 \\ 0 & -\gamma_{32} & \gamma_{33}+j\omega & -\gamma_{34} \\ -\gamma_{51} & -\gamma_{52} & -\gamma_{53} & -\gamma_{54} \end{bmatrix}$$

$$M_{\phi-\phi} = \det\begin{bmatrix} \gamma_{11}+j\omega & -\gamma_{12} & 0 & 0 \\ -\gamma_{21} & \gamma_{22}+j\omega & -\gamma_{23} & 0 \\ 0 & -\gamma_{32} & \gamma_{33}+j\omega & -\gamma_{34} \\ 0 & 0 & -\gamma_{43} & \gamma_{44}+j\omega \end{bmatrix}$$

Then, the FN spectrum can be calculated as follows:

$$\begin{aligned}
\mathrm{FN} =& \frac{\omega^2}{|\Delta|^2}\left|F_{\mathrm{RS}}M_{\mathrm{RS}-\phi} - F_{\mathrm{ES}}M_{\mathrm{ES}-\phi} + F_{\mathrm{GS}}M_{\mathrm{GS}-\phi} - F_S M_{S-\phi} + F_\phi M_{\phi-\phi}\right|^2 \\
=& \frac{\omega^2}{|\Delta|^2}\Big\{\langle F_{\mathrm{RS}}F^*_{\mathrm{RS}}\rangle M_{\mathrm{RS}-\phi}M^*_{\mathrm{RS}-\phi} - \langle F_{\mathrm{RS}}F^*_{\mathrm{ES}}\rangle M_{\mathrm{RS}-\phi}M^*_{\mathrm{ES}-\phi} \\
&+ \langle F_{\mathrm{RS}}F^*_{\mathrm{GS}}\rangle M_{\mathrm{RS}-\phi}M^*_{\mathrm{GS}-\phi} - \langle F_{\mathrm{RS}}F^*_{S}\rangle M_{\mathrm{RS}-\phi}M^*_{S-\phi} + \left\langle F_{\mathrm{RS}}F^*_{\phi}\right\rangle M_{\mathrm{RS}-\phi}M^*_{\phi-\phi} \\
&+ \langle F_{\mathrm{ES}}F^*_{\mathrm{ES}}\rangle M_{\mathrm{ES}-\phi}M^*_{\mathrm{ES}-\phi} - \langle F_{\mathrm{ES}}F^*_{\mathrm{RS}}\rangle M_{\mathrm{ES}-\phi}M^*_{\mathrm{RS}-\phi} - \langle F_{\mathrm{ES}}F^*_{\mathrm{GS}}\rangle M_{\mathrm{ES}-\phi}M^*_{\mathrm{GS}-\phi} \\
&+ \langle F_{\mathrm{ES}}F^*_{S}\rangle M_{\mathrm{ES}-\phi}M^*_{S-\phi} - \left\langle F_{\mathrm{ES}}F^*_{\phi}\right\rangle M_{\mathrm{ES}-\phi}M^*_{\phi-\phi} + \langle F_{\mathrm{GS}}F^*_{\mathrm{GS}}\rangle M_{\mathrm{GS}-\phi}M^*_{\mathrm{GS}-\phi} \\
&+ \langle F_{\mathrm{GS}}F^*_{\mathrm{RS}}\rangle M_{\mathrm{GS}-\phi}M^*_{\mathrm{RS}-\phi} - \langle F_{\mathrm{GS}}F^*_{\mathrm{ES}}\rangle M_{\mathrm{GS}-\phi}M^*_{\mathrm{ES}-\phi} - \langle F_{\mathrm{GS}}F^*_{S}\rangle M_{\mathrm{GS}-\phi}M^*_{S-\phi} \\
&+ \left\langle F_{\mathrm{GS}}F^*_{\phi}\right\rangle M_{\mathrm{GS}-\phi}M^*_{\phi-\phi} + \langle F_S F^*_S\rangle M_{S-\phi}M^*_{S-\phi} - \langle F_S F^*_{\mathrm{RS}}\rangle M_{S-\phi}M^*_{\mathrm{RS}-\phi} \\
&+ \langle F_S F^*_{\mathrm{ES}}\rangle M_{S-\phi}M^*_{\mathrm{ES}-\phi} - \langle F_S F^*_{\mathrm{GS}}\rangle M_{S-\phi}M^*_{\mathrm{GS}-\phi} - \left\langle F_S F^*_{\phi}\right\rangle M_{S-\phi}M^*_{\phi-\phi} \\
&+ \left\langle F_\phi F^*_{\phi}\right\rangle M_{\phi-\phi}M^*_{\phi-\phi} + \langle F_\phi F^*_{\mathrm{RS}}\rangle M_{\phi-\phi}M^*_{\mathrm{RS}-\phi} - \langle F_\phi F^*_{\mathrm{ES}}\rangle M_{\phi-\phi}M^*_{\mathrm{ES}-\phi} \\
&+ \langle F_\phi F^*_{\mathrm{GS}}\rangle M_{\phi-\phi}M^*_{\mathrm{GS}-\phi} - \langle F_\phi F^*_{S}\rangle M_{\phi-\phi}M^*_{S-\phi}\Big\}
\end{aligned}$$

Interband Cascade Lasers

The diffusion coefficients among the carrier noise source F_N, the photon noise source F_S, and the phase noise source F_ϕ of ICLs are derived as [2, 3]:

$$\begin{aligned}
\langle F_N F_N^*\rangle &= 2D_{NN} = \eta\frac{I}{q} + G_0(N - N_0)S + \frac{N}{\tau_{sp}} + \frac{N}{\tau_{aug}} \\
\langle F_S F_S^*\rangle &= 2D_{SS} = mG_0(N - N_0)S + \frac{S}{\tau_p} + m\beta\frac{N}{\tau_{sp}} \\
\left\langle F_\phi F_\phi^*\right\rangle &= 2D_{\phi\phi} = \frac{2D_{SS}}{4S^2} \\
\langle F_N F_S^*\rangle &= 2D_{NS} = -\left(m\beta\frac{N}{\tau_{sp}} + mG_0(N - N_0)S\right) \\
\left\langle F_N F_\phi^*\right\rangle &= 2D_{N\phi} = 0 \\
\left\langle F_S F_\phi^*\right\rangle &= 2D_{S\phi} = 0
\end{aligned}$$

with G_0 being $G_0 = \Gamma_p v_g a_0 / A$. When considering small-signal modulation, the linearized rate equations with Langevin terms can be derived in the frequency domain as follows:

$$\begin{bmatrix} \gamma_{11} + j\omega & \gamma_{12} & 0 \\ -\gamma_{21} & \gamma_{22} + j\omega & 0 \\ -\gamma_{31} & 0 & j\omega \end{bmatrix} \begin{bmatrix} \delta N(\omega) \\ \delta S(\omega) \\ \delta\phi(\omega) \end{bmatrix} = \begin{bmatrix} F_N(\omega) \\ F_S(\omega) \\ F_\phi(\omega) \end{bmatrix} \tag{B.5}$$

with

$$\gamma_{11} = G_0 S + \frac{1}{\tau_{sp}} + \frac{1}{\tau_{aug}}; \quad \gamma_{12} = G_0 (N - N_0);$$

$$\gamma_{21} = m G_0 S + m \frac{\beta}{\tau_{sp}}; \quad \gamma_{22} = \frac{1}{\tau_p} - m G_0 (N - N_0);$$

$$\gamma_{31} = \frac{1}{2} m \alpha_H G_0;$$

Following the same analysis in the previous section, the

$$\mathrm{RIN}(\omega) = \frac{|\delta S(\omega)|^2}{S^2}$$

By applying Cramer's rule and substituting the second column, which represents the effects of all three equations on the photon number S, one obtain

$$\Delta_S = \det \begin{bmatrix} \gamma_{11} + j\omega & F_N & 0 \\ -\gamma_{21} & F_S & 0 \\ -\gamma_{31} & F_\phi & j\omega \end{bmatrix}$$

To continue the analytical approach, the Laplace expansion is used, which translates Δ_S to:

$$\Delta_S = -F_N M_{\mathrm{N-S}} + F_S M_{\mathrm{S-S}} - F_\phi M_{\phi-\mathrm{S}}$$

where the three minors M are defined as follows

$$M_{\mathrm{N-S}} = \det \begin{bmatrix} -\gamma_{21} & 0 \\ -\gamma_{31} & j\omega \end{bmatrix}; \; M_{\mathrm{S-S}} = \det \begin{bmatrix} \gamma_{11} + j\omega & 0 \\ -\gamma_{31} & j\omega \end{bmatrix}; \; M_{\phi-\mathrm{S}} = \det \begin{bmatrix} \gamma_{11} + j\omega & 0 \\ -\gamma_{21} & 0 \end{bmatrix} = 0$$

Once again, following the same analysis in the previous section, the

$$\mathrm{FN}(\omega) = |\delta\omega_S(\omega)|^2 = |j\omega\delta\phi|^2 = \left| j\omega \frac{\Delta\phi}{\Delta} \right|^2$$

where Δ_ϕ is

$$\Delta_\phi = \det \begin{bmatrix} \gamma_{11} + j\omega & \gamma_{12} & F_N \\ -\gamma_{21} & \gamma_{22} + j\omega & F_S \\ -\gamma_{31} & 0 & F_\phi \end{bmatrix}$$

Therefore again,

$$\Delta_\phi = F_N M_{N-\phi} - F_S M_{S-\phi} + F_\phi M_{\phi-\phi}$$

where the three minors M are defined as follows

$$M_{N-\phi} = \det\begin{bmatrix} -\gamma_{21} & \gamma_{22} + j\omega \\ -\gamma_{31} & 0 \end{bmatrix};\ M_{S-\phi} = \det\begin{bmatrix} \gamma_{11} + j\omega & \gamma_{12} \\ -\gamma_{31} & 0 \end{bmatrix};$$

$$M_{\phi-\phi} = \det\begin{bmatrix} \gamma_{11} + j\omega & \gamma_{12} \\ -\gamma_{21} & \gamma_{22} + j\omega \end{bmatrix}$$

References

1. McCumber D (1966) Intensity fluctuations in the output of CW laser oscillators. I. Phys Rev 141(1):306. https://link.aps.org/doi/10.1103/PhysRev.141.306
2. Ahmed M, Yamada M, Saito M (2001) Numerical modeling of intensity and phase noise in semiconductor lasers. IEEE J Quantum Electron 37(12):1600–1610
3. Coldren LA, Corzine SW, Mashanovitch ML (2012) Diode lasers and photonic integrated circuits. Wiley
4. Duan J, Wang X-G, Zhou Y-G, Wang C, Grillot F (2018) Carrier-noise-enhanced relative intensity noise of quantum dot lasers. IEEE J Quantum Electron 54(6):1–7

Appendix C
Density Operator and Phase-Space Distribution in a (Very Small) Nutshell

Density Operator

Quantum-mechanical state vectors $|\psi\rangle$ carry the maximum possible information about a system in accordance with the principles of quantum mechanics. This information generally comprises quantum numbers linked to a collection of commuting observables. Moreover, if $|\psi_1\rangle$ and $|\psi_2\rangle$ are two possible quantum states, their coherent superposition also represents a valid quantum state:

$$|\psi\rangle = c_1 |\psi_1\rangle + c_2 |\psi_2\rangle$$

if the coefficients c_1 and c_2 are known. In the scenario where the states $|\psi_1\rangle$ and $|\psi_2\rangle$ are orthogonal ($\langle\psi_2|\psi_1\rangle = 0$), the condition $|c_1|^2 + |c_2|^2 = 1$ must hold. However, there are frequent cases, and in fact, more often than not, where the state vector is not precisely determined. Such situations arise, for example, when the system of interest interacts with another system, possibly a significantly large one such as a reservoir, resulting in entanglement between the two systems. Quantum states described by state vectors are referred to as *pure states*. On the other hand, states that cannot be fully characterized by state vectors are classified as *mixed states*. Mixed states are represented by density operators $\hat{\rho}$:

$$\hat{\rho} = \sum_i |\psi_i\rangle\, p_i\, \langle\psi_i| = \sum_i p_i\, |\psi_i\rangle\langle\psi_i| \tag{C.1}$$

Here, the summation is performed across an ensemble, akin to the concept in statistical mechanics. In this context, p_i signifies the probability of the system occupying the ith state of the ensemble, denoted by $|\psi_i\rangle$, where the condition $\langle\psi_i|\psi_i\rangle = 1$ holds. The probabilities satisfy the evident relationships:

S. Zhao, *Noise, Dynamics and Squeezed Light in Quantum Dot and Interband Cascade Lasers*, Springer Theses, https://doi.org/10.1007/978-3-031-66818-0

$$0 \leq p_i \leq 1, \quad \sum_i p_i = 1, \quad \sum_i p_i^2 \leq 1$$

For the specific case in which all p_i except the jth one are equal to zero, i.e., $p_i = \delta_{ij}$, one can derive:

$$\hat{\rho} = \left|\psi_j\right\rangle\left\langle\psi_j\right|$$

the density operator corresponding to the pure state $\left|\psi_j\right\rangle$ is obtained. Notably, in this specific instance, the density operator is essentially the projection operator onto the state $\left|\psi_j\right\rangle$. In the broader context presented by (C.1), the density operator takes the form of a summation involving projection operators across the ensemble, with each projection operator weighted by the respective probabilities associated with each ensemble member.

Now, let's introduce a complete, orthonormal basis $|\varphi_n\rangle$, $(\sum_n |\varphi_n\rangle \langle\varphi_n| = \hat{I})$, which represents eigenstates of a certain observable. Then for the ith member of the ensemble:

$$|\psi_i\rangle = \sum_n |\varphi_n\rangle \langle\varphi_n \mid \psi_i\rangle = \sum_n c_n^{(i)} |\varphi_n\rangle$$

where $c_n^{(i)} = \langle\varphi_n \mid \psi_i\rangle$. The matrix element of $\hat{\rho}$ between the n and n' eigenstates is

$$\left\langle\varphi_n|\hat{\rho}|\varphi_{n'}\right\rangle = \sum_i \langle\varphi_n \mid \psi_i\rangle\, p_i\, \langle\psi_i \mid \varphi_{n'}\rangle = \sum_i p_i c_n^{(i)} c_{n'}^{(i)*}$$

The quantities $\left\langle\varphi_n|\hat{\rho}|\varphi_{n'}\right\rangle$ form the elements of the density matrix. Taking the trace of this matrix

$$\begin{aligned} \operatorname{Tr}\hat{\rho} &= \sum_n \left\langle\varphi_n|\hat{\rho}|\varphi_n\right\rangle = \sum_i\sum_n \langle\varphi_n \mid \psi_i\rangle\, p_i\, \langle\psi_i \mid \varphi_n\rangle \\ &= \sum_i\sum_n p_i \langle\psi_i \mid \varphi_n\rangle \langle\varphi_n \mid \psi_i\rangle = \sum_i p_i = 1 \end{aligned} \tag{C.2}$$

Since the density operator $\hat{\rho}$ is Hermitian (as indicated by its construction in (C.1)), the diagonal elements $\langle\varphi_n|\hat{\rho}|\varphi_n\rangle$ must be real. Consequently, it follows from (C.2) that $0 \leq \langle\varphi_n|\hat{\rho}|\varphi_n\rangle \leq 1$. For one of the states of the ensemble $|\psi_i\rangle$, by itself pure, the expectation value of some operator $\hat{O}$ is given by

$$\left\langle\hat{O}\right\rangle_i = \left\langle\psi_i|\hat{O}|\psi_i\right\rangle$$

For the statistical mixture, the ensemble average is given by

$$\left\langle \hat{O} \right\rangle = \sum_i p_i \left\langle \psi_i | \hat{O} | \psi_i \right\rangle$$

which is the average of the quantum-mechanical expectation values weighted by the probabilities p_i. Formally, this can be expressed as

$$\left\langle \hat{O} \right\rangle = \mathrm{Tr}(\hat{\rho}\hat{O}) \tag{C.3}$$

since

$$\begin{aligned}
\mathrm{Tr}(\hat{\rho}\hat{O}) &= \sum_n \left\langle \varphi_n | \hat{\rho}\hat{O} | \varphi_n \right\rangle \\
&= \sum_n \sum_i p_i \left\langle \varphi_n \mid \psi_i \right\rangle \left\langle \psi_i | \hat{O} | \varphi_n \right\rangle \\
&= \sum_i \sum_n p_i \left\langle \psi_i | \hat{O} | \varphi_n \right\rangle \left\langle \varphi_n \mid \psi_i \right\rangle \\
&= \sum_i p_i \left\langle \psi_i | \hat{O} | \psi_i \right\rangle
\end{aligned}$$

Phase-Space Distribution

Since the density operator $\hat{\rho}$ can be expanded in terms of the photon number states:

$$\hat{\rho} = \sum_n \sum_m |n\rangle \langle n| \, \hat{\rho} \, |m\rangle \langle m| = \sum_n \sum_m \hat{\rho}_{nm} \, |n\rangle \langle m|$$

All the matrix elements $\rho_{mn} = \langle m|\hat{\rho}|n\rangle$ are needed to fully determine the operator $\hat{\rho}$. The diagonal elements, $p_n = \rho_{nn}$, correspond to the probabilities of observing n photons in the field. Likewise, the expansion can be made in terms of the coherent states as

$$\hat{\rho} = \int \int \frac{\mathrm{d}^2\alpha}{\pi} \frac{\mathrm{d}^2\beta}{\pi} \, |\alpha\rangle \langle\alpha| \, \hat{\rho} \, |\beta\rangle \langle\beta| \tag{C.4}$$

However, there exists an alternative method to express $\hat{\rho}$ in terms of coherent states, resulting in $\langle\alpha| \, \hat{\rho} \, |\beta\rangle = P(\alpha, \alpha^*)\delta^2(\alpha - \beta)$, which is:

$$\hat{\rho} = \int P(\alpha, \alpha^*) \, |\alpha\rangle \langle\alpha| \, \mathrm{d}^2\alpha \tag{C.5}$$

where $P(\alpha, \alpha^*)$ is a weight function often referred to as the Glauber-Sudarshan P function. The expression on the right-hand side of (C.5) represents the "diagonal" representation of the density operator, and the P function is analogous to the phase-space distributions encountered in statistical mechanics. In this context, the real and imaginary components of α function are the variables of the phase space. Notably, $P(\alpha, \alpha^*)$ is constrained to be real due to the Hermitian nature of the operator $\hat{\rho}$. Then,

$$\begin{aligned}\operatorname{Tr}\hat{\rho} &= \operatorname{Tr}\int P(\alpha, \alpha^*)|\alpha\rangle\langle\alpha|\mathrm{d}^2\alpha \\ &= \int\sum_n P(\alpha, \alpha^*)\langle n \mid \alpha\rangle\langle\alpha \mid n\rangle\mathrm{d}^2\alpha \\ &= \int P(\alpha, \alpha^*)\sum_n\langle\alpha \mid n\rangle\langle n \mid \alpha\rangle\mathrm{d}^2\alpha \\ &= \int P(\alpha, \alpha^*)\langle\alpha \mid \alpha\rangle\mathrm{d}^2\alpha = \int P(\alpha, \alpha^*)\mathrm{d}^2\alpha = 1,\end{aligned}$$

$P(\alpha, \alpha^*)$ can also serve to calculate any normal-ordered operators in the quantum averages (with annihilation operators to the right and creation operators to the left).

$$\left\langle(\hat{a}^\dagger)^m\hat{a}^n\right\rangle = \operatorname{Tr}\left[\hat{\rho}(\hat{a}^\dagger)^m\hat{a}^n\right] = \int(\alpha^*)^m\alpha^n P(\alpha, \alpha^*)\mathrm{d}^2\alpha$$

However, for certain quantum states of light, the properties of $P(\alpha, \alpha^*)$ can deviate significantly from those of a conventional probability distribution, where one would naturally anticipate $P(\alpha, \alpha^*) \geq 0$. There exist quantum states for which $P(\alpha, \alpha^*)$ becomes negative or exhibits highly singular behaviour. In such instances, the corresponding quantum states are termed "nonclassical". This may be seen by the following explicit calculation:

$$\begin{aligned}\left\langle\Delta X^2\right\rangle &= \frac{1}{4}\left\langle(\hat{a} + \hat{a}^\dagger)^2 - \left\langle\hat{a} + \hat{a}^\dagger\right\rangle^2\right\rangle \\ &= \frac{1}{4}\left[\left\langle\hat{a}^2 + (\hat{a}^\dagger)^2 + 2\hat{a}^\dagger\hat{a} + 1\right\rangle - \left\langle\hat{a} + \hat{a}^\dagger\right\rangle^2\right] \\ &= \frac{1}{4}\left\{1 + \int\mathrm{d}^2\alpha P(\alpha, \alpha^*)\left[(\alpha + \alpha^*)^2 - \left\langle\alpha + \alpha^*\right\rangle^2\right]\right\}\end{aligned} \tag{C.6}$$

which shows that $\left\langle\Delta X^2\right\rangle < \frac{1}{4}$ only if $P(\alpha, \alpha^*)$ is not positive definite. The constant term $\frac{1}{4}$ appears on the right-hand side of (C.6) coming from the commutator of $\hat{a}$ and $\hat{a}^\dagger$ during the process of normal-ordering the original expression. It essentially represents a manifestation of the quantum character of the light, often referred to as "shot noise."

When considering anti-normal ordering, the Husimi Q representation can be readily acquired:

$$\left\langle \hat{a}^m (\hat{a}^\dagger)^n \right\rangle = \frac{1}{\pi} \int \alpha^m (\alpha^*)^n \langle \alpha | \hat{\rho} | \alpha \rangle \, d^2\alpha$$

so that

$$Q(\alpha) = \frac{1}{\pi} \langle \alpha | \hat{\rho} | \alpha \rangle$$

It is clear that this representation is always positive since $\hat{\rho}$ is a positive-definite operator. Also, $|Q(\alpha)| \leqslant 1/\pi$. For a coherent state $|\beta\rangle$, one gets that $Q(\alpha)$ is a Gaussian in phase space:

$$Q(\alpha) = \frac{1}{\pi} e^{-|\alpha-\beta|^2}$$

It's worth noting that, with regard to the Q representation, the equation is equivalent to (C.6) is:

$$\begin{aligned} \left\langle \Delta X^2 \right\rangle &= \frac{1}{4} \left[\left\langle \hat{a}^2 + (\hat{a}^\dagger)^2 + 2\hat{a}\hat{a}^\dagger - 1 \right\rangle - \left\langle \hat{a} + \hat{a}^\dagger \right\rangle^2 \right] \\ &= \frac{1}{4} \left\{ -1 + \int d^2\alpha \, Q(\alpha) \left[(\alpha + \alpha^*)^2 - \left\langle \alpha + \alpha^* \right\rangle^2 \right] \right\} \end{aligned} \quad \text{(C.7)}$$

Employing symmetrical ordering of the operators gives rise to the Wigner representation denoted as $W(\alpha)$ [1]. This representation can be demonstrated to be always nonsingular, although it might still exhibit negative values. It should be noted that the equation equivalent to (C.6) is now:

$$\begin{aligned} \left\langle \Delta X^2 \right\rangle &= \frac{1}{4} \left[\left\langle \hat{a}^2 + (\hat{a}^\dagger)^2 + \hat{a}\hat{a}^\dagger + \hat{a}^\dagger \hat{a} \right\rangle - \left\langle \hat{a} + \hat{a}^\dagger \right\rangle^2 \right] \\ &= \frac{1}{4} \int d^2\alpha \, W(\alpha) \left[(\alpha + \alpha^*)^2 - \left\langle \alpha + \alpha^* \right\rangle^2 \right] \end{aligned} \quad \text{(C.8)}$$

Reference

1. Wigner E (1932) On the quantum correction for thermodynamic equilibrium. Phys Rev 40(5):749

Appendix D
Balanced Homodyne Detection

Figure D.1 illustrates the schematic of balanced homodyne detection. Mode "$\hat{a}$" comprises the single field that could potentially be squeezed, while mode "$\hat{b}$" holds a strong coherent classical field, which can be considered as a coherent state with amplitude β. The beam splitter is assumed to be of equal division (50:50). According to the quantum theory of beamsplitter,

$$\hat{c} = \frac{1}{\sqrt{2}}(\hat{a} + i\hat{b}); \quad \hat{d} = \frac{1}{\sqrt{2}}(\hat{b} + i\hat{a}) \tag{D.1}$$

The detectors placed in the output beams measure the intensities $I_c = \langle \hat{c}^\dagger \hat{c} \rangle$ and $I_d = \left\langle \hat{d}^\dagger \hat{d} \right\rangle$, and the difference in these intensities is

$$I_c - I_d = \langle \hat{n}_{cd} \rangle = \left\langle \hat{c}^\dagger \hat{c} - \hat{d}^\dagger \hat{d} \right\rangle = i \left\langle \hat{a}^\dagger \hat{b} - \hat{a}\hat{b}^\dagger \right\rangle \tag{D.2}$$

Assuming the mode $\hat{b}$ to be in the coherent state $\left|\beta e^{-i\omega t}\right\rangle$, with $\beta = |\beta| e^{-i\Psi}$, then:

$$\langle \hat{n}_{cd} \rangle = |\beta| \left[\hat{a} e^{i\omega t} e^{-i\theta} + \hat{a}^\dagger e^{-i\omega t} e^{i\theta} \right] \tag{D.3}$$

Here, $\theta = \Psi + \pi/2$. Assuming that the light in mode "$\hat{a}$" is also at frequency ω (and, in practice, both the "$\hat{a}$" and "$\hat{b}$" mode light fields originate from the same laser), one can assign $\hat{a} = \hat{a}_0 e^{-i\omega t}$ so that

$$\langle \hat{n}_{cd} \rangle = 2|\beta| \left\langle \hat{X}(\theta) \right\rangle \tag{D.4}$$

where

$$\hat{X}(\theta) = \frac{1}{2}\left[\hat{a}_0 e^{-i\theta} + \hat{a}_0^\dagger e^{i\theta} \right]$$

S. Zhao, *Noise, Dynamics and Squeezed Light in Quantum Dot and Interband Cascade Lasers*, Springer Theses, https://doi.org/10.1007/978-3-031-66818-0

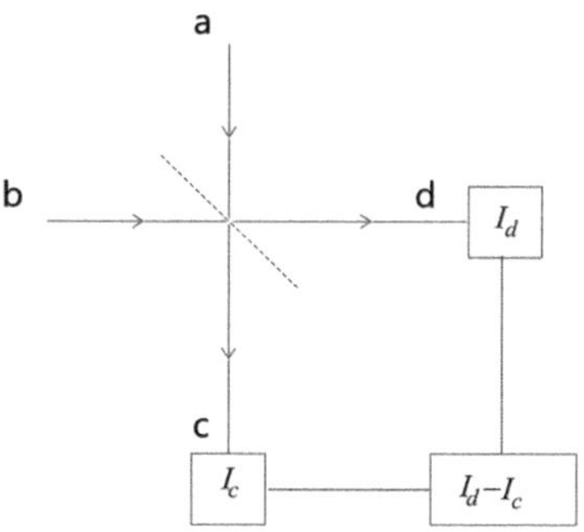

Fig. D.1 Schematic of the balanced homodyne method for the detection squeezing. The field to be detected enters along $\hat{a}$ while a strong coherent field is injected along $\hat{b}$. The boxes in the lower left and upper right represent photodetectors measuring the respective photocurrents. The box on the lower right represents a correlation device that subtracts the photocurrents

is the field quadrature operator at the angle θ. By varying θ, achievable by altering the phase Ψ of the local oscillator (LO), one can effectively measure any desired quadrature of the signal field. In the case of a strong LO (much larger than the input signal), the variance of the output photon number difference operator $\langle \hat{n}_{cd} \rangle$ can be stated as:

$$\langle \Delta \hat{n}_{cd}^2 \rangle = 4|\beta|^2 \left\langle \Delta \hat{X}^2(\theta) \right\rangle \tag{D.5}$$

The dominant term now is only due to the interference between the input signal noise and the LO power and the LO noise is eliminated completely. Under the input squeezing condition $\left\langle \Delta \hat{X}^2(\theta) \right\rangle < \frac{1}{4}$, one ca deduce that $\langle \Delta \hat{n}_{cd}^2 \rangle < |\beta|^2$. In practical experiments, the phase of the LO is adjusted to yield the maximum possible quadrature squeezing. The shot-noise level is calibrated by blocking the signal field, ensuring that only the LO field reaches the detector. The measurement outcomes are then subjected to spectral analysis, allowing the determination of the amount of squeezing as a function of the frequency.

By incorporating additional beam splitters, it becomes possible to perform simultaneous balanced homodyne measurements on both quadratures of the input signal. Such a configuration requires a minimum of four beam splitters. However, it's essential to acknowledge that this arrangement is consistently influenced by vacuum fluctuations that enter through the unused ports. In practice, balanced homodyne detection is the most commonly employed scheme for detecting squeezed states of light.

GPSR Compliance

The European Union's (EU) General Product Safety Regulation (GPSR) is a set of rules that requires consumer products to be safe and our obligations to ensure this.

If you have any concerns about our products, you can contact us on ProductSafety@springernature.com

In case Publisher is established outside the EU, the EU authorized representative is:

Springer Nature Customer Service Center GmbH
Europaplatz 3
69115 Heidelberg, Germany

Batch number: 10732090

Printed by Printforce, the Netherlands